Evolutionary Perspectives
on Environmental Problems

Dustin J. Penn
& Iver Mysterud, editors
Foreword by E.O. Wilson

AldineTransaction
A Division of Transaction Publishers
New Brunswick (U.S.A.) and London (U.K.)

Library of Congress Catalog Number: 2005041816
ISBN: 978-0-202-30754-1 (cloth); 978-0-202-30755-8 (paper)
Printed in the United States of America

Library of Congress Cataloging-in-Publication Data

Evolutionary perspectives on environmental problems / Dustin J. Penn and
     Iver Mysterud, editors
          p. cm.
     Includes bibliographical references and index.
     ISBN 0-202-30754-9 (cloth : alk. paper)—ISBN 0-202-30755-7 (pbk. :
     alk paper)
       1. Human evolution. 2. Human ecology. 3. Nature--Effect of human
     beings on. 4. Soical evolution. 5. Social ecology. 6. Sociobiology.
     7. Hunting--Environmental aspects. I. Penn, Dustin J. II. Mysterud,
     Iver.

GN281.E897 2005
599.93'8--dc22

                                                    2005041816

For Edward O. Wilson
and his outstanding contributions to sociobiology,
conservation, and scientific integration

Contents

Acknowledgements   xi

Foreword   xiii
    Edward O. Wilson

Introduction: 1
    The Evolutionary Roots of Our Ecological Crisis
    Dustin J. Penn and Iver Mysterud

Part 1: Human Nature and Resource Conservation

1.   Human Behavioural Ecology and Conservation   9
    Joel T. Heinen and Bobbi S. Low

2.   The Evolved Psychological Apparatus of Human   31
    Decision-Making is One Source of Environmental Problems
    Margo Wilson, Martin Daly, and Stephen Gordon

Part 2: The Ecological Noble Savage Hypothesis

3.   Game Conservation or Efficient Hunting?   53
    Raymond Hames

4.   Behavioral Ecology of Conservation   67
    in Traditional Societies
    Bobbi S. Low

5.   Evolutionary Ecology and Resource Conservation   81
    Michael S. Alvard

Part 3: The Tragedy of the Commons

6.   The Tragedy of the Unmanaged Commons   105
    Garrett Hardin

7.   Closing the Commons: Cooperation for Gain or Restraint?   109
    Lore M. Ruttan

8. Revisiting the Commons: Local Lessons,   129
   Global Challenges
   Elinor Ostrom, Joanna Burger, Christopher B. Field,
   Richard B. Norgaard, and David Policansky

9. Grassland Conservation and the Pastoralist Commons   141
   Monique Borgerhoff Mulder and Lore M. Ruttan

Part 4: The Evolution of Discounting
      and Conspicuous Consumption

10. Conserving Resources for Children   157
    Alan R. Rogers

11. Two Truths about Discounting   165
    and Their Environmental Consequences
    Norman Henderson and William J. Sutherland

12. Sex Differences in Valuations of the Environment?   169
    Margo Wilson, Martin Daly, Stephen Gordon,
    and Adelle Pratt

13. The Evolutionary of Magnanimity: When is It Better to   183
    Give than to Receive?
    James L. Boone

Part 5: Overpopulation and Fertility Declines

14. Evolutionary Economics of Human Reproduction   201
    Alan R. Rogers

15. More Status or More Children? Social Status, Fertility   219
    Reduction, and Long-Term Fitness
    James L. Boone and Karen L. Kessler

16. The Demographic Transition: Are We Any Closer to an   237
    Evolutionary Explanation?
    Monique Borgerhoff Mulder

Part 6: Biophilia

17. Biophilia and the Conservation Ethic   249
    Edward O. Wilson

18. Human Behavioral Ecology: 140 Years without Darwin   259
    is Too Long
    Gordon H. Orians

Conclusion: Integrating the Biological and Social Sciences 281
to Address Environmental Problems
Iver Mysterud and Dustin J. Penn

References   297
Contributors   343
Author Index   347
Subject Index   357

Acknowledgements

We want to acknowledge all of the contributors to this volume and especially E. O. Wilson and Monique Borgerhoff Mulder for their help and encouragement.

We regret that Garrett Hardin passed away (14 Sept. 2003), and we would like to acknowledge that he supported and encouraged our work on this book. He played a central role in inspiring us and many others to think about human behavior, overpopulation and our environmental impact from an evolutionary perspective. We thank Mary Curtis, and Laurence Mintz and others at Transaction, and Richard Koffler, Mai Shaikhanuar-Cota, Sara Jackson at Aldine de Gruyter, our original publisher for helping to bring this Reader to fruition. We would like to acknowledge the support of the Department of Biology, University of Utah, the Konrad Lorenz Institute for Ethology, the Austrian Academy of Sciences and the Department of Biology, University of Oslo, Norway. Finally, we especially wish to thank our editorial assistants, David Wood and especially Renate Hengsberger, who worked tirelessly to help us put this reader together. We could not have done it without their help!

Foreword

Edward O. Wilson

This collective work is an attempt to overcome one of the oldest and most obdurate biases of civilization—namely the strong feeling that *Homo sapiens* has catapulted itself above and beyond nature. In this largely unexamined worldview, ten million or more species composing natural ecosystems are now outside, and we are safely—more or less—inside. The corollary is that the remaining outside exists to be brought inside and converted to civilized use.

As the authors of the following chapters show in illuminating and often disturbing detail, this traditional dichotomy of humanity-versus-nature is false and dangerous. On the one hand, it perpetuates our destructive mishandling of the biosphere. On the other hand, it scants the self-understanding that *Homo sapiens* needs to settle down on our home planet, hence as a prerequisite to survival. Nature, to put the matter as succinctly as possible, is part of us, and we are part of nature. In order to grasp the benefits of that inseverable union, we need to make it the focus of a great deal of new research, on the intersection of ecology, behavioral biology, evolutionary theory, and the social sciences. Self-interest, if nothing else, say the authors, dictates that we move toward a consilience of these subjects with all deliberate speed.

The human impact on the natural world is obvious over most of the planet's surface. The global population has exploded from under 2 billion to over 6 billion in less than a century. The rate of growth is now slowing. The global fertility rate has dropped from 4.3 children per woman in 1960 to 2.6 children per woman at the present time. The breakpoint, at which zero population growth will be attained is 2.1 and is likely to be reached within several decades. In 1999 the United Nations Population Division of the United Nations Department of Economic and Social Affairs estimated that if present demographic trends continue, the global fertility rate will reach breakpoint and then drop below it sometime late in the century. The total population will peak soon afterward at 8-10 billion, then commence a slow decline. Meanwhile, per-capita production and consumption are also rising worldwide, with no peak yet in sight. The global ecological footprint, the amount of productive land required for food, water, and all services for each person on average, is about 2.1 hectares. In the United States the footprint is approximately 9.6

acres, and in the developing countries only one-tenth as much; and in almost all countries it is increasing.

So what is happening to nature as the human biomass and consumption move toward their climax? Natural ecosystems, appropriated and converted to accommodate our growth, are disappearing at an accelerating rate. The world's rainforests, for example, where most of the known plant and animal species live, are down to 55 percent of their original cover and are being clear-cut at about one percent of remaining cover each year. In most parts of the world, species are declining and vanishing due (in rank order of their impact) to habitat destruction, invasive species, pollution, and overharvesting. Even in the relatively well-protected United States, one-third of the species are now classified as vulnerable or endangered, with one percent already officially declared extinct. Some specialists on biodiversity agree that if existing levels of decline continue, half the species of plants and animals will be gone or "committed" to early extinction by the end of this century.

An extreme anthropocentrist might argue: So what? Is it such a bad thing to totally transform the planet to the service of humanity? The answer is emphatically yes, it is a bad thing—even for anthropocentrists. In 1997 a team of ecologists and economists estimated that the total services provided free by Earth's natural ecosystems is about U.S. $33 trillion, more than the gross world product (i.e., all the gross national products combined). These services include, in addition to wood, food, and other products they yield, basic ecosystems functions such as soil formation and regeneration, waste recycling, water purification, and climate regulation. To replace them with artificial substitutes is very expensive and growing more so. A prosthetic Earth, its native biosphere adulterated and largely replaced, its physical and life-bearing skin micromanaged like a spaceship, is a world humanity cannot afford.

It has been my experience, confirmed by trends in the U.S. polls, that people easily understand these simple matters of cost and benefit when the data and estimates are laid out before them. Moreover, they are strongly inclined to agree that erasing the Creation is morally wrong. It is in fact a sin, as Patriarch Bartholomew I recently declared to his 250 million Orthodox Christian followers. The defilement of Earth's biosphere must seem a permanent abomination to adherents of every religious and secular belief. It cuts humanity off from our roots, and it obscures our million-year-old genetic history.

Homo sapiens may someday choose to crowd onto spacecraft and emigrate to the Stars. But not now. For the indefinite future and possibly forever, the natural Earth that gave us birth is our only possible home. We are exquisitely adapted in body and mind to live within its razor-thin biosphere, and cosmically naive to think otherwise. Ours is a biological species designed by evolution to live in a biological world, and this one in particular. The more we learn about our true deep origins and present-day relation to the rest of life, the sooner we will be able to settle down on Earth and be at peace with it and one another.

Introduction: The Evolutionary Roots of Our Ecological Crisis

Dustin J. Penn and Iver Mysterud

A stiffer dose of biological realism appears to be in order.... The only way to make a conservation ethic work is to ground it in ultimately selfish reasoning...
An essential component of this formula is the principle that people will conserve land and species fiercely if they foresee a material gain for themselves, their kin, and their tribe.—E. O. Wilson, 1984 (131-132)

We are facing an increasing number of environmental problems, including toxic pollution, global warming, destruction of tropical forests, extinction of biological diversity, and depletion of natural resources (Wilson, 2001). Our environmental problems are undermining our planet's capacity to sustain our growing populations. The cause of our environmental problems is no mystery; they are due to human actions, namely over-populating, over-consuming, and misusing technology. Therefore, efforts to reduce our environmental impact would benefit by a better understanding of human behavior. The problem is that biological and social sciences lack the integration necessary to understand human behavior or ecology. Biologists (ecologists, conservation biologists, toxicologists) study environmental problems, but they are not usually cross-trained in the social sciences (psychology, anthropology, and economy) and do not generally address human behavior. Social scientists study humans, but they are not usually trained in ecology or other biological sciences. They are generally more interested in economic and social issues than the environment, and many fail to see the connections. Thus, integrating the biological and social sciences to address our environmental problems is one the most important challenges in science (Wilson, 1998a, b).

The main obstacle to integrating the social and biological sciences is the widely assumed but erroneous division between humans versus nature. Fortunately, this ancient dichotomy is being dismantled, especially as more social scientists are beginning to recognize the importance of the environment for humanity. Indeed, one of the most important scientific revolutions has been the integration of ecology into the social sciences. Newly emerging interdisciplinary sciences, such as ecological economics (Costanza, 1991) and environmental psychology (Gardner & Stern, 1996; Gifford, 1997), are helping to recognize our species' *ecological constraints*, and demolish the false divisions between humans and nature. However, the synthesis is incomplete.

1

Our concern is that environmental researchers have not yet integrated their assumptions about humanity with evolutionary biology (Penn, 2003; Ridley & Low, 1994; Wilson, 1998a). Many still hold romantic notions about people in non-western cultures (Rousseau's mythical noble savages), and obsolete notions that the human mind is a nonmaterial force (a ghost in the machine) or an infinitely plastic device shaped only by experience and culture (a blank slate) (Pinker, 2002). Few consider the *biological constraints* on human behavior, or how our biology enables our flexibility and potential for change (Ridley, 2003). Environmental thinkers recognize that humans are ecologically part of nature, but they often assume other aspects of the human/nature dichotomy, such as the nature/nurture and culture/biology dichotomy (Tooby, 2001; Tooby & Cosmides, 1992). Consequently, many advocate policies that are as unrealistic as Marxists' assumptions that people might one day behave just like altruistic worker bees, sacrificing for the general good (Low & Ridley, 1994; Ridley & Low, 1994). Sustainability is an admirable goal, *but our policies need to be sustainable themselves*, and therefore, we need policies that are compatible with human nature.

Fortunately, another scientific revolution has been brewing as an increasing number of biologists and social scientists are applying principles from evolutionary biology to understand human behavior (Barkow et al., 1992; Betzig, 1997; Pinker, 2002). These evolutionary sciences are providing another assault on the false human/nature dichotomy, and the artificial division between the biological and social sciences (Wilson, 1998a). These disciplines have their roots in the biological study of behavior (ethology and behavioral ecology), and especially social behavior (sociobiology) (Dawkins, 1976; Wilson, 1975). Today, there are several evolutionary approaches to studying human behavior (Mysterud, 2004), in principle complementary, but to a certain extent competing (Smith, 2000). Human *behavioral ecology* aims to understand the evolutionary functions of behavior, especially of people in traditional societies (Smith & Winterhalder, 1992b). *Human ethology* is an integrative approach, considering both the proximate and evolutionary bases for behavior, and especially species-typical (universal) behaviors (Eibl-Eibesfeldt, 1989). *Evolutionary psychology* integrates evolution and cognitive sciences to understand how the human mind and decision-making has been "designed" by natural selection for solving problems of survival and reproduction faced by our ancestors (Barnett et al., 2002; Buss, 1999). One of the most exciting developments is *gene-culture co-evolutionary theory*, which aims to understand how genes and culture influence each other over evolutionary time (Boyd & Richerson, 1985; Durham, 1991; Lumsden &Wilson, 1981; Richerson & Boyd, 1992). Taken together, these disciplines are providing revolutionary insights into many aspects of human behavior, including reproductive behavior, cooperation, conflict, language, politics, medicine, history, culture, morality, and religion (Ridley, 1996; Wright, 1994). Moreover, they provide a major step towards integrating the biological and social sciences (Pinker, 2002; Wilson, 1998a).

The findings of evolutionary research on human behavior have profound implications for the environmental sciences, and the aim of our Reader is to bring together a variety of papers that show how and why. Part 1, "Human Nature and Resource Conservation," contains general reviews that address environmental problems from different evolutionary perspectives. The first chapter by Heinen and Low (1992) was the first to our knowledge to address environmental conservation from the perspective of the human behavioral ecology (Heinen & Low, 1992). The second chapter, by Margo Wilson, Daly, and Gordon (1998), was the first to address environmental problems from the perspective of evolutionary psychology (Wilson et al., 1998). Together, these chapters provide an excellent introduction to show how evolutionary analyses can be applied to environmental problems (for another more recent comprehensive overview see Penn, 2003).

Part 2, "The Ecological Noble Savage Hypothesis," contains chapters that address the notion that our environmental problems are due to Western culture, and that our ancestors and people in indigenous societies live in harmony with nature until the corrupting influences of Western culture (White, 1967). This idea, known as the ecological noble savage hypothesis (Redford, 1991a), has been widely assumed and promoted by various environmental philosophies, such as Deep Ecology and Ecofeminism (Merchant, 1982; Sessions, 1995). However, human behavioral ecologists that have investigated this idea have failed to find any evidence that indigenous hunter-gatherers live as reverent conservationists or make special conservation efforts (Alvard, 1993; Hames, 1987b, 1991; Kay, 1994; Low, 1996; Stearman, 1994; Vickers, 1988, 1994). Moreover, traditional peoples, like Westerners, sometimes deplete their natural resources in "ostentatious displays of resource-accruing potential and success in social competition" (Wilson et al., 1998). Furthermore, an increasing number of archaeological studies indicate that as humans spread around the planet they caused massive extinctions and other ecological problems (Alroy, 2001; Anderson, 1989; Diamond, 1998; Flannery, 2001; Holdaway and Jacomb, 2000; Martin and Klein, 1984; Miller et al., 1999; Olson and James, 1984; Roberts et al., 2001; Steadman, 1995; Steadman et al., 2002). These findings should be a wake-up call for our species' potential for destroying our own environment, and undermining stability social (Diamond, 1995, 1998).

We have included one of the first studies on this important topic by Hames (1987b), a review by Alvard (1999) (Alvard, 1999), and a less direct though more encompassing study by Low (1996). To our knowledge, all other studies on this topic to date have reached similar conclusions (Krech, 1999; Ruttan & Borgerhoff Mulder, 1999; Smith & Wishnie, 2000), though see Yochim, 2001. They generally conclude that people in non-Western and traditional societies may profess a mystical reverence for nature, but their actual behavior is not necessarily consistent with what they say they believe (Krech, 1999; Low, 1996; Tuan, 1970). They argue that the low environmental impact of aboriginal peoples is probably due to their small numbers, migratory lifestyle, and lack of modern technology and markets. This does not imply that humans are incapable of conservation, or that we have nothing to learn from indigenous peoples (or that they do not deserve custody of their lands). It turns out that many popular notions about the cultural practices of non-Western and traditional peoples are exaggerated or myths (Brown, 1991; Edgerton, 1992; Freeman, 1983, 1998; Kauffman, 1995; Shermer, 1997). Therefore, we need to be more skeptical about claims about human behavior, and find out more about what people actually do before making prescriptions for change.

Part 3, "The Tragedy of the Commons," contains chapters that address the conservation of common-pool or open-access natural resources, such as fisheries, forests, grazing lands, freshwater, and clean air. In his classic paper on the topic, Garrett Hardin (1968) pointed out that when individuals are free to exploit many natural resources, such as a commons pasture, then there is no incentive to show prudent restraint (Hardin, 1968). On the contrary, each user has an incentive to exploit such resources before others do, and consequently they become over-exploited by everyone. This phenomenon is also known as the fisherman's problem (Gordon, 1954), a collective action problem (Hawkes, 1992; Olson, 1965), a social trap (Costanza, 1987; Platt, 1973), a social dilemma (Dawes, 1980, 1988), and a multi-person prisoner's dilemma (Kollock, 1998). Hardin's elegant idea helps to explain why people often abuse natural resources, and why common-pool resources are particularly vulnerable to overexploitation and pollution. It has been supported by numerous studies, and has become the central paradigm for addressing sustainability and ecological problems (Adams et al., 2003; Agrawal, 2003; Dietz et al., 2003; McCay & Acheson, 1987; Ostrom, 1990).

Hardin's paper was criticized initially because it failed to point out that communal resources are often managed successfully through cooperative management of stakeholders (at least until there is a threat of outsiders capturing the resource) (Berkes et al., 1989). In the paper that we included here, Hardin (1994) laments not entitling his paper "The tragedy of the *unmanaged* commons" to avoid this misunderstanding (Hardin, 1994). His paper was also criticized for failing to consider how individual interests can be constrained by social norms within a group, though this is inaccurate since Hardin (1968) argued that the tragedies can be averted by social pressure, or in his words, "mutual coercion, mutually agreed upon." Research on managing common-pool resources is beginning to incorporate evolutionary perspectives of human cooperation (Penn, 2003; Ridley, 1996; Ridley & Low, 1994). We included a paper by Ruttan (Ruttan, 1998) and another by Borgerhoff Mulder and Ruttan (Borgerhoff Mulder & Ruttan, 2000) that have rather thoughtful approaches investigating the use and management of natural resources in traditional societies. We also included a paper by Ostrom et al. (1999), which provides an excellent example of how perspectives from the biological and social sciences can be integrated to address the problem of managing common-pool resources.

Part 4, "The Evolution of Discounting and Conspicuous Consumption," addresses the problem of explaining why people are so ecologically short-sighted, and why people in developed countries consume so many resources. Social scientists blame our environmental problems on the human mind's short time-horizons, i.e., our tendency to discount the future and needs of future generations (Costanza, 1991). Such proximate explanations, however, are only part of the story. Evolutionary analyses are also necessary to explain *why* humans discount the future. Discounting is not a pathological feature of human behavior, nor one easily "cured" by an environmental education (though sustainainability advocates often seem to think so). It is surely an adaptation "designed" by natural selection to enable individuals to make decisions that enhance their survival and reproductive success. It is unlikely that we can eliminate discounting, but there is room for optimism. Whereas some economists argue that discounting should not be applied beyond an individual's own expected lifetime, an evolutionary perspective suggests that "one's descendants are an extension of one's self, and organisms may be expected to have evolved to act in ways that will promote their fitness both before and after their deaths" (Wilson et al., 1998: 516). In this part, Rogers (1991) explores how natural selection is expected to favor discounting rates to conserve resources for offspring. Henderson & Sutherland (1996) suggest that using discount rates based on evolutionary analyses will improve environmental policies. Margo Wilson and her colleagues (1996) show evidence that men have higher discount rates than women, which suggests that discounting rates may be shaped by sexual selection (Henderson & Sutherland, 1997; Rogers, 1991; Wilson et al., 1996) Although achieving zero discounting is unrealistic, evolutionary theory does not imply that institutional and other social changes cannot alter people's time preferences (e.g., as in using social pressure to avoid the tragedy of the commons).

One of the greatest challenges is to explain why people in developed countries consume so many resources, and more than are necessary to survive and reproduce. Contrary to the assumptions of many economists, increases in absolute levels of wealth and consumption do not make people happier; above a minimal threshold, it is relative rather than absolute wealth that matters (de Graaf et al., 2001; Durning, 1992; Frank, 1999). This makes sense from an evolutionary perspective: people's concern for social status and other "positional goods" become more intense once their immediate survival concerns are satisfied (Hirsch, 1978). People appear to be on a runaway consumption treadmill, always trying to keep up with the Joneses, and with the constant exposure to extraordinarily wealthy families on television, both actual and virtual, they may be trying to keep up with the Gateses (de Graaf et al., 2001; Durning, 1992; Frank, 1999). Mass media appears to be designed to exploit our instinctive concerns for keeping up in this consumption race (Penn, 2003), and recent evidence indicates that we are being manipulated (e.g., watching television is addictive, and watching television is linked to developing attention disorders [Christakis el al., 2004]). To explain why people engage in extravagant displays of resources and wealth (and why they pay so much attention to the Joneses), Veblen (1899) (Veblen, 1899/1981) suggested that "conspicuous consumption" functions as an honest (cheat-proof) competitive strategy to demonstrate wealth and social status. His hypothesis is a special case of the handicap principle, which Zahavi suggested to explain the evolution of extravagant displays in nature (Zahavi, 1975; Zahavi & Zahavi, 1997). Evolutionary researchers have begun to combine Veblen's and Zahavi's ideas to understand conspicuous consumption (Bliege Bird et al., 2001; Boone, 1998; Frank, 1999; Miller, 1999; Neiman, 1997; Penn, 2003; Smith and Bird, 2000; Sosis, 2000). We have included a paper by Boone (1998), which was one of the first to take this approach.

Part 5, "Overpopulation and Fertility Declines," contains chapters that address the evolution of human reproductive decisions. Efforts to stabilize population growth would benefit by a better understanding of why family size has been declining in some but not other societies. Demographers have long thought that fertility declines were due to lower child mortality that accompanies economic development (demographic transition). However, this idea was based on the erroneous assumption that natural selection favors altruistic population-regulation to benefit the species (Bates & Lees, 1979; Hawks & Charnov, 1988; Turke, 1989). Recently, evolutionary researchers have been trying to understand why people in developed countries are having few children, and other aspects of reproductive decisions (evolutionary demography) (Bock, 2000; Borgerhoff Mulder, 1998; Kaplan & Lancaster, 2000a; Low, 1993; Mace, 2000a; Rose, 1997; Turke & Betzig, 1985; Voland, 1998). They point out that natural selection favors optimizing rather than maximizing the number of offspring due to life history tradeoffs between the parent survival and reproduction (Strassmann & Gillespie, 2002), and quantity and quality of offspring (Borgerhoff Mulder, 2000). Therefore, reducing fertility is not necessarily maladaptive if this enables parents to invest more resources into their progeny to enhance their fitness (Rogers, 1990b). In this part, we have included papers by Rogers (1990b), Boone (1999), and a review by Borgerhoff Mulder (1998) that explore how these life history tradeoffs might help to explain human fertility patterns (Boone & Kessler, 1999; Borgerhoff Mulder, 1998; Rogers, 1990b).

Of course, low fertility may be a maladaptive response to life in the modern world, but this would still require an evolutionary explanation. Low fertility and childlessness may be due to exposure to modern chemical pollutants (Jensen et al., 2002) or modern mass media designed to manipulate us into pursuing ever more consumer goods (Penn, 2003). Reproductive decisions are not based on a single factor, and several factors are associated with fertility declines, including women's empowerment (Penn, 1999), and population density (Lutz and Qiang, 2002), though these have only begun to be explored from an evolutionary perspective. The challenge is to find incentives that help stabilize population growth, and yet do not require counter-productive increases in consumption.

Part 6, "Biophilia," contains chapters that aim to understand the evolution of environmental aesthetics, which should help to explain why people cherish nature as well as destroy it. E. O. Wilson's (1984; 1993) biophilia hypothesis suggests that humans have an instinctive preference for green spaces and natural landscapes, just as other species have evolved adaptive habitat preferences. This hypothesis has some empirical support (see Gardner & Stern, 1996; Kellert & Wilson, 1993). For example, people are attracted to images of habitats that provide food and protection (Kaplan, 1987, 1992), and savannah-like landscapes, similar to where our ancestors evolved in Africa (Gardner & Stern, 1996; Orians, 1986; Orians & Heerwagen, 1992). The most important implication of the biophilia hypothesis is that people obtain emotional and other psychological benefits from preserving natural habitats (Kellert & Wilson, 1993). This seems quite likely given our species' amazing fondness for plants, pets, zoos, parks, and gardens, though research on the evolution of environmental aesthetics is only beginning. We have included papers on this topic by E. O. Wilson (1993) and Orians (1998) that will help interested readers get started (Orians, 1998; Wilson, 1993).

In the concluding chapter, we discuss how increasing the integration between the biological and social sciences is improving our understanding of how we are altering our environment–and how these changes also affect us. The rapid environmental changes affect the economy (Costanza, 1991), social stability (Homer-Dixon, 1999), and human health (Colborn et al., 1996). The recent rise in some health problems, such as allergies, obesity, and heart disease, for example, appear to be a consequence of our not being evolutionarily adapted to the modern world (Nesse & Williams, 1994; Trevathan, 1999). Research that applies evolutionary analyses to medical problems, or "Darwinian medicine," has much in common with the approaches in this Reader. Therefore, evolutionary analyses to address environmental problems might be called "Darwinian ecology" (Penn, 2003). The integration of evolutionary analyses into the social sciences is also improving our understanding of human cooperation (Hauert et al., 2002; Milinski et al., 2001; Ridley, 1996; Vogel, 2004), which has particular relevance for efforts to address our environmental problems (Penn, 2003).

We hope that bringing the research of these authors together into a single volume will help to increase an interest in the evolution of human behavior. Evolutionary analyses are unlikely to provide any cures for our environmental problems, and, in fact, they often only help us to see why it is difficult to achieve our goals. Still, knowing our species' limitations is crucial for designing *sustainable* environmental and population policies. Our planet is too small to continue our current rates of population growth and consumption. As resources become scarce, due to unequal distribution as well as over-exploitation, we are likely to see increased social unrest and violent conflicts (Homer-Dixon, 1999). The growing disparity of resources in rich versus poor countries is a sure recipe for disaster. As E. O. Wilson (2001a: 150) points out, "The income gap is the setting for resentment and fanaticism that causes even the strongest nations, led by the American colossus, to conduct their affairs with an uneasy conscience and a growing fear of heaven-bound suicide bombers" (Wilson, 2001). We can not just help to increase the standard of living of poor countries, without reducing ours, because we would need two more Earths to bring the rest of the world up to our level of consumption of the developed world (Wilson, 2001). Addressing our environmental problems is complex and difficult work, although bringing evolutionary perspectives into the debates will help to confront and resolve differences among environmental thinkers, who often have pre-Darwinian worldviews. This is important because holding an incorrect worldview about humans and our place in nature often blocks our ability to solve our problems effectively (Brown, 2001).

1

Human Behavioural Ecology and Environmental Conservation

Joel T. Heinen and Roberta ("Bobbi") S. Low

Introduction

Today we face global environmental problems of unprecedented—and often unforeseen—proportions. Much apparent human *"progress"* may be an illusion (Brown, 1990, 1991) because such high prices accompany it: examples include deforestation (*e.g.*, Hecht & Cockburn, 1990), desertification (Chiras, 1988; Revelle & Revelle, 1988), pollution and other problems of intensified agriculture (Revelle & Revelle, 1988), massive extinctions (N. Myers, 1984; Wilson, 1988), and growing economic disparities (*e.g.*, Smith, 1982).

Elsewhere (Low & Heinen, 1993), we are arguing that the *"conventional wisdoms"* on which we base our strategies for conservation may be wrong, and that if they are indeed wrong, our strategies will not work. These *"conventional"* wisdoms include our perception that people in pre-industrial (*"traditional"*) societies, being more directly and immediately dependent on the ecology of the natural systems around them, were deliberately more conserving and respectful of those resources (*e.g.*, Bodley, 1990). "Conventional wisdoms" suggest that, as we have developed technological insulation against ecological fluctuations, we have, in important ways, *"lost touch"* with ecological realities and constraints, and have, at least to some degree, lost our respect for them. Another *"conventional wisdom"* is that all of us, as ethical individuals, pay deference to the common good; thus, because none of us wishes to cause destruction of resources, each of us will accept some level of personal cost to ensure the common good.

Heinen, Joel T. and Low, Roberta ("Bobbi") S.: Human Behavioural Ecology and Environmental Conservation: Environmental Conservation, Vol. 19, No. 2, Summer 1992. Reprinted with the permission of Cambridge University Press.

As we have noted elsewhere (Low & Heinen, 1993), if these conventional wisdoms were true, ecological information would be sufficient to solve ecological problems. Yet, by the 1970s, we had more than enough information to attack problems that nevertheless remain, and even grow worse, today. What is wrong? We can see that our current efforts are discouragingly minimal at least in their effects. If we are correct, we need a new paradigm or pattern of behaviour for analysing these problems. The examples given above, for all their variety, have in common an underlying theme: *conflicts of interests over resources*. This theme is central in the field of behavioural ecology, which examines how environmental conditions shape organisms' behaviour and life-history (*e.g.,* Cronk, 1991; Krebs & Davies, 1991). Behavioural ecological theory generates testable predictions about resource-use patterns. We argue that it can also suggest specific strategies to promote wise resource-use.

The behavioural ecological approach argues that humans, in the manner of other living organisms, evolved to get resources in order to survive and reproduce, and that the well-being of the individual, of his/her family, and of his/her social group (comprising those people with whom the individual most regularly interacts) takes precedence, while the good of the larger group (the population or species) has never been relevant. *We argue that natural selection has shaped all living organisms to exploit resources effectively, in competition with each other, and that our human problem is that, through our cleverness, we have created a novel evolutionary circumstance—we now have such technology that the very behaviours which we evolved to do well, outpacing others, are those most likely to ruin us* (Low & Heinen, 1993). Using this new paradigm, we can integrate information from other fields (economics, psychology, etc.), and suggest new strategies, showing which strategies are most likely to be effective under particular conditions.

A Behavioural Ecological Approach

The evolution and ecology of resource-use in other species are well-studied (*e.g.*, Daly & Wilson, 1983; Alcock, 1984; Trivers, 1985; Dawkins, 1986, 1989; Krebs & Davies, 1991 and specific studies cited therein). Resource acquisition has reproductive costs and benefits; is influenced by environmental extremeness, predictability, and patchiness, and can show sexual dimorphism. This proposition, stemming from Charles Darwin's (1859, 1871) explicit statement to that last effect, seems simple—but can give rise to complex and profound results. The important point to variation is its heritability; however, in social animals, there can be significant cultural heritability (Cavalli-Sforza & Feldman, 1981; Lumsden & Wilson, 1981; Boyd & Richerson, 1985; Dawkins, 1986).

All living things have evolved to acquire and use resources to survive and reproduce, though in ways often constrained by ecological conditions. The most effective resource strategies in any particular environment should become the most common—involving, for example, those individuals using efficient strategies leaving more offspring in the next generation than their competitors. In this paradigm, there are two central concepts: (1) genetically selfish behaviours, *i.e.* those which enhance an individuals' reproduction, tend to be favoured by natural selection; and (2) successful reproductive strategies vary ecologically with the richness, controllability, and predictability, of important resources.

Successful reproduction, however, does not necessarily mean producing the most offspring; it's never that simple. Producing the maximum number of offspring in any reproductive bout seldom, in fact, leads to maximum lifetime reproduction. There are many examples of seemingly destructive behaviours which actually increase lifetime reproductive success: for example, infanticide, lethal conflict, delayed reproduction, and sterility (*e.g.*, non-reproductive helpers at the nest, *cf.* Woolfenden & Fitzpatrick, 1984; sterile honeybee workers, *cf.* Seeley, 1985; and infanticide, *cf.* Hausfater and Hrdy, 1984), are all phenomena that one would think would decrease, rather than increase, reproductive success. Detailed analyses have shown that these behaviours are found in specific ecological and social circumstances and that their impact is an increased net reproductive success in those environments.

Each individual has reproductive interests, but these interests are shared by other individuals which share common genes—genes identical by descent—and thus several avenues are open to enhance net reproductive success (inclusive fitness maximization, or kin selection, *cf.* Hamilton, 1964). We expect organisms, including humans, to engage in activities that benefit relatives if they can recognize and interact with them; the extent to which this is true will tend to depend on the degree of relatedness. Thus helping relatives, even at some cost to oneself, can be genetically profitable.

Human intelligence probably evolved in the context of resource- and mate-competition (Alexander, 1971, 1979, 1987; Humphrey, 1983). Cooperation (Trivers, 1971) can be a highly effective competitive (*i.e.* reproductive) strategy. Reciprocity occurs only in long-lived, social species-species in which individuals recognize each other and are likely to interact repeatedly. Some examples of animals, other than humans, which actively reciprocate and form coalitions, are Chimpanzees (*Pan troglodytes*) (Goodall, 1986), Lions (*Panthera leo*) (Packer & Pusey, 1982), and African Elephants (*Loxodonta africana*) (Moss, 1988). If individuals interact only rarely or occasionally, indirect reciprocity is unlikely: individuals will mirror the behaviour of others in a "tit-for-tat" manner (reciprocating until the other actor defaults; *see* Axelrod and Hamilton, 1981). When risks are high, helping behaviours are likely to occur only or primarily among kin. However, we expect organisms in long-lived, social species (including humans), to do things which benefit potential reciprocators without immediate profit, depending on the probability that there will be future interactions between/among the individuals concerned.

The relevance of these patterns to resource problems should be immediately obvious. There are clear patterns to what we do (Table 1.1). Many behaviours that we call "altruistic" in fact have evolved because they were likely to benefit the inclusive fitness of those who followed them; such behaviours are genotypically "selfish". Examples include parental and nepostic behaviours as well as friendship. "Genotypically altruistic" behaviours, *i.e.*, those which benefit reproductive competitors at a cost to the doer (examples are rare, *see* Table 1.1), cannot evolve through natural selection, and while

Table 1.1

Categories of Behaviours, and their Phenotypic and Genotypic Effects. Many behaviours that we call "altruistic" are in fact genetically profitable, as we can see when we separate behaviours according to their phenotypic (apparent) effect, *versus* their genotypic effect (effect on genetic lineage). Behaviours above the box can evolve by natural selection (genotypically selfish), and those below the box cannot (genotypically altruistic). We thus expect the latter types of behaviours to be rare (adapted from Alexander, 1974). *See text* for explanation.

PHENOTYPIC EFFECT?	GENOTYPIC EFFECT?	EXAMPLES
PROFITABLE ("selfish")	PROFITABLE ("selfish")	numerous; all overt competition
COSTLY ("altruistic")	PROFITABLE ("selfish")	parental car, nepotism, reciprocity
		BECAUSE NATURAL SELECTION FAVOURS ONLY GENO- TYPICALLY PROFITABLE BEHAVIOURS. UNDER NATURAL SELECTION, BEHAVIOURS ABOVE THE LINE SHOULD BE COMMON BELOW THE LINE, RARE
PROFITABLE ("selfish")	COSTLY ("altruistic")	? A rich miser who disinherits his family, leaving an anonymous gift to a home for unwed mothers
COSTLY ("altruistic")	COSTLY ("altruistic")	Mother Theresa

they may occur, are always vulnerable to invasion from genotypically self-ish (lineage-enhancing) behaviours (Alexander, 1974). Accordingly, over time, genetic altruists will comprise an ever-decreasing proportion of the population (*e.g.*, the last Shaker died a few years ago). Only if selection worked to favour the group, rather than genetic lineages, would genetic altruism be common.

In behavioural ecology, this is called the "levels of selection" problem. Considering this evolutionary background, it is not surprising that certain ecological conditions change the costs and benefits of helping. Minnis (1985 p. 38), for example, noted that as food stress increases, sharing of food in human groups increases to a point, and then decreases, at least in pre-industrial societies. In really extreme cases, dependent kin (children) can be abandoned (*e.g.*, Turnbull, 1972; Boswell, 1988). Even when resources are not scarce, sharing tends to be directed towards kin, and towards individuals from whom the giver might receive benefits (Chagnon, 1982; Turke & Betzig, 1986; Hill & Kaplan, 1988).

Levels of Selection and Conservation Issues

Typically, when many, unrelated, individuals simultaneously have common access to resources, whether the resources are grazing lands or whale populations, they tend to exploit the resource more than is wise from a long-term sustainability perspective, in order to gain individually. As we have noted (Low & Heinen, 1993) this "commons" problem (Hardin, 1968) is probably the most frequent "levels of selection" problem seen by conservationists.

Why are these problems common? If selection maximized the interests of the group, they should not be. Yet many find it tempting to think of selection as acting "for" groups, without specifying relationships among individuals, or the impact of behaviours on individual inclusive fitness. Perhaps this is because kin selection involves individuals other than direct lineal descendants, and reciprocity may involve completely unrelated individuals—thus, individual selection results in groups that look "fit". This approach is flawed: natural selection does not favour individuals which give up inclusive fitness for group good.

Yet this does not mean that we are simply harsh and fierce competitors, with no redeeming moral features. First, helping our family and friends, favoured by individual-level selection, is common and favoured by natural selection (phenotypically altruistic but genotypically selfish behaviour, *see* Table 1.1). Sometimes, too, the group may appear to benefit, incidentally, as a result of the cumulative selection of individuals (*e.g.*, Williams, 1966). Interestingly, Wynne-Edwards (1962) argued that all species except humans were groupselected, because human populations seemed not to be "regulated". Even then, the conflicts between individual profit and group good were already clear. In fact, humans alone may show any evidence of group selection at all. Laws, for example, are inflictions of constraint on individual behaviour by coalitions of others in the group (Alexander, 1987). There is little evidence, however, that we can convince individuals to change their behaviour solely because potential group benefits will follow.

This phenomenon is central to understanding human resource-use. If we suggest, as in the environmental movement of the 1970s, that everyone should pay an immediate, relatively small cost (taking shorter and fewer showers, recycling extensively, etc.), in the interests of gaining long-term global benefits which will be shared with non-relatives and competitors, we are asking for behaviours that have no evolutionary precedent. When we ask people to do things that cost them individually, with no benefit in the short term, and no matter how sensible they may be, we see defection; it hasn't worked as a widespread strategy in the past, and we see no convincing evidence that it will in the future; we review several examples elsewhere (Low & Heinen, 1993).

Novel Evolutionary Environments

Individuals which have more resources than others, typically have greater reproductive success (Low, 1989b, 1989c, 1990a, 1990b). In modern societies, however, humans may have broken the link between resource accumulation and inclusive fitness (Low & Heinen, 1993). In the manner of other organisms, we humans apparently evolved to strive for resources, using them for ourselves and our families. We also typically derive proximate rewards of satisfaction and pleasure from success in that struggle.

When environmental conditions change, previously advantageous behaviour can continue to be driven by proximate cues (that, in the past, correlated with reproductive advantage), even when the proximate cues are unhinged from the (past) functional advantage. This is most common with environmental changes that represent evolutionarily novel events; we have reviewed several examples (Low & Heinen, 1993). In fact, the subjective assessment of "generalist" life-history traits as being somehow superior is, in fact, related to the fact that humans can alter other species' habitats rapidly; most animals which have become rare or endangered are ecological specialists of which the habitat was altered by human action more quickly than the species could adapt to the alteration by natural selection (*e.g.*, Schaller et al., 1985; *see also* Ehrlich & Ehrlich, 1981).

Because no organism, including humans, has evolved to be aware of ultimate costs and benefits, but only of proximate rewards or punishments, we have evolved to find pleasurable those things which enhance our survivorship or reproduction. Novelty complicates this process (*e.g.*, sugar, *cf.* Low & Heinen, 1993). There is evidence that the human eye grows to compensate for blurred vision in children and adolescents, and this evolved mechanism, which tends to correct visual defects as a child grows, can overcompensate in children who read a great deal (an evolutionarily novel event); the result is a high incidence of near-sightedness in such children.

We argue that humans are in an evolutionarily novel situation with regard to the use and conservation of resources. Humans have evolved to strive to acquire, use, and control, resources, and to derive pleasure and pride from these processes; furthermore, we are able to do this at unprecedented rates and to unprecedented extents, due to technological advances (evolutionarily novel events). Due to these novel events and an evolutionary history, we thus have the ability and proclivity to cause environmental catastrophe; due to the same evolutionary history, we have far less ability or proclivity to correct the situation.

Natural Selection and Human Resource-Use

Our goal, then, is to design conservation programmes which are in accordance, and not in conflict, with our history as a long-lived, social mammal. If human resource-use had followed the same behavioural ecological rules as that of other species, very different predictions would have emerged about how to influence future patterns of resource use, compared with our "conventional wisdoms" (Low & Heinen, 1993): people are unlikely to give up short-term individual or familial benefits for long-term societal or global gains. This is likely to have been true throughout our history as a species, rather than constituting a new phenomenon associated with technological innovation.

We humans, in the manner of allied species, evolved to be aware only of proximate rewards and punishments, not of ultimate costs and benefits. We have evolved to maximize short-term rewards that, in our history, correlated with reproductive success. In new and future evolutionary environments, these strategies may be harmful, not helpful, to humans as a whole. If we could simply set aside our evolutionary past and use logic, perhaps we could easily act as if the Earth were our family, regardless of personal rewards. But that seems to be difficult (*see* Hawkes & Charnov, 1988), and today in developed countries we have enormous technology for resource exploitation, ability to export costs, and high proximate rewards of status and power for exploitation.

This gives us an ability to cause massive extinctions, deforestation, global warming, desertification, and many other derogations. Even the crucial problem of overpopulation (*e.g.*, Ehrlich & Ehrlich, 1990) arises in part from us as individuals satisfying our (evolved) proximate desires to have children. We literally can destroy the Earth to satisfy our proximate goals. Appeals to people to make relatively small sacrifices for the ultimate good of all people everywhere have had only limited success; we have not yet evolved sufficiently to consider the global population as our family. This is the situation which Boulding (1977) discussed when he spoke of the problems of building a "global public"—a large group willing to bear individual costs for a common good.

We humans don't often think of ourselves as having evolved in response to pressures on our reproductive success, yet elsewhere (Low & Heinen, 1993) and here we argue that much evidence supports this thesis. A behavioural ecological approach makes several predictions about resource use, fertility, mortality, male-female differences, and about how these are predicted to differ in various environments. If humans, like other organisms, did evolve to garner resources that enhance their reproductive interests, then these general patterns lead to specific predictions about human resource-use and reproductive patterns (*idem*). We accordingly tested five hypotheses that seemed to us to have central implications for effective environmental conservation.

If enhancement of individual and familial survival and reproduction is the trait favoured by natural selection, we argued (*idem*) that, in traditional societies, resources should be most conservatively used when there is rapid and clear feedback regarding the impact on family and individual welfare (when overexploitation carries clear individual and familial costs). A review of existing data indeed suggested that individuals use what they can find out about the effects of their behaviour on the resource, to modulate the level of use.

We also argued (*idem*) that deliberate overexploitation in traditional societies is likely when (a) it yields individual genetic profit (and/or its correlates of wealth, power, and social status), and (b) technology is sufficient to accomplish overexploitation. Thus the impact of introduction of more efficient technologies will vary, depending on whether their use will result in greater (short-term) individual and familial benefit. A review of traditional societies indeed suggested that often the technology has been more-than-sufficient to devastate resources, but only when it became profitable to exploit (*e.g.*, the introduction of markets), did individuals become exploitative. Other examples suggested that, when payoffs existed but technology was inefficient, we saw behaviour which we incorrectly attributed to conservative philosophy.

Indirect or incidental damage (*e.g.*, through habitat destruction) is most likely when information or feedback about the resource is limited (*idem*). Even when we move from considering traditional societies to modern industrial conditions, those resources that are most likely to be overexploited should be those with slow feedback cycles—those on which it is hard to see the impact of resource-use. This is especially critical for resources in which the size of the resource pool is difficult to measure and there are numerous alternative potential users: examples are many non-renewable resources such as coal, oil, natural gas, and Pleistocene water-deposits, and renewable resources such as whales and large terrestrial mammals (*e.g.*, many fur-bearers). If the resource is immediately profitable and used by many, non-related, non-interacting individuals, the most profitable individual strategy is rapid exploitation. If property rights and market economies allow, some individuals may be able to follow a monopoly strategy, gaining control over the resource, and then regulating prices to maximize profit.

Because of the different shape of mating *versus* parental return curves (Low, 1990a; also Clutton-Brock, 1988 and entries therein), male and female humans, in the manner of other male and female mammals, differ in the amount of resources which they strive to acquire, and in the risks they are willing to take. Because women's reproductive value (age-related) is important in mate selection, while for men, resource control is likely to be more important than age (Low, 1989a, 1989b, 1990a, 1990b), women's reproduction correlates poorly with wealth, power, or status. Even in cultural milieux in which marriage ages were late, women who married earlier had, simply, higher lifetime fertility than women who married later. Men's fertility, on the other hand, tends to be related to their wealth and/or status, not to their age of marriage. Thus in most societies, men can benefit more directly from resource exploitation, and are likely to be the driving force in risky, high-stakes resource garnering, and hence in overexploiting resources.

If enhancement of individual and familial survival and reproduction is the evolutionary context of resource striving, then our strategies for convincing people to shift resource-use to more conserving patterns must appeal to their perceived short-term, familial and local, interests (Low & Heinen, 1993). No other species has evolved to behave in the long-term interests of the larger group (unless that group comprises only close relatives), at the expense of short-term individual and familial interests, and we propose that this is also true of humans.

Our other review (*idem*) accordingly suggested that humans have evolved to use resources in reproductively selfish ways, and thus, if benefits of conservation can be made to outweigh costs for people—for example through a system of economic or other incentives which confer immediate or very short-term benefits on individuals and/or their families and potential reciprocators—then effective conservation strategies are likely to persist and spread. If this is the case, governments and organizations may find it productive to implement policies that create systems of incentives to conserve: *the more immediate the benefit, the more successful should be the outcome.*

Discussion

If "conventional wisdoms" were true, resource abuse would be likely to increase with the degree of technology, however measured; and information about the effects of our actions would be sufficient to solve ecological problems, as individuals would accept costs for the common good. Neither of the predictions from "conventional wisdom" have been supported (Low & Heinen, 1993): resource abuse sometimes accompanies technological innovation, but only if individuals can benefit in the short term. When shifts in cultural values occur, they appear to be secondary. The principal alternative to our "conventional wisdom", the behavioural ecological paradigm, suggests that our complex human intelligence evolved in cooperating social groups in competition with other human groups, and has always been directed towards our own familial and reciprocal short-term benefit—even when we could see clearly that long-term societal, or even global, detriment would result. How, then, can we foster strong normative conservation ethics, if we wish to do so?

Williams (1989) referred to evolution as *"immoral"* (we would prefer *"amoral"*) because it fosters the traits which we have described. He makes an important point: we cannot overlook the fact that selection *uniformly* favours genotypically selfish behaviour if we have any interest in understanding the process and in manipulating, for any purpose, strategies that are influenced by this process. We contend that this process had had a seminal influence on all aspects of human behaviour, including those of import relating to conservation issues. Yet human societies are complex, and differ in their ethnic and religious heterogeneity, degree of sociality and mobility, and degree of relatedness of individuals. All of these factors modify predictions which one might make about the relative usefulness of various conservation strategies; for example, while we predict that, the shorter-term, more individual and local, the reward is, the greater will be the likelihood of success (Low & Heinen, 1993), nonetheless, if a society is small and comprises non-mobile, similar individuals, longer-term strategies may well be successful.

We suggest that behavioural ecology predicts a pattern to the sorts of strategies that are most likely to be effective in different situations (*see* Figure 1.1 for a graphical representation). For example, problems such as local recycling efforts (which have some relatively immediate and personal costs and benefits) would be more easily tackled with educational efforts and appeals to help the community by *"doing the right thing"* than, for example, drought in Ethiopia or the price of gasoline (*cf.* Figure 1.1). Gasoline (namely *"gas"* or *"petrol"*) is currently much more costly (from three to six times) in Europe compared with the USA; as one would expect, use *per caput* in Europe is much lower, as a consequence of price, distances between living and working places, and availability of alternative transportation. In the US, we are certainly aware of the problem—we have all the information we need—but it is obviously politically difficult to pass new taxes in the US (for precisely the reasons we have outlined). Cowhey's (1985) analysis of energy policy is a sterling example of *"crisis"* policy management in the face of conflicts of interest at a number of levels: government, individual politician, company CEOs (Corporate Executive Officers), and consumers. For the reasons which Cowhey gives in more detail than we can here, it would be difficult to raise gasoline taxes even though it is clear that, if this were done, incentives for use of alternative energy-sources and alternative (mass) transit would be created.

The difficulty in this case would be to convince voters and Members of Congress from the industrialized states to pass such a tax, owing to the immediate costs of taxation to consumers and to the automobile industry; the immediate individual costs to a politician promoting tax increases are seldom seen

Figure 1.1

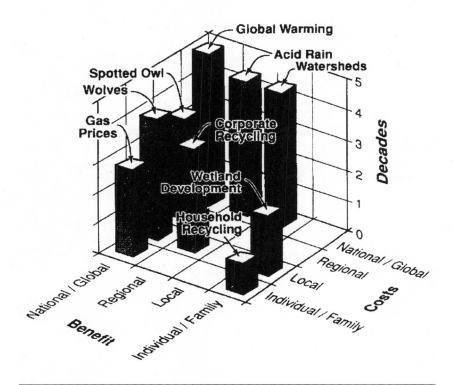

The degree of difficulty in solving environmental problems will differ with a number of factors, most of which are related to scale and discounting issues. The easiest problems are those in which the costs are paid, and the benefits derived, by the same individuals, and the benefits come quickly (*e.g.* household recycling); in such cases, simply having the correct information may lead to a solution. When costs or benefits accrue at higher-than-individual levels, there is a potential for conflicts of interest within a level (*e.g.* gasoline prices: we all agree that something must be done, but we loathe to inflict taxes). When costs are incurred at one level, but benefits are derived at another level, resolution is also difficult; for example, in the conservation of the Spotted Owl, preservation of old-growth forest might lead to regional economic depression, but have the national/international benefit of saving an endangered species. In such cases, actors are likely to discount the costs or benefits to others. Finally, the longer the elapsed time between benefits and costs, the greater the likelihood of discounting. The most difficult cases are those such as global warming: the costs are incurred from the individual to the national levels (with great potential for conflicts of interests), and the costs must be paid now for potential benefits in, say, 20 to 50 years. In such cases, anyone asked to pay a cost now is predicted to discount the value of information, the benefits to be derived, etc. In sum, the farther any case is from the "individual costs, individual benefits" corner, and the longer the time-scale, the more difficult it will be to solve.

as counterbalanced by the long-term societal gain (*cf.* Table 1.1). We suggest that smaller-scale economies, with less-heterogeneous populations, may be more effective at the sort of coalition efforts required, for example, to raise the price of petrol. As Daly & Cobb (1989) pointed out, the rational behaviour of an individual, given the current incentive system, is not necessarily a rational policy for society as a whole.

With these complex problems, the question then becomes: how to change incentives, such that individuals are likely to behave in ways which benefit society? As we note elsewhere (Low & Heinen, 1993), the most difficult resource-use problems have one or more of the following characteristics: inadequately-known resource-base: slow feedback; externalized costs; and use by many, unrelated non-interacting, individuals (*e.g.*, global warming, *cf.* Figure 1.1). Even in such extreme cases, if appropriate incentives could be provided, we would expect use-patterns to change. Non-monetary incentives can be effective, especially if they play on our evolved psychological mechanisms. Here, then, we suggest some patterns for major, current strategies: defining problems at an effective level ("scale problems"), information, social incentives, economic incentives, coalition-building to force compliance, or government action. Our review of theory suggests that information and education alone will be insufficient to solve many resourcerelated problems, though they may be instrumental in solving some.

Some Characteristics of Resource-related Problems

There are certain characteristics which appear to be common to any resource-related problem: inherent costs and benefits of any solution, and temporal and spatial discounting of the problem. We contend that humans are likely to discount some kinds of environmental problems in additional ways: "sensory" and "species" discounting. Thus the farther away a problem is, or the more distant in the future, the more we discount it. This is a "scale" issue. We contend that scale issues in one form or another will determine the nature of appropriate solutions, given the way humans have evolved to respond to environmental circumstances (i.e. extremes and unpredictability). Because humans evolved in relatively small groups of reciprocating individuals, to varying degrees genetically related to other group members, they are likely to respond most immediately to relatively small-scale, short-term problems which affect the individual and his/her family and friends directly. Notice that, if humans were in fact "group selected" (*sensu* Wynne-Edwards, 1962), this would not be true; we would then be able to convince people to do things to their own short-term disadvantage, for the long-term good of humanity.

Defining manageable portions of any problem to be tackled seems to be an important strategic device (*e.g.*, Weick, 1984), and one that is related to our evolutionary past in small groups with shared interests (as we have argued, shared interests are the key). It appears too overwhelming for the human psyche to think of major problems, such as are listed in our introduction. Weick (1984) noted that such litanies of problems may preclude any solution, for people may become too overwhelmed to take action. Instead, he argues, building a series of "small wins" creates a sense of control, reduces frustration and anxiety, and fosters enthusiasm. Small, specific, and visible, objectives are most obtainable, generating the energy to attack another small goal. The sooner the visible reward is in coming, the more effective will be the strategy. One example which Weick gives is that of William Ruckelshaus, who announced, on the first day of his tenure as head of the US Environmental Protection Agency, that the Agency was pursuing five major lawsuits against large American cities. Other sources of pollution were more visible; other sources of pollution were perhaps even more widespread and damaging. But Ruckelshaus's move was dramatic, well-researched, well-orchestrated—and winnable: it established a foundation for success and a sense of effectiveness.

There are inherent costs and benefits to any environmental problem. At what scale these costs and benefits are manifested will, we contend, in part determine potential solutions. For example, the economic benefits of maintaining national parks and protected areas are frequently estimable at the national level (*e.g.*, Dixon and Sherman, 1990): they may include increased foreign exchange and general revenues for a variety of service industries (airlines, hotels, restaurants, etc.). Nature tourism is now a major industry world-wide, and the major industry for some developing nations (Richter, 1989; Whelan, 1991). However, some of the costs of maintaining such areas are frequently very localized, and include limited access to formerly unrestricted resources, crop damage, livestock depredation, and human deaths from wildlife (*e.g.*, Heinen & Kattel, 1992*a*). It is now widely accepted that management strategies for protected areas must be altered—to allow local people greater access to resources, and allow them to benefit more directly than hitherto from the tourism industry (*e.g.*, MacKinnon *et al.,* 1985; Hough & Sherpa, 1989; McNeely, 1988).

We contend that this is so because humans have evolved to be concerned primarily with short-term costs and benefits on a restricted temporal and spatial scale. Any long-term, national benefits are secondary, and programmes which rely on them (in the manner of many national conservation strategies) are therefore not likely to be stable. Heinen (1993), for example, found that attitudes of residents around one Nepalese protected area were generally poor, despite some measurable economic benefits of adoption, and that respondents overestimated their costs and underestimated their benefits from the reserve, as judged by the results of corroborative surveys. Nepalese conservation legislation has shifted in recent years towards allowing more local uses and benefits to local people, and allowing more participation in management actions by local people, though conflicts are still common (Heinen & Kattel, 1992*b*).

The World Conservation Union (IUCN, 1990) recognizes ten types of protected areas; we predict that protected areas (all of which have inherent local

costs) can be maintained into the distant future only if their management programmes include adequate local benefits to compensate for these costs. Benefits
may include providing employment and training for residents, compensating for
crop damage and livestock depredation, and allowing some consumptive uses
of forest products. Similarly, some endangered species recovery programmes in
the USA are locally costly but nationally beneficial (because most Americans
want them, but local residents suffer some costs). Recovery programmes
for large mammalian predators such as Wolves (*Canis lupus*) (*cf.* Figure 1.1;
Dunlap, 1988) and Grizzly Bears (*Ursus arctos horribilis*) (Glick *et al.*,
1991), and for the Spotted Owl (*cf.* Figure 1.1) in the Pacific Northwest
(Liverman, 1990), come to mind. It is interesting to note that the recovery
plan for the Kirkland's Warbler (*Dendroica kirklandii*), a highly-endangered
bird which nests in early-successional pine in north-central Michigan, has
never been controversial; we suspect that this is because the programme
includes regular cutting and sale of the pine (Byelich *et al.*, 1985), and hence
is in accordance with local, short-term economic interests—unlike many other
endangered species recovery plans.

Inherent in problems of scale with regard to environmental issues and solutions, is the ability of humans to discount. We contend that this ability is an
evolved response to living with environmental unpredictability (*e.g.*, Colwell,
1974; Low, 1978). The longer or farther something is away from our immediate reproductive interests, the less it will pay us to invest time, money, and/ or
energy, in its consideration. Discounting theory, widely applied in economics
(Gramlich, 1990), is applicable here. The nuclear accident at Chernobyl was
obviously a global ecological disaster, but it was much more immediate to
people of the Ukraine and Central and Eastern Europe (who saw the effects
of acute radiation sickness), followed by people of Western Europe (who
were receiving fallout for several weeks; *see* Wolfson, 1991), than it was to
North Americans. Granted, there were widespread expressions of concern
and offers of help from the US and Canada, but more immediate panic (and
action) in Europe; the risk was simply not as great to North Americans. The
opposite was true for the nuclear accident at Three Mile Island in Pennsylvania,
in which the radioactivity was almost completely contained and hence posed
little in the way of health-risk (so far). We may expect discounting to influence human perceptions on a temporal scale as well; long-term costs (*e.g.*,
potential for global warming due primarily to the consumption of fossil fuels;
Schneider, 1989), would probably not, in and of itself, be a sufficient reason
to convince Western (especially North American) motorists not to drive, given
their immediate benefits of driving and immediate costs of finding alternative
transportation.

In many pollution issues, our sensory limitations may also come into play.
We refer to this as "sensory" discounting—"if you can't see it or smell it, it's
simply not perceived as being a problem". Examples may be carcinogenic

substances that humans are not able to sense at low concentrations (due probably to the lack of an evolutionary history with the substance), such as benzene; we would suggest that humans may be unable to perceive a risk in such cases until it's too late. This may be compared with a garbage dump which is likely to elicit complaints due to odour, though it does not pose as great a health-risk as others do. Another form of discounting which we see with regard to some ecological issues, especially relating to toxics, is that of "species discounting" (our phrase). We believe there is a tendency not to take action to remedy a toxic situation until there are direct threats to human health—even in cases where threats to wildlife or ecosystem health are quite apparent, and despite our knowledge of the movement of toxic materials through ecological food webs, and hence ultimate threats to human health in almost all cases of toxics in the environment, *e.g.*, PCB "revisionism" (Stone, 1992).

Some Potential Solutions to Resource-Related Problems

In summary, the characteristics we have outlined are all scale-issues, and include the costs and benefits of solving any environmental problem, where (spatially and temporally) these costs and benefits are manifested, and how this is influenced by the evolved human propensity to engage in spatial, temporal, and possibly also in sensory and species, discounting (in certain cases). Given this theoretical framework, we can make some predictions about what types of solutions are likely to work with environmental problems in any particular set of characteristics. The general categories of solutions which we consider here are: information/education programmes, social incentives (through direct reciprocity), economic incentives (through small fines, etc.), the formation of coalitions, and broad governmental regulations. In general, we predict that solutions will be easier for problems which are smaller in scale, and more immediate to the actor, than for those that are larger and farther away. Such limited environmental problems may be solved through education/information programmes alone. We predict that in general, as problems become larger in scale, more difficult and drastic solutions will be required. Our theoretical framework is depicted in Figure 1.1. Some potential solutions, and the conditions in which they may be important, follow.

Information/Education

Information about the effect of our behaviour is obviously important; ultimately, self-interest dictates avoiding our own destruction. Yet because of the great changes in the scope of time and space involved, we suggest that information alone will seldom be sufficient to change people's behaviour. If any change is seen as costly in individual time, money, or even "attention", more than simple information may be required to effect it. Many solutions to resource problems may lie in adopting conservation strategies that provide direct individual or familial benefits, or that serve as advertisements to potential reciprocators, as well as giving monetary advantages.

We argue elsewhere (Low & Heinen, 1993) that information will probably work most effectively at the local and short-term level. Information should work in those situations that are most analogous to people's management, in traditional societies, of resources which give quick feedback. Thus, when we are discussing the fact that, if local landfills become full, we must impose a cost on each garbage bag collected from our house weekly, we are likely to use information about how to recycle quickly, though, in such a case, an economic incentive is also implicit (*cf.* Figure 1.1). Information about any direct ill-effects of toxics such as garden pesticides, may be sufficient for some people to stop using them. However we contend, for reasons given above, that information alone is unlikely to be important in solving any major environmental issues; but information along with other factors may be appropriate, or even necessary, in solving most or all environmental problems.

Social Incentives: Reciprocity

We humans evolved as a highly social species, and reciprocity is a powerful force—one that we have probably underestimated in our attempts to encourage resource-conserving behaviours. Cost and benefit currencies were not originally, and need not be, monetary. Our costs and benefits as a social primate are older than the invention of barter and money, though not older than family structure and reciprocity. If we are right, some solutions may come without economic levers. Potentially important rewards include advertising one's status as a good citizen and cooperator. Even our brief review above suggests this (*see also* Low & Heinen, 1993).

It may be possible to manipulate extension of our definitions of "family" or shared interest in real ways, to the benefit of conservation programmes. These techniques have been extraordinarily successful in the service of ends that many of us consider less than ideal (*e.g.*, training for warfare: Holmes, 1985). Throughout our evolutionary history, most conflicts were fought by individuals or small groups of related men, and most conflicts were fought over reproductively important matters (Manson & Wrangham, 1991; Low, 1993a).

We contend that the basic reasons for warfare have not changed—today, most modern military conflicts are again fought by different ethnic groups competing for resources, *e.g.*, in the Himalayan region (Ives & Messerli, 1989)—although the scope and lethality of warfare have changed through technological innovation (*e.g.*, the Persian Gulf War). It is not coincidental that successful military training eliminates many differences in appearance with haircuts and uniforms; that it uses kinship and reciprocity terms ("son", "brother", "buddy", or their equivalents); that units are trained and moved together, so that one's life depends on men one knows and can presumably trust. There have been notable failures, such as in Vietnam, where individuals were rotated (Hackworth, 1989); in short it is evident that modern military training mimics our evolutionary past when men who fought together were kin and friends (*e.g.*, Manson & Wrangham, 1991). If we can call more widely on analogous strategies in the service of conservation (*e.g.*, Peace Corps, Conservation Corps, etc.), perhaps we can add a powerful approach to conservation programmes as well.

Because of our inescapable history as a long-lived, social primate, we possess evolved proximate mechanisms to maximize our functioning in the social context. If we can play upon these evolved mechanisms again, we may be able to promote conservation strategies more effectively than if we require conscious cost-benefit calculations, or rely primarily upon economic costs and benefits. Because cultural transmission is so very important (though not unique) in humans (*e.g.*, Hamilton, 1975; Cavalli-Sforza & Feldman, 1981; Lumsden & Wilson, 1981; Boyd & Richerson, 1985), we have evolved to be "docile" but disposed to be taught (Simon, 1990).

Though there are sex differences in the intensity of training to be obedient, even boys in highly polygynous societies, in which ferocity and aggression abound, are taught to obey their elders (Low, 1990a). Thus children raised in a conservation ethic may well, as adults, find it less onerous than did their parents to perform slightly (phenotypically) costly behaviours which will have a longer-term benefit for the group. Similarly, the promotion of conservation ethics as a social norm plays on our evolved sociality—if one "ought" to do it, it can be socially costly to defect (*e.g.*, D.G. Myers, 1990). The question remains: how much time do we have? If it takes a generation or two to create socially altruistic, group-benefit-oriented individuals, what will remain for them to save? We need to work in both the short and long term, to use both economic and social (familial, reciprocal) costs and benefits, and to design education and social-norm strategies on the most local scale we can. It is interesting to note that most traditional swidden (cut-and-burn) agriculture systems in tropical forests are managed by individual family units, and are highly stable over time (Dove, 1983).

Among non-relatives, our neighbours are those with whom we are most likely both to have had (socially) profitable reciprocal relationships, and those on whom we are least likely to defect (*e.g.*, in tit-for-tat games, repeated interaction is important; Axelrod & Hamilton, 1981; Axelrod, 1984; Axelrod & Dion, 1988). If we have promoted a normative standard regarding conservation, compliance is likely to be higher the better we know, and the more we interact with, our neighbours. Latane & Rodin (1969) described the well-known "bystander effect": when there are sufficient bystanders, people defect—their chance being greater of remaining anonymous, and hence less likely to be charged with failure to act. Thus information (recycling, above), reciprocal interactions, and promotion of social norms, are all likely to be more effective when translated into local scenarios than when remaining in wider ones.

We described open-access, common-pool resources above as among the most difficult to manage for the long-term group-level good (*e.g.*, global warming in Figure 1.1). There are examples of effective *"commons"* management, and while they differ dramatically in many ways, they have important similarities (Ostrom, 1990). The populations are stable and relatively non-mobile. As Ostrom (1990 p. 88) noted: *"Individuals have shared a past and expect to share a future. It is important for individuals to maintain their reputations as reliable members of the community."* Thus, repeated reciprocal interactions among members of a small and stable community can be a powerful force for generating long-term group-beneficial behaviour.

Economic Incentives

The two previous strategies, information about effects and social rewards, are likely to work best with smaller-scale problems—local landfills, for example—for obvious reasons. Behavioural ecology predicts that an effective way to change patterns of resource use more broadly and effectively than otherwise is to examine people's perceived short-term benefits and play on those, emphasizing individual (and corporate) short-term benefits (for compliance), or costs (for defection) in cases in which the behaviour will have long-term societal effects. As McIntosh (1985) pointed out, it is only recently that there has been acceptance of the idea that humans are governed by much the same physical, physiological, and ecological, rules as are other mammalian species; we extend this statement to evolutionary principles as well. At one level, the logic we propose, namely that individuals *"calculate"* (though not necessarily consciously) their benefits and costs in any action, has been said or implied by others (*e.g.*, McNeely, 1988, and many entries in Hardin and Baden, 1977).

Economic incentives can work in situations in which information alone is unlikely to work: in larger, more heterogeneous groups with little willingness to pay individual costs for the good of the group, *e.g.*, due to lack of social interactions and/or familial ties. For example, in Figure 1.1, *"wetland development"* is seen as a condition in which a local (and sometimes even regional) ecocomplex or component ecosystems may be degraded by an individual developer; economic incentives, enforced by coalitions, might be effective. Thus, as we note elsewhere (Low & Heinen, 1993), economic incentives will work in a large number of situations; for example, in those areas in which firms offer five or ten cents for returned bottles and cans, they have a vastly speeded rate of recycling and should prosper over those in which there are no incentives (although there may be more than enough information about landfills).

Coalitions and Negotiations

The most successful coalitions comprise individuals who have convergent interests: *e.g.*, kin (who have fewer reproductive conflicts than competitors), monogamous mates (who have convergent reproductive interests), or frequently reciprocating individuals (as in Ostrom's [1990] examples of successful "commons"). While all of us are loath to inflict costs upon ourselves, there is a possibility that we can use coalitions—which will sometimes accept small costs in order to impose greater costs on others or on society at large—to help us to "leapfrog" from small-scale successes to larger-scale successes. At the state level, for example, additional gasoline taxes for road repairs are not uncommon. The elimination of externalities at local and regional levels strikes us as a likely basis for the success of such strategy.

Negotiation is a standard strategy in the face of conflicts of interest in, for example, labour relations, and a large literature about successful strategies exists. Incentives, regulations, and taxes for environmental problems, however, are currently not negotiated in the standard way. Within existing legal constraints, is there a way in which we can bring what we know about negotiation to bear on these problems? Currently, the main negotiations concerning environmental issues are conducted at the international level—with the attendant problems of involving numerous interests and levels, and with conflicts of interest at each level (*e.g.*, Putnam, 1988).

Government Regulations

So long as "government" is considered as a unified, external force to solve environmental problems, we will not solve anything. It is only too easy to suggest that "the government" should raise gasoline prices, for example: but as we noted above, the real problem is how to make it worth while to individuals to act in ways that will cause and support such a governmental response. Thus in reality we cannot create the fantasy of an external governing body, as each of us is, even if trivially, involved in the actual choice of representatives, and each of us can be involved in influencing the voting behaviour of our representatives by letter-writing, etc. We humans have a long history of voting against taxing ourselves, even when the information which we have makes it clear that some additional taxes would help to ameliorate a situation causing considerable social or environmental concern.

Fortunately, coalitions and negotiations can significantly influence government decisions. Ostrom (1990) suggested that the basic premises of much sociological research in this area is flawed because of the assumption that government is a singular entity acting in isolation from, and above, other actors. Groups of individuals with common interests can affect governmental regulations by planning strategies that include gaining knowledge of the voting record and constituency of individual members of parliaments or congresses, and by lobbying heavily the members who are the most powerful and/or are undecided (and hence may be swayed to vote in the direction of the lobby). Many (or most) environmental regulations in the USA, for example, have been influenced by the lobbying efforts of environmental NGOs, whose membership is made up largely of people with personal interests in conservation (bird-watchers, anglers, hunters, campers, and other recreational users).

Conclusions and Summary

We contend that humans, as living organisms, evolved to sequester resources to maximize reproductive success, and that many basic aspects of human behaviour reflect this evolutionary history. Much of the environment with which we currently deal is evolutionarily novel, and much behaviour which is ultimately not in our own interests, persists in this novel environment. Environmentalists frequently stress the need for "sustainable development", however it is defined (*see* Redclift, 1987), and we contend that a knowledge of how humans are likely to behave with regard to resource use, and therefore a knowledge of what kinds of programmes are likely to work in any particular situation, is necessary to achieve sustainability. Specifically, we predict that issues which are short-term, local, and/or acute, such as an immediate health-risk, will be much easier to solve than issues which are broad, and which affect individuals other than ourselves, our relatives, and our friends. The bigger the issue is, the less effective is likely to be the response. Hence, the biggest and most troublesome ecological issues will be the most difficult to solve—*inter alia* because of our evolutionary history as outlined above.

This may not appear to bode well for the future of the world; for example, Molte (1988) contends that there are several hundred international environmental agreements in place, but Carroll (1988) contends that, in general, none of them is particularly effective if the criterion for effectiveness is a real solution to the problem. There are countless examples of "aggressors" (those nations causing the problem) not complying with an agreement, slowing its ratification, or reducing its effectiveness (*e.g.*, the US *versus* Canada, or Great Britain *versus* Sweden, with regard to acid rain legislation: Figure 1.1, *cf.* Bjorkbom, 1988). The main problem in these cases is that the costs are externalized and hence discounted by those receiving the benefits of being able to pollute. Any proposed change is bound to conflict with existing social structures, and negotiations necessarily involve compromise in a *quid pro quo* fashion (Brewer, 1980). We contend, along with Caldwell (1988) and Putnam (1988), that nations are much too large to think of as individual actors in these spheres. Interest groups within nations can affect ratification of international environmental treaties; for example, automobile industry interests *versus* those of environmental NGOs in the USA on the acid rain issue. It may even be that our evolutionary history is inimical to the entire concept of the modern nation state.

Barring major, global, socio-political upheaval, we suggest that a knowledge of the evolution of resource use by humans can be used to solve at least some resource-related problems in modern industrial societies. In some cases, these can probably be solved with information alone, and in other cases, the problems can probably be solved by playing on our evolutionary history as social reciprocators; environmental problems which tend to be relatively local and short-term may be solvable in these ways. Economic incentives can provide solutions to many other types of problems by manipulating the cost and benefits to individuals. We suggest that broader, large-scale environmental problems are much more difficult to solve than narrower, small-scale ones, precisely because humans have evolved to discount such themes; stringent regulations and the formation of coalitions, combined with economic incentives to use alternatives and economic disincentives (fines) not to do so, may be the only potential solutions to some major, transboundary environmental issues.

In preparing this argument, we have reviewed literature from many scholarly fields well outside the narrow scope of our expertise in behavioural ecology and wildlife conservation. Our reading of many works from anthropology, economics, political science, public policy, and international development, will doubtless seem naïve and simplistic to practitioners of those fields, and solving all environmental problems will ultimately take expertise from all of these fields and more. In general, however, we have found agreement for many of our ideas from these disparate disciplines, but much of their literature does not allow for a rigorous, quantitative hypothesis-testing approach to analysing the main thesis presented here—an approach that we, as scientists, would encourage. We hope to challenge people interested in environmental issues from many perspectives, to consider our arguments and find evidence, *pro* or *con*, so that we (collectively) may come closer to a better analysis of, and ultimately to solutions for, our most pressing environmental problems.

Acknowledgements

We are particularly indebted to William B. Stapp, Beverly Strassman, Garry Brewer, Kenneth Lockridge, Alice Clarke, David Zaber, David White, Shannon Sullivan, Paul Turke, and the members of the Human Behavior and Evolution Program at the University of Michigan; to George Williams, of SUNY, Stony Brook, to R. Paul Shaw of the World Bank and Stanton Braude of the University of Missouri, St Louis, and to Martin Daly and Margaret Wilson of McMaster University, Henry Harpending of Pennsylvania State University, John Baden of the University of Washington, and especially one anonymous reviewer.

2

The Evolved Psychological Apparatus of Human Decision-Making is One Source of Environmental Problems

Margo Wilson, Martin Daly, and Stephen Gordon

Resource Exploitation by *Homo Sapiens*

It has become increasingly difficult to ignore or deny the fact that the Earth's biota are in crisis. The abundance and diversity of flora and fauna have been and are being diminished at an accelerating pace, both as a direct result of human exploitation and as an indirect result of habitat loss and environmental degradation (Wilson, 1992). Despite the efforts of parties with economic interests antagonistic to conservation, it is no longer possible for informed citizens to doubt the reality of these trends, nor is there reason to doubt that the diversity and abundance of species will continue to decline for some time as a result of human numbers and activities. What is controversial is what to do about it (Clark, 1991).

The accumulation and dissemination of information about the crisis and its roots in human action are clearly not all that is required to bring about an effective remedial response. Yet, according to Ridley and Low (1994), many conservationists have assumed, at least implicitly, that if people were fully informed of the problems and their causes, they would change their priorities and activities in order to conserve resources for the future, and by relying on that assumption conservationists have implicitly embraced an unrealistic model of human beings as rational collectivists. Education is not sufficient, Ridley and Low argue, because natural selection has not designed human psychology to give priority to either the common good or the distant future, but to relatively short-term gains and positional advantages in a zero-sum intraspecific competition. According to this argument, the forces that have shaped human nature over evolutionary time have been forces that favor rapid, thorough exploitation of our resource base rather than stewardship. The human animal is not exceptional in this regard: because selection is predominantly a matter of within-species differentials in reproductive success, the phenotypes that proliferate are precisely those that enable organisms to exploit resources sooner and more effectively than their competitors, especially conspecifics, and to externalize or pass on to future others the costs of that resource exploitation.

The popular notion that aboriginal people who are uncorrupted by "western" values are reverent conservationists appears to be a romantic myth. The evidence from present-day hunting and foraging societies (Hames, 1987, 1991; Alvard and Kaplan, 1991; Alvard 1993, 1994a, Chapter 17, this volume), from ethnographic accounts of nonstate societies (Low, 1996), and from studies of human history and prehistory (Diamond, 1992; Kay, 1994) lends scant support to the idea that nonindustrialized foragers abide by a conservation ethic, nor to the proposition that greedy modern westerners are exceptional in their reluctance to subordinate their present wants to the future or to the common good. Moreover, although the conflict between human wants, on the one hand, and conservation goals, on the other, is often discussed in terms of human survivorship and comfort, human resource exploitation goes beyond these "essentials": nonindustrialized peoples, like westerners, deplete resources in ostentatious displays of resource-accruing potential and success in social competition (for industrialized societies, see Kaplan and Hill, 1985; Hawkes, 1993; and for western societies, see Frank, 1985; Ng and Wang, 1993; Howarth, 1996).

In our view, the Ridley and Low argument is overstated in the extent to which they suggest that current understanding of the natural selective process implies a "selfish" as opposed to a more collectivist evolved social psychology (Daly and Wilson, 1994). *Homo sapiens* is, after all, a social species with many psychological adaptations for social actions (e.g., Daly and Wilson, 1988; Cosmides, 1989; Simpson and Kenrick, 1997). Nevertheless, Ridley and Low's general point seems to be well taken: both theory and the available data on human behavior support the thesis that *Homo sapiens* is not by nature a conservationist, and hence that recognizing environmental problems, deploring them, and gaining a sophisticated understanding of their sources in our actions, may still not be enough to motivate the behavioral changes required to rectify them.

In this chapter, we argue for a more evolutionarily and psychologically informed model of *Homo economicus,* since economics is possibly the most relevant discipline to guide the development of incentive structures which will alleviate the current conservation crisis. This more realistic economic model will necessarily have to consider variations in human preferences and decision-making in relation to variables such as sex, age, and parental status that behavioral ecologists and other evolutionists consider fundamental. To illustrate our argument, we focus primarily on sex differences and age. The possibility that men and women "value" environmental goods somewhat differently is a topic that has hitherto received surprisingly little attention (Low and Heinen, 1993), despite an obvious selectionist rationale for predicting evolved sex differences in such domains as the subjective acceptability of various sorts of risks in the pursuit of status and resources. We also briefly discuss how a selectionist perspective on life history suggests that preferences and decision-making are also likely to have evolved to vary systematically with age.

Toward an Evolutionarily Informed Model of *Homo Economics*

The social science with an obvious role to play in remediating the current global crisis is economics. It is economic forces that drive technological innovations with their associated risks of contamination, despoliation, and expropriation. The developing field of ecological economics (e.g., Costanza, 1991) has much to say about common pool resource use and conservation incentives, consumer practices, monetary valuation of environmental goods, and the processes and consequences of externalizing costs, including pricing costs of foregone future resource use.

Economic ways of thinking make sense to evolutionary ecologists, who for decades have borrowed concepts like cost-benefit analysis, marginal values, investment, and profitability. Recently, the flow of ideas between these disciplines has become bidirectional. Several economists are now considering how past selection pressures have designed psychological processes underlying preferences, cooperation, and other aspects of economic transactions (e.g., Becker, 1976; Rubin and Paul, 1979; Frank, 1985, 1988; Bergstrom and Bagnoli, 1993; Samuelson, 1993; Simon, 1993; Bergstrom, 1995; Binmore et al., 1995; Mulligan, 1997; Sethi and Somanathan, 1996; Ben-Ner and Putterman, 1998; Romer, 1995), and some have begun to attend to variations in preferences and utility functions in relation to variables that evolutionists would consider central, such as sex, age, and parental and other kinship statuses (e.g., Rubin and Paul, 1979; Becker, 1981; Bergstrom, 1995; Eckel and Grossman, 1996; Mulligan, 1997). However, the dominant model of *Homo economicus* continues to be a folk psychological one in which preferences are translated into action by "rational" processes of deliberative decision-making that do not necessarily correspond to the psychological machinery that has actually evolved (Daly and Wilson, 1997).

Traditional economic analysis has assumed not only that actors are rational utility maximizers, but also that there is a unitary currency of utility in which all "goods" can be valued. (See Sunstein, 1994, for a critique of the assumption of a unitary currency of utility.) These assumptions make the application of economic decision theory to the behavior of nonhuman animals seem meta-phorical. But the application of cost-benefit models to *Homo sapiens* is really no less metaphorical. All complex animals confront the problem of how to value seemingly incommensurate goods in a common "currency." How many prospective calories will cover the predation risk cost of foraging activity X? Is mating opportunity Y sufficiently valuable to warrant accepting prospective injury risk Z by competing for it? From this comparative perspective, the real innovation in the invention of money was not that it reduces disparate utilities to one, but that it facilitates otherwise difficult reciprocal exchange. Money permits the elaboration of economic transactions by eliminating the necessity that one party trust the other to reciprocate in future, as well as by enabling exchanges in which the "buyer" does not otherwise have a commodity presently desired by the "seller." Unfortunately, this fungibility of assets in modern economic systems increases the appeal of destructive resource exploitation because exploiters can take their profits and invest elsewhere.

Some readers may protest that the costs, benefits, and trade-offs that we invoke in explaining risky decision-making by animals are only statistical characterizations of the natural selective past, whereas for human actors prospective costs and benefits are actually calculated and considered and hence are proximate determinants of behavioral choices. Perhaps so, but the model of decision makers as conscious and rational deliberators is, in fact, just as problematic when applied to people as when applied to kangaroo rats or starlings. Experimental psychologists have shown that people do not have the sort of privileged insight into the determinants of their own decisions that rational actor models presume and that the sense of having engaged in conscious deliberation and reasoned choice is largely illusory and after the fact (e.g., Nisbett and Wilson, 1977; Nisbett and Ross, 1980; Kahneman et al., 1982; Marcus, 1986). Although there are controversies about how best to characterize the psychological processes that produce human choice behavior, the evidence is unequivocally contrary to the assumption that people engage in the sort of simple rational calculus of utility maximization customarily attributed to *Homo economicus* (e.g., Kahneman and Tversky, 1979, 1984; Nisbett and Ross, 1980; Loewenstein and Thaler, 1989; Shafir, 1993; Gigerenzer and Hoffrage, 1995; Cosmides and Tooby, 1996; Hoffman et al., 1996).

Consider, for example, the classic demonstration by Kahneman and Tversky (1979) that people weigh alternatives very differently when exactly the same end states are framed as gains versus losses. Most people prefer a sure $1,500 gain over letting a coin toss determine whether they would get $1,000 or $2,000, and this "risk aversion" is not hard to rationalize: it apparently reflects the diminishing marginal utility of money, presumably because each successive dollar's incremental effect on our expected well-being really is smaller than the last. (The difference between being penniless or a millionaire is much greater than the additional impact of a second million.) However, if people are presented with exactly the same alternative outcomes framed as an initial award of $2,000 followed by a choice between relinquishing $500 or taking a 50% chance on being obliged to relinquish $1,000, most switch to "risk acceptance" (preferring the gamble). This is very much harder to rationalize in terms of the curvilinear utility of money. Losing any ground whatever from a state already attained apparently has a strong negative emotional valence.

How are the mental processes that produce such apparent inconsistencies of preference to be understood? Adaptationist thinking suggests several testable hypotheses. One is that voluntarily relinquishing prior gains has evolved to be aversive in the specific context of social bargaining because in ancestral environments, to relinquish prior gains was to advertise weakness, inviting future demands for additional concessions. Another hypothesis is that people may be averse to alternatives that take more time or require more steps, ultimately because delay and complexity have entailed risk of defection or duplicity. Even those decision theorists who have been critical of the assumption that people are rational utilitarians with full conscious knowledge of their own preferences (e.g., Kahneman and Tversky, 1984; Loewenstein and Thaler, 1989; Shafir, 1993; Knetsch, 1995) and who have thus attempted to model the psychological processes that produce these "irrational" effects have yet to consider such possibilities. In addition to its value as a cautionary tale against simple rational-actor models, Kahneman and Tversky's gain-loss framing effect is potentially interesting with respect to decisions about how to pitch conservation efforts to the public: the emotional appeal of a campaign to avoid the loss of what we already possess may be more powerful than the appeal of promised gains through remediation.

Another area in which economic analysis might benefit from considering how the evolved human psyche works is in efforts to attach prices to nonmarket resources. Certain "goods," such as air, have not ordinarily been monopolizable, exchangeable, or partible, and have not traditionally been treated as property, nor even thought of as resources. Other "goods," such as the tranquility or beauty of a setting, are clearly threatened by various sorts of economic exploitation and must somehow be valued in decisions about whether the gains from that exploitation are sufficient to offset the losses in these nonmarket resources. Armed with a unitary currency (money) and the conception of human decision-makers as capable of articulating veridical, rational preferences, economists interested in placing values on nonmarket goods have invented the "contingent valuation method" (CVM; e.g., Carson and Mitchell, 1993; Goodwin et al., 1993; Willis and Garrod, 1993; Cummings and Harrison, 1994; Smith, 1994; Heyde, 1995).

In a CVM study, a sample of people are asked how much they would be willing to pay to retain or attain some benefit. Ideally, respondents in a CVM study are given sufficient relevant information to permit a meaningful answer to some question such as how much would you be willing to pay in order to engage in a recreational activity X at place Y under conditions Z on a total of N days in the next year, or what is the maximum additional amount that you would pay before deciding that X is too expensive (e.g., Cummings et al., 1986; Carson and Mitchell, 1993). Critics of this method have been alarmed by the growing use of CVM studies in policy making and in legal decisions concerning compensation and have decried the presumption that it is appropriate or even possible to place dollar values on such goods as human health, aesthetic worth, or species survival (e.g., Sunstein, 1994; Heyde, 1995). Moreover, when CVM survey data are used to determine the damages to be paid by environmental despoilers, as they have been and are being used, then the incentive structures for decision-makers planning environmentally hazardous endeavors may become such that damaging even the recreational resources of the wealthy will be more costly (and hence more to be avoided) than damaging resources that are crucial to the lives and health of much larger numbers of people of lesser means (see also Boyce, 1994).

But the problems with the CVM are not limited to the questionable justness of its policy applications. There are good reasons to doubt that people are capable of giving meaningful, valid answers to CVM questions (e.g., Fischoff, 1991; Kahneman and Knetsch, 1992; Kahneman et al., 1993; Cummings and Harrison, 1994; Guagnano et al., 1994; Binger et al., 1995; Gregory et al., 1995; Loewenstein and Adler, 1995). Answers to CVM questions regularly violate the expectation that increments in the quantity of a good will increase its subjective value, for example, as may be illustrated by Kahneman's (1986) demonstration that different groups of people attached almost the same average dollar value in extra taxes to preserving the fish stocks of lakes in a small area of the province of Ontario as they were willing to pay for all the lakes in Ontario. Professed willingness to pay is also apt to be greatly exaggerated until respondents are reminded of the many possible demands on their limited means. For example, Hamilton, Ontario, residents who were asked how much they would be willing to pay to improve boating conditions in the local harbor gave a mean answer that was 30-fold higher if this was the first such CVM question in the interview than if it was the second (Dupont, 1996).

Being asked to put a price on certain environmental goods may be so out of the normal context in which a preference would be elicited that it is impossible to give a meaningful response. Indeed, it is questionable whether the sorts of preferences that the CVM obliges interviewees to articulate even exist prior to the questioning or are instead constructed in ways affected not only by the stable attributes of the respondent (as the CVM assumes), but also by the circumstances of the interview and the contextual framing of the task (Fischoff, 1991; Boyce et al., 1992; Kahneman and Knetsch, 1992; Irwin et al., 1993; Baron and Greene, 1996). Ajzen et al. (1996), for example, showed that respondents who had been "primed" by the inclusion of do-gooder bromides (e.g., "It's better to give than to receive") in an ostensibly unrelated word-unscrambling task committed almost twice as much to a public good from which they would derive no personal benefit as did respondents who had unscrambled only neutral control sentences.

It is also questionable whether even cooperative respondents are able to predict what they would really do or pay if the situation ceased to be hypothetical (Bohm, 1994; Loewenstein and Adler, 1995), and it is even more questionable whether they have conscious access to the determinants of their choices. Nevertheless, CVM researchers ask people to articulate just these things and accept the answers at face value. When Kahneman and Knetsch (1992) proposed, for example, that professions of willingness to pay for environmental protection or remediation might represent "the purchase of moral satisfaction" rather than the specific environmental benefit's value to the respondent, several CVM researchers announced that they had disconfirmed this hypothesis by showing that respondents who were instructed to choose "the reason" for their choice of dollar values from a menu mainly picked something else (e.g., Loomis et al., 1993; MacDonald and McKenney, 1996).

If we are going to price nonmarket goods in making tough decisions among alternatives that all have negative aspects, as it seems we must, then we need to move beyond these simplistic conceptions of decision makers as rational and decision criteria as consciously accessible. Recent efforts (e.g., Gigerenzer et al., 1988, 1991; Cosmides, 1989; Gigerenzer and Hoffrage, 1995; Cosmides and Tooby, 1996; Wang, 1996) have begun to incorporate evolutionary psychological models into explanations for the seemingly irrational aspects of the ways in which people process information and order their priorities. Success in this endeavor partly depends on correctly hypothesizing the nature of the adaptive problems that emotional reactions and other psychological processes were designed to solve in order to clarify the functional organization of complex psychological phenomena involved in decision-making under uncertainty, risk-taking, discounting the future, collective action, cooperating in use of common pool resources, and many other aspects of decision-making relevant to conservation of resources, species, and habitats.

Risk as Variance of Expected Payoffs

An adaptationist perspective on human psychology and action could contribute to understanding of several aspects of the contemporary ecological crisis. The need to elucidate the psychological adaptations of most direct and remediable relevance to the continuing population explosion is one obvious example. Another area in which evolutionary theorizing has already contributed is in identifying the circumstances under which the restraint of selfish consumption in cooperative ventures is realizable and those under which opportunities for "cheating" make cooperation unstable (Axelrod, 1984; Cosmides and Tooby, 1989; Boone, 1992; Hawkes, 1992). But in addition to the much-discussed problems entailed by the natural selective advantages enjoyed by the most prolific and selfish phenotypes, the ways in which selection has shaped such subtle specifics as time preferences, social comparison processes, and sex differences may also have important implications for conservation and environmental remediation efforts. If we are to mitigate the ills caused by human reluctance to reduce resource accumulation and consumption, for example, it seems important to elucidate the precise ways in which human decision-making discounts the future and how this discounting responds to uncertainty, both in ontogeny and in facultative responsiveness to variable aspects of one's immediate situation. The perceived costs of giving up present consumption depend on one's material circumstances, but little is known about subjective valuations and perceptions of uncertainties as a function of material and social circumstances.

Experimental studies of nonhuman animal foraging decisions have established the ecological validity of a risk-preference model based on variance of expected payoffs. Rather than simply maximizing the expected (mean) return in some desired commodity such as food, animals should be, and demonstrably are, sensitive to variance as well (Real and Caraco, 1986). Whereas seed-eating birds generally prefer to forage in low variance microhabitats as compared to ones with a similar expected yield but greater variability, for example, they switch to preferring the high variance option when their body weight or blood sugar is so low as to predict that they will starve unless they can find food at a higher than average rate (Caraco et al., 1980). Although the high variance option increases the bird's chances of getting exceptionally little, a merely average yield is really no better, and the starving birds accept the risk of finding even less in exchange for at least some chance of finding enough to survive. Such experiments have produced essentially similar results in several species of seed-eating birds (Caraco and Lima, 1985; Barkan, 1990), as well as in rats (Kagel et al., 1986; Hastjarjo et al., 1990).

It may be possible to understand risk acceptance by human explorers, adventurers, and warriors in analogous terms. Even taking dangerous risks to unlawfully acquire the resources of others might be perceived as a more attractive option when safer, lawful means of acquiring material wealth yield a pittance, although the expected mean return from a life of robbery may be no higher and the expected life span shorter. Interestingly, variations in robbery and homicide rates between places are better explained by variance in income than by absolute values of poverty (e.g., Hsieh and Pugh, 1993).

There is also experimental evidence that human decision-making is sensitive to variance as well as to expected returns. Psychologists and economists, using various hypothetical lottery or decision-making dilemmas, have documented that people's choices among bets of similar expected value are affected by the distribution of rewards and probabilities (e.g., Lopes, 1987, 1993). They are also influenced by whether numerically equivalent outcomes are portrayed as gains or losses as discussed above (Kahneman and Tversky, 1979). The underlying psychological dimension governing these choices among alternative, uncertain outcomes has been conceptualized as one ranging from "risk-averse" to "risk-seeking" (or "risk-prone" or "risk-accepting"). In the experimental nonhuman studies described above, the starving below-weight animal preferring the high variance option would be deemed risk-seeking. Diversity in risk aversion or risk seeking could be mediated psychologically by ei-ther variation in the subjective utilities of the outcomes or variation in perceptions of the probabilities associated with each outcome or both (Real, 1987).

Sex Differences in Risk Acceptance and Resource Use?

Consideration of the ways in which sexual selection differentially affects the sexes suggests that women and men confronted by uncertainty might have different subjective utilities or subjective probabilities and that these psychological determinants of risk acceptance or aversion might also vary in relation to life-history variables and cues indicative of expected success in intrasexual competition. Psychologists studying risk acceptance have documented sex differences and age effects but have focused mainly on stable individual differences (e.g., Trimpop, 1994; Zuckerman, 1994) and have scarcely addressed how risk preferences may be affected by social and material cues of one's life prospects and by one's relative social and material success.

The rationale for anticipating sex differences in the way people value and exploit the environment, as well as differences in willingness to risk damaging one's health, is an argument that has been applied to other aspects of risk taking and to sexually differentiated adaptations for intrasexual competition (e.g., Wilson and Daly, 1985, 1993). Its premise is that ancestral males were subject to more intense sexual selection (the component of selection due to differential access to mates) than were ancestral females, with resultant effects on various sexually differentiated attributes.

Successful reproduction, in *Homo* as in most mammals, has always required a long-term commitment on the part of a female, but not necessarily on the part of a male. Female fitness has been limited mainly by access to material resources and by the time and energy demands of each offspring, whereas the fitness of males, the sex with lesser parental investment, is much more affected by the number of mates (Trivers, 1972; Clutton-Brock, 1991). It follows that the expected fitness payoffs of increments in "mating effort" (by which term we encompass both courtship and intrasexual competition over potential mates) diminish much more rapidly for females than for males, and it is presumably for this reason that such effort constitutes a larger proportion of total reproductive effort for men than for women. One hypothesis inspired by these considerations is that men may find rapid resource accrual, resource display, and immediate resource use somewhat more appealing than women and that men may be more inclined to disparage risks and discount the future in their decisions about acquiring and expending resources (Low and Heinen, 1993).

Following Bateman (1948), Williams (1966), and Trivers (1972), sex differences in the variance in reproductive success are widely considered indicative of sex differences in intrasexual competition. Relatively high variance generally entails both a bigger prize for winning and a greater likelihood of failure, both of which may exacerbate competitive effort and risk acceptance. Bigger prizes warrant bigger bets, and a high probability of total reproductive failure means an absence of selection against even life-threatening escalations of competitive effort on the part of those who perceive their present and probable future standing to be relatively low. Although it is worth cautioning that fitness variance represents only the potential for selection and that variations in fitness could in principle be nonselective (Sutherland, 1985), intrasexual fitness variance appears to be a good proxy of the intensity of sexual selection because it is a good predictor of the elaboration of otherwise costly sexually selected adaptations. In comparative studies, sex differences in such attributes as weaponry for intraspecific combat are apparently highly correlated with the degree of effective polygamy of the breeding system—that is, with sex differences in fitness variance (e.g., Clutton-Brock et al., 1980). It is also worth cautioning that there can be other evolutionary explanations for sex differences in risk acceptance besides the Bateman–Williams–Trivers theory of sexual selection (see, for example, Regelmann and Curio, 1986), but this theory currently appears to be the one of greatest relevance to mammals in general and humans in particular.

All evidence suggests that the human animal is and long has been an effectively polygynous species, albeit to a lesser degree than many other mammalian species. Successful men can sire more children than any one woman could bear, consigning other men to childlessness, and this conversion of success into reproductive advantage is ubiquitous across cultures (Betzig, 1985). Of course, great disparities in status and power are likely to be evolutionary novelties, no older than agriculture, but even among relatively egalitarian foraging peoples, who make their living much as most of our human ancestors did, male fitness variance consistently exceeds female fitness variance (Howell, 1979; Hewlett, 1988; Hill and Hurtado, 1995). Moreover, in addition to the evidence of sex differences in the variance of marital and reproductive success in contemporary and historically recent societies, human morphology and physiology manifest a suite of sex differences consistent with the proposition that our history of sexual selection has been mildly polygynous: size dimorphism with males the larger sex, sexual bimaturism with males later maturing, and sex differential senescence with males senescing faster (Harcourt et al., 1981; Møller, 1988).

If the fitness of our male ancestors was more strongly status dependent than that of our female ancestors, as seems likely, then from the perspective of sexual selection theory, men may be expected to be more sensitive than women to cues of their status relative to their rivals. If intrasexual competition among men has largely depended on acquisition of resources (both material and so-cial), which were converted into reproductive opportunities, and if there has been a history of high variance in the distribution of resources and reproductive opportunities, then the masculine psyche is likely to have evolved to accept greater risk in its efforts to acquire, display, and consume resources, especially when accepting a small payoff has little or no more value than no payoff, as, for example, when a small payoff leaves a poor man still unmarriageable. This argument treats risk as variance in the magnitude of payoffs for a given course of action. In life-threatening circumstances people often take the riskier (higher variance) course of action. But people also take great risks when present cir-cumstances are perceived as "dead ends." For example, history reveals that suc-cessful explorers, warriors, and adventurers have often been men who had few alternative prospects for attaining material and social success. Later-born sons of aristocratic families were the explorers and conquerors of Portuguese colonial expansion, for example, while inheritance of the estate and noble status went to first-born stay-at-home sons (Boone, 1988). Similarly, later-born sons and other men with poor prospects have been the ones who risked emigration among more humble folk, too (e.g., Clarke, 1993), a choice which sometimes paid off handsomely, as in European colonial expansion, but must surely have more often led to an early death.

Sex Difference in Disdain for Health Risks?

One of the many domains within which men manifest greater risk acceptance than women is in health monitoring and preventive health care. Apparently, the average number of physician contacts per year is greater for males than females before puberty, but between 15 and 45 years of age, women visit physicians almost twice as often as men (Woodwell, 1997), even after one has accounted for birth-related visits and sex differences in rates of accident and illness. We hypothesize that men will also disregard the health hazards of various environ-mental contaminants more than women. And if men are relatively insensitive to the risks that they themselves incur, it seems likely that they will also be relatively insensitive to the risks that their activities entail for other people and for other fauna and flora.

One way to test these ideas is to ask people how they would behave in hypo-thetical dilemmas. As an example of this approach, we asked 173 introductory psychology students (90 women and 83 men) at McMaster University in Ham-ilton, Ontario, to consider the following hypothetical situation and then answer questions as if the situation applied to them.

> Imagine that you presently live in a mid-sized southern Ontario city of 300,000 people, where you were born and where most of your family and friends still reside. You have been looking for work and you suddenly find yourself with two job offers to choose between.

If you accept Baylor & Wilson's offer of employment at $30,000 per annum, you can continue to live and work in your home town. If you accept Smithers and Company's offer of $35,000 [$50,000] instead, you will be relocated to a city of 600,000 people in another province. From what you've heard, this city sounds like an interesting and beautiful place to live, but air pollution levels and respiratory disease rates are twice [ten times] what they are in the city where you now live.

Which offer do you accept?

Baylor & Wilson _____ Smithers & Company _____

The alternatives in square brackets were presented to distinct sets of subjects, making a 2x2x2 between-groups experimental design: male versus female subjects x the magnitude of the incentive to move ($5,000 versus $20,000 higher salary) x the magnitude of the deterrent costs in air quality and attendant health hazard (2-fold versus 10-fold).

Although all subjects were university students, at the same life stage and almost unanimously unmarried and childless, women and men responded somewhat differently to the experimental variables (Figure 2.1). As we predicted on the basis of the arguments above, men were attracted by an extra financial incentive more than were women, although not significantly so. More striking, and statistically significant, was the differential response to environmental risk: women were substantially deterred by higher costs in air quality and health hazards, but men were completely unaffected by this variable, choosing identically regardless of whether the stated costs were 2-fold or 10-fold. Other evidence also indicates that women may be more concerned about environmental health hazards than men (e.g., Flynn et al., 1994; Sachs, 1996, 1997). In a previous study involving a similar dilemma (but no variation of financial incentives and health risks), Wilson et al. (1996) found that men were significantly more likely than women to say they would accept a promotion "which would significantly boost your career" but would require moving to a city where the respiratory health risk was 10% higher than that of the hometown. In this earlier version, there were many parents among the subjects, and 41% of those who were parents said they would accept the promotion, compared to 81% of those without children, a difference that remained significant when the age of respondents was controlled.

Earlier in this chapter, we criticized "contingent valuation" studies for asking people how much they would be willing to pay for a particular benefit and taking their answers at face value, and we must acknowledge that the results we report here may have similar validity problems. Unlike CVM studies of nonmarket goods, however, we have asked people to consider a situation that is likely to be a common experience of most people: deciding to take one job rather than another, with benefits and costs associated with both. In principle, data from people's actual decisions between different employment opportunities can be compared with our results (as sometimes can be done and some-

Figure 2.1

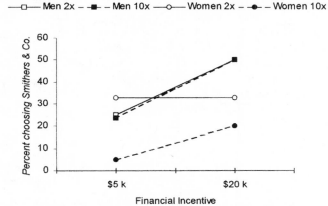

Percentage of men and women choosing the job at Smithers & Company, which would entail moving far away, either a $5,000 or $20,000 incentive above the home-town job, and either a 2 times or 10 times greater risk of respiratory problems than that of the hometown. Women were significantly more deterred by the health risk than males ($p = .03$ by logit loglinear analysis). The tendency for men to be more attracted by a financial incentive was not significant ($p = .14$).

times has been done in validating CVM results with "revealed preference" analyses of what people have actually paid for different goods or benefits; Smith, 1994; Carson et al., 1996). We attach no significance to the specific percentages of men and women choosing the Smith & Co. employment opportunity, but only to the sex difference in the impacts of an imagined financial incentive and an imagined health hazard. The apparently greater willingness of men than of women to treat health hazards as acceptable costs of opportunities for financial benefit should be further tested with real-world data on choices among different job opportunities.

Sex Difference in Disregard of Environmental Degradation?

In addition to the expectation that men are more likely than women to disdain personal health risks in their pursuit of economic and status advantages, we hypothesize that men are more likely than women to disregard or downplay environmental degradation. Support for this proposition already exists (e.g., Mohai, 1992; Sachs, 1997), but the possibility that it is a reflection of the male's psyche's greater prioritizing, of present profits as a result of dif-ferential histories of sexual selection has not been articulated or explored.

The rate at which one "discounts the future" is the rate at which the subjective value of future consumption diminishes relative to the alternative of present consumption (or, the "interest rate" required to motivate foregoing consumption). If *A* discounts the future more steeply than *B*, then *A* will value a given present reward relative to expected future rewards more highly than *B* and will be less tolerant of what psychologists call "delay of gratification." Hence, variable willingness to engage in nonsustainable modes of resource exploitation such as clearcutting or otherwise expending one's capital may be construed, at least in part, as variation in the rates at which decision makers discount the future.

Do men discount the future more steeply than women in the specific realm of conservation decisions? Wilson et al. (1996, p. 154) asked another set of 104 McMaster University people (36 men and 68 women ranging in age, from 17 to 24) to consider the following dilemma:

> Imagine you are farming a tract of land. Your father, like his father before him, lived off the profits from the farm without taking additional wage work elsewhere. You were fortunate to earn a scholarship to university to study agriculture, and now that you have inherited the farm you are considering changing the techniques of farming to be more specific and business-like. Prior to inheriting the farm you had a successful career as a broker specializing in agricultural commodities. [After your wife died suddenly, you've decided to leave that job to return to the farm. Your two children are delighted about the prospect of living on the farm.] Presently, you are pondering whether to follow one course of action (Plan A) or another (Plan B).

> Plan A: Convert the farm entirely to hybrid corn production for livestock feed. Corn is extremely profitable to grow, but it requires heavy chemical fertilization which over time will percolate into the water table with a very high probability that the land will not be usable in 60 years without heavy chemical supplements.

> Plan B: Convert the farm entirely to hay for livestock feed. Hay in good years can bring a good market price, but generally hay yields a modest profit. On the other hand, hay production does not diminish the quality of the soil and chemical supplements are not needed.

> Which plan did you choose? A or B? _____

Men were significantly more likely to choose the soil-degrading option (39% of men and 16% of women, Figure 2.2). In order to determine whether these "decision makers" were utilizing sound economic logic, we asked them to rate their agreement (on seven-point Likert scales) with propositions that might reflect the reasoning behind their choices. As expected, the proposition that "because you can always invest the profits from farming in other economic ventures including other farmland, you should weight profit over damage to the land" was endorsed significantly more strongly ($p < .0001$) by those who chose corn than by those who chose hay, but there was no significant effect of sex of subject.

In this scenario, one factor that might be expected to influence decisions that may have long-term negative effects on the quality of your farm is whether your children are likely to continue farming. This was the rationale for adding the two bracketed sentences ["After your wife died . . ."] for half the subjects.

Figure 2.2

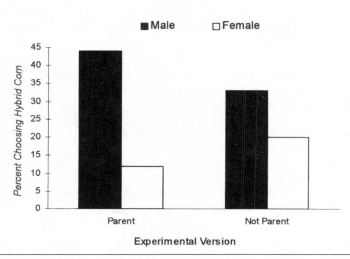

Percentage of men and women choosing the soil-degrading option (plan A: hybrid corn) according to the experimental version of the hypothetical dilemma they considered (being a widowed parent of two children, versus parental status unmentioned). Men were significantly more likely to choose the soil-degrading option ($\chi^2 = 6.6$, $p < .01$), but neither parental status ($p = .91$) nor the interaction of sex of the subject by parental status ($p = .27$) was significant by logit loglinear analysis.

We had anticipated that parental status would increase the likelihood that sub-jects of both sexes would be deterred from planting corn due to the possible long-term costs, but inclusion of this sentence did not result in any detectable difference in the choice of crops (Figure 2.2). Perhaps imagining that one has children cannot evoke the mindset of actual parenthood. (In this sample, only four people were married and only two had children.) Another possibility is that some subjects interpreted the existence of children as a source of increased demand for imminent cash flow. (And it may be relevant that the experiment was conducted in a region where it has become the norm that farmland is retained only until suburban real estate developers are prepared to pay the farmer's ask-ing price.)

We also anticipate that the percentage of people choosing corn versus hay might vary with expertise or other characteristics of the sample, so, for example, economics majors rather than psychology majors may be more likely to choose corn, and conservation biology majors may be more likely to choose hay than our sample of psychology majors. However, we predict that, in general, a sex difference in choice will remain; departures from this expectation may reveal interesting insights into the determinants of decision-making relevant to con-servation efforts.

What other factors might we expect to influence the steepness of discounting functions? If we assume that there is an evolved, facultative decision process behind such discounting, then obvious candidates are life expectancy and other sources of variable, subjective probability that one will retain control of the resources in question in the future. Wilson and Herrnstein (1985) have argued, on the basis of diverse evidence, that men who engage in predatory violence and other risky criminal activity have different "time horizons" than law-abiding men, weighing the near future relatively heavily against the long term. What these authors failed to note is that facultative adjustment of one's personal time horizons could be an adaptive response to predictive information about one's prospects for longevity (Daly and Wilson, 1990; Rogers, 1991, 1994; Hawkes, 1992; Gardner, 1993; Wilson and Daly, 1997) and the stability of one's social order and ownership rights.

Much of the social science literature on discounting and time horizons treats tolerance of delayed gratification as a proxy for intelligence. We see this as an anthropomorphic stance predicated on the claim that the capacity to plan far ahead and adjust present behavior to long-term future expectancies is a hallmark of complex cognitive capacity in which the human animal is unmatched. From an evolutionary adaptationist perspective, however, discounting and delay of gratification represent essentially the same issue as that addressed by Fisher (1930) and all subsequent life-history theorists: how is the future optimally weighted in deciding present allocations of effort (e.g., Roitberg et al., 1992; Clinton and LeBoeuf, 1993). The answers depend on the expected present and future reproductive payoffs associated with each alternative, expectations that may vary facultatively in response to available cues, and these issues are as germane to nonhuman animals (and plants) as to sophisticated cognizers. From this perspective, what selects for willingness to delay gratification is a high likelihood that present somatic effort can be converted to future reproduction. Thus, rather than reflecting stupidity, short time horizons are likely to characterize those with short life expectancies, those whose sources of mortality are not strongly or predictably dependent on their actions, and those for whom the expected fitness returns of present striving are positively accelerated rather than exhibiting diminishing marginal returns.

How human beings and other animals discount the future has been described in considerable detail by experimental psychologists, but a fuller understanding of these processes awaits the infusion of evolutionary adaptationist insights (Bateson and Kacelnik, 1996; Benson and Stephens, 1996; Kacelnik and Bateson, 1996; McNamara, 1996). The most noteworthy conundrum concerns the shape of discount functions, which are often, perhaps typically, hyperbolic rather than "rationally" exponential (Kirby and Herrnstein, 1995; Green and Myerson, 1996). The puzzling thing about hyperbolic discount functions is that they engender predictable reversals of preference between alternative futures with different time depths and hence predictable regret of what will become bad decisions in retrospect (e.g., Hoch and Loewenstein, 1991; Roelofsma, 1996). Suppose, for example, that a large reward two weeks hence is preferable to a smaller reward one week hence. If future discounting is hyperbolic, then as time passes the appeal of the more imminent reward rises more steeply than that of the more distant, until it may come to be preferred when almost at hand. One consequence is that people and other animals may even invest effort in erecting impediments to their own anticipated future lack of "self-control" or capacity to delay gratification (Kirby and Herrnstein, 1995). Understanding why the psychological underpinnings of time preference have evolved to produce such seemingly maladaptive internal struggles and why the evolved human psyche defies normative economic theory by discounting different utility domains at different rates (Chapman, 1996) may provide impor-tant clues for understanding why "waste" and inefficiency are so hard to eradicate. (See Kacelnik, 1997, for a possible adaptationist explanation for hyperbolic discounting).

Life-Stage Patterns of Risk Preference: Young Men as the Most Risk-Accepting Demographic Group

One may also hypothesize that sexually differentiated valuations of natural resources may be especially conspicuous in those life stages in which males have been selected to compete for reproductive opportunities most intensely. By this reasoning, the life-stage in which laying claim to resources and expending rather than conserving them should be most attractive is that in which such behavior would have had the greatest expected fitness payoff for our ances-tors. There is reason to believe that that lifestage for men is and has long been young adulthood (Daly and Wilson, 1990). Once men are husbands, they have something to lose in intrasexual competition, and once they are fathers, concern for their offspring's well-being may result in alterations of their valuations of the environment, especially if the resources would be those of recurring value from one generation to the next, such as land or water rights. Remark-ably, however, effects of parenthood on environmental attitudes and behavior are virtually unstudied.

Several lines of evidence about life-span development support the idea that young men constitute a demographic class specialized by a history of selec-tion for maximal competitive effort and risk-taking. Young men appear to be psychologically specialized to embrace danger and confrontational competition (e.g., Gove, 1985; Jonah, 1986, Lyng, 1990, 1993; Bell and Bell, 1993). Risk of death as a result of external causes (accidents, homicides, and suicides) is greater in men than in women and is maximally sexually differentiated in young adulthood, both in the modern west (Wilson and Daly, 1985, 1993; Holinger, 1987; Daly and Wilson, 1990), and in nonstate, foraging societies more like those in which we evolved (Hewlett, 1988; Hill and Hurtado, 1995).

The fact that men senesce faster and die younger than women even when they are protected from external sources of mortality suggests that these sex differences in mortality have prevailed long enough and persistently enough that male physiology has evolved to discount the future more steeply than female physiology. In the case of homicides, young men are not only the principal victims but also the principal perpetrators; indeed, men's likelihood of killing is much more peaked in young adulthood than is the risk of being killed (Daly and Wilson, 1988, 1990). All of these facts can be interpreted as reflections of an evolved life span schedule of risk proneness.

An alternative to this hypothesis, however, is that age patterns reflect responses to changes in relevant circumstances that happen to be correlated with age. Mated status, for example, would be expected to inspire a reduction in dangerous risk-taking because access to mates is a principal issue inspiring competition, and married men have more to lose than their single counterparts. Marital status is indeed related to the probability of committing a lethal act of competitive violence, but age effects remain conspicuous when married and unmarried men are examined separately (Daly and Wilson, 1990). Similarly, men are most likely to be economically disadvantaged in young adulthood, and poverty, too, is a risk factor in intrasexual competitive homicide, but young adulthood and unemployment status are again separable risk factors for homicide (Daly and Wilson, 1990).

Dangerous acts are adaptive choices if the positive fitness consequences are large enough and probable enough to offset the costs (Daly and Wilson, 1988). Disdain of danger to oneself is especially to be expected where available risk-averse alternatives are likely to produce a fitness of zero: if opting out of dangerous competition maximizes longevity but never permits the accrual of sufficient resources to reproduce, then selection will favor opting in (Rubin and Paul, 1979; Enquist and Leimar, 1990).

From a psychological point of view, it is interesting to inquire how age- and sex-specific variations in effective risk-proneness are instantiated in perceptual and/or decision processes. As we noted above, one possible form of psychological mediation entails flexible time horizons or discount rates. Other psychological processes with the effect of promoting risk-taking can also be envisaged. One could become more risk prone as a result of one or more of the following: intensified desire for the fruits of success, intensified fear of the stigma of nonparticipation, finding the adrenalin rush of danger pleasurable in itself, underestimating objective dangers, overestimating one's competence, or ceasing to care whether one lives or dies. As drivers, for example, young men both underestimate objective risks and overestimate their own skills in comparison to older drivers (Finn and Bragg, 1986; Matthews and Moran, 1986; Brown and Groeger, 1988; Trimpop, 1994). There is also some evidence that the pleasure derived from skilled encounters with danger diminishes with age (Gove, 1985; Lyng, 1990, 1993). In general, sensation-seeking inclinations, as measured by preferences for thrilling, dangerous activities, are higher in men than in women and decrease with age in a pattern quite like that of violent crime perpetration (Zuckerman, 1994).

Youths are especially unlikely to seek medical assistance or other health-enhancing preventive measures (Millstein, 1989; Adams et al., 1995), and young men are the demographic group most willing to take risks with drugs and intoxicants and to risk contracting sexually transmitted diseases (Irwin, 1993; Millstein, 1993). Relative disdain for their own lives can also be inferred from the fact that men's suicide rates maximally surpass women's in young adulthood (Holinger, 1987; Gardner, 1993).

In this context, it may be worth noting that the data in Figures 2.1 and 2.2 were collected almost entirely from young adults, in whom risk acceptance and sex differences therein may be most pronounced. However, the subjects were also people with good economic prospects and life expectancies, and these factors should have diminished risk acceptance. Because of their demographic uniformity, these samples were unsuitable for assessing the possibility of differential responses according to age, marital, and parental status. Whether this artificial technique is suitable for exploring, life-span developmental changes and differences between economic classes and other life circumstances remains to be seen.

It is clear that the most risk-prone demographic classes accept risk in diverse domains, and it seems likely that the same association would hold in comparing individuals within demographic categories. But the degree to which risk proneness is domain general is still largely an open question. Zuckerman (1994) has argued that sensation-seeking is a stable personality characteristic: a domain-general mindset which is highly correlated with individual differences in neuron membrane physiology, and he has developed a "sensation-seeking scale," on which men score significantly higher than women, and both sexes (but especially men) score highest in young adulthood. We asked subjects who participated in the hypothetical job choice dilemma (Figure 2.1) to complete Zuckerman's "thrill and adventure seeking" scale, and we, too, found a significant sex difference (average score for males was 7.6 and for females 6.1; $t = 3.54$, df $= 119$, $p < .001$). However, sensation-seeking scores were not associated with subjects' choice responses to the dilemma, and we are currently conducting research aimed at assessing the degree to which risk acceptance is consistent within individuals across different contexts and alternative operationalizations of risk.

Rogers (1994, 1997) has brought evolutionary reasoning to bear on the issue of optimal age-specific rates of future discounting, given the age-specific mortality and fertility schedules of human populations. His analysis suggests that people of both sexes should have evolved to have the shortest time horizons and to be maximally risk accepting in young adulthood. More specifically, his theoretical curve of age-specific optimal discount rates looks much like the human life-span trajectory of reckless risk proneness that may be inferred from data on accidental death rates and homicide perpetration. The claim that optimal discount rates decline as one ages may seem paradoxical, given the argument that indicators of a short or uncertain expected future life span should be cues favoring risk acceptance. The factors responsible for Rogers' counterintuitive result are certain peculiarities of human life history and sociality, namely, gradually diminishing fertility long before death and a shifting allocation of familially controlled resources between personal reproductive efforts and descendants' reproductive efforts.

Economists such as Norgaard and Howarth (1991) and Common (1995) consider it a conceptual error to extend the concept of future discounting beyond the individual actor's reasonably expected life span and argue that conserving resources for future generations is an issue of resource allocation and equity, instead. But to behavioral ecologists, one's descendants are an extension of one's self, and organisms may be expected to have evolved to act in ways that will promote their fitness both before and after their deaths. Thus, appropriate modeling of the factors affecting optimal discount rates requires consideration of the psychology of human kinship and lineage investment (Rogers, 1991; Kaplan, 1996).

Conclusions and Recommendations

We believe that effective solutions to environmental and conservation problems require a sophisticated understanding of their sources in human desires and actions. Answering this challenge will surely require an integration of conceptual and empirical contributions from several disciplines. Our thesis has been that the use of the Darwinian/Hamiltonian selectionist paradigm of behavioral ecology as metatheory for psychology and economics may constitute one particularly promising route toward productive interdisciplinary synthesis.

A strength of bringing this behavioral ecological perspective to bear on the study of human decisionmaking that impacts conservation and environmental degradation is that it has drawn attention to the likelihood of variations with respect to sex, life stage, parenthood, social status, inequity, and life expectancy cues, and unites these variables in a theoretical framework capable of generating predictions. This perspective has also contributed to the growing realization that research and education are insufficient to stem the tide of environmental degradation without sophisticated attention to modifying incentive structures as argued by Ridley and Low (1994). And as we argued in criticizing some CVM studies, thinking evolutionarily draws attention to the fact that the functional organization of the human mind is not designed to produce accurate introspections, but rather to produce effectively reproductive action in ancestral environments, an understanding that sensitizes the researcher to the potential pitfalls of opinion polling.

A weakness is that the individualistic focus of evolutionary psychologists and behavioral ecologists has yet to shed much light on political processes, especially in state-level societies with complex governmental and other institutions, with the result that the implications drawn from evolutionists' insights are still likely to be rather far removed from practical policy recommendations.

The suggestion that our evolved "human nature" is a source of environmental exploitation and degradation is not a claim that nothing can be done, but a warning that effective conservation and remediation strategies will have to incorporate an understanding of relevant evolved psychological processes in order to modify human action.

Summary

The serious reduction in abundance and diversity of the earth's flora and fauna is a fact, but what can be done about it remains controversial. We argue that the use of the Darwinian/Hamiltonian selectionist paradigm of behavioral ecology as metatheory for psychology and economics offers a promising route to a sophisticated understanding of human desires and actions which are the sources of and solutions to conservation problems.

Our critique of the "contingent valuation method" widely used by economists centers on the point that the functional organization of the human mind is not designed to produce accurate introspections but rather to produce effectively reproductive action in ancestral environments.

A strength of the behavioral ecological perspective in developing hypotheses relevant to exploitation and despoliation is that it has drawn attention to the likelihood of variations in human decision-making with respect to sex, life stage, parenthood, social status, inequity, and life expectancy cues. In two experimental studies concerning sex differences in hypothetical decisions, men were significantly more likely than women to prefer a crop with higher profit but higher risk of soil degradation, and the men were more willing to treat personal health hazards as acceptable costs of opportunities for financial benefits.

Acknowledgments

Our studies of environmental attitudes, perceptions, and decision-making have been supported by grants from the Great Lakes University Research Fund (GLURF grant 92-102), the Tri-Council Eco-Research Programme of Canada (TriCERP grant 92293-0005), and by a John D. and Catherine T. MacArthur Foundation grant to the Preferences Network. We thank Tim Caro, Monique Borgerhoff Mulder, and two anonymous readers for comments on a previous draft.

3

Game Conservation or Efficient Hunting?

Raymond Hames

Cultural ecologists often argue that hunter-gatherers have developed a number of cultural institutions and practices designed to allow them to achieve equilibrium with their environments. The equilibrium is found in an evident lack of resource degradation in the home ranges of the groups in question (Sahlins 1968) and the apparent long-term stability of their populations (see, for example, Birdsell 1968). It results, in part, from conservation practices motivated by and reflected in, for example, reverence for animal life shown in religion, ritual, and mythology (Martin 1978; Nelson 1979, 1982), limited needs (Sahlins 1968), or game taboo systems (McDonald 1977) which deter overexploitation. Destruction of natural resources and severe environmental degradation—tragedies of the commons—are said to be characteristic of state systems in which desire for profit or personal aggrandizement motivate uncontrolled exploitation. When native peoples engage in similarly destructive behaviors and opt for short-term maximization, these actions are attributed to their linkages to state systems through trade or colonization (Leacock 1954).

The question of whether none, all, or some hunting peoples are conservationists has not been clearly answered for even one group. An adequate answer to this question requires behavioral data on patterns of resource use and acquisition. Although religious beliefs, for example, may motivate individuals to behave, conservation can be achieved only through their behavior and its material consequences. There have been few attempts to demonstrate the behavioral correlates and environmental impacts of cultural practices believed to have conservation functions. An exception is H. Feit's analysis (1973) of Waswanipi Cree hunting-ground rotation. However, even in this excellent work it is impossible to demonstrate whether hunting-ground rotation is designed to

Hames, Raymond, 1987. *Game Conservation or Efficient Hunting? In* The Question of the Commons. The culture and ecology of communal resources (McCay, B. & Acheson, J., eds). pp. 92-107. Tucson: The University of Arizona Press. Reprinted with the permission from Raymond Hames.

maintain animal populations ("conservation") or to permit hunters to hunt as efficiently as possible.

Amazonian hunters and fishers may regulate their predatory impact on game populations in at least two related ways. The first, conservation, involves reducing the overall intensity of hunting, decreasing its intensity in different habitats or "patches," or tabooing the hunting of particular species or age and sex classes within species. This behavior requires short-term restraint for long-term benefits. The second, resource management, may be defined as activities that enhance the environment of game animals with the effect of increasing their abundance. Examples include burning scrublands in native North America and Australia to maintain forage for bison and kangaroo, respectively (Lewis 1982), and in Amazonia, the organic enrichment of lakes through dumping practices (Stocks 1983a,b). Management in this sense has not often been reported in Amazonia, perhaps because it has not been looked for and is more difficult to detect than conservation. I thus focus on conservation.

Using data on hunting and fishing among lowland Amazonian peoples, I deduce and test three pairs of predictions of hunters' responses to game depletion from two competing perspectives: conservation and short-term efficiency. At the end of the chapter I suggest some conditions under which conservation is likely to evolve. Any persuasive account of conservation as a human adaptation requires a theory that shows that conservation is by design (*sensu* Williams 1964b) and not a side-effect of some other process, specifies the conditions under which conservation will evolve, and predicts how individuals will systematically regulate their behavior to conserve resources.

Responses to Game Depletion

Conservation Hypotheses

I pose three conservationist predictions about how hunters respond to game scarcity. The first is to decrease the overall rate of hunting as game density declines. This assumes that depletion is, in large part, a function of the intensity of hunting and that the relaxation of hunting effort allows game to rebound from depleted levels. This is the conservation version of a time-allocation model. The second, or patch-choice method, is similar but involves the diminution of hunting effort in patches of the environment (habitats) that have been severely depleted of game species. That is, hunters allocate less time to depleted patches and more to rich patches, allowing the former to recover. The third, or taboo method, involves total abstention from particular game species, or even particular age and sex classes within species, through cultural rules. McDonald (1977) argues that certain game species are tabooed in lowland South America to prevent overexploitation or local extinction.

Efficiency Hypotheses

These conservationist predictions can be compared to those of short-term optimization models about how foragers should respond to game depletion. Optimal foraging theory (Pyke *et al.* 1976) predicts that hunters will attempt to maximize their net rates of return while hunting. In the simplest of these models a hunter is assumed to be indifferent to the long-term effects of his hunting behavior and attempts to maintain the highest rate of return over the short term—that is, to hunt as efficiently as possible. In regard to time allocation, if game becomes depleted (indexed by decreasing hunting efficiency) an optimal forager will allocate more time to hunting—the exact opposite of the conservation hypothesis. The efficiency hypothesis assumes that the value of animal protein is relatively high and inelastic so that hunters will attempt to make up for decreased rates of return by increased hunting. This is not to suggest that a hunter is attempting to maintain a particular level of consumption (although there may be levels below which he does not want to fall), but rather that increased time allocated to hunting has greater utility than alternative uses of time (Winterhalder 1983).

The patch method relates to a concrete issue in optimal-foraging theory known as the marginal-value theorem (Charnov 1976a,b). This model predicts that a hunter will leave or ignore patches that have low rates of return and will spend time in patches that have higher rates of return, as long as the extra travel time to richer patches does not lower his overall rate of return. In doing so a hunter will maximize *his* hunting efficiency. Therefore, we should expect a positive correlation between the amount of time spent in a patch and the rate of return for that patch. Unfortunately, this prediction is identical to the one made for the patch-method version of the conservation hypothesis.

The tabooing method of conservation relates to the model of diet-breadth expansion and contraction in optimal-foraging theory. Hunters will start to take game species they previously ignored, due to their low rates of return, as they deplete highly preferred game species (those which have the highest rates of return upon encounter). By doing so, hunters will maximize their rates of return in the face of game depletion. The main difference between this prediction and the conservation model is that in the diet-breadth model high-ranked, depleted game species are *always* taken when they are encountered in the search, whereas in the conservation model they are *never* taken when encountered during a hunt.

To my knowledge, few Amazonian human ecologists have gone to the field to develop or test theories of conservation or optimal foraging (but see Hawkes *et al.* 1983). As a result, available data lacks the precision needed to evaluate conclusively the competing hypotheses. It can, however, be used for preliminary quantitative evaluations of the time-allocation and patch-choice hypotheses. Anecdotal and qualitative analysis must suffice for the taboo hypothesis.

Time-Allocation Hypotheses

The conservation prediction is that hunters will decrease hunting time alloca-tion as game depletion increases. The optimization or efficiency prediction is that hunters will increase hunting time as game depletion increases. Vickers's (1980) longitudinal study of Siona-Secoya hunting contains the most appropri-ate evidence for evaluating these hypotheses. Vickers collected data on mean length of hunt and hunting efficiency in the village of Shushufindi during the first, second, and fourth years of its existence (Table 3.1). Assuming that hunting efficiency is a fair index of game depletion, we find that as hunting efficiency decreases (increasing game depletion) in the fourth year, hunting time increases. This suggests that the Siona-Secoya are not conserving game. Instead, game has a sufficiently high value to induce greater efforts toward its acquisition compared to alternative uses of time.

Table 3.1
Siona-Secoya Hunting Time and Efficiency

SETTLEMENT DURATION	TIME (HR/HUNT)	EFFICIENCY (KG/HR)
First and second years	7.56	2.48
Fourth Year	8.48	1.40

Source: Vickers 1980

Comparative data collected by John Saffirio (Saffirio and Hames 1983) on Yanomama "forest" villages and "highway" villages on both hunting and fishing time-allocation and efficiency provides another evaluation of the time-allocation hypotheses.[1] Highway villages are Yanomama villages that have been established along a portion of the Trans-Amazon system. During the construction of the highway, game was heavily depleted by highway workers who hunted with rifles and shotguns, and a great deal of game was evidently scared off by the disruptive influence of the highway itself. In addition, the fishing productivity of the area declined when streams that formerly crossed the land now occupied by the high-way were diverted. Forest villages, by comparison, occupy relatively undisturbed forest and are located five days' distance from the highway. Table 3.2 shows that highway hunters hunt a third less efficiently than forest villagers and that they hunt more than twice as intensively as forest villagers. Table 3.2 also shows that highway villagers fish half as efficiently as forest villagers and allocate twenty percent more time to it than do forest villagers.

Both the Siona-Secoya and Yanomama examples directly contradict the conservation hypothesis. Hunters and fishers who have depleted game re-

Table 3.2
Hunting and Fishing Time and Efficiency in Yanomama
Forest and Highway Villages

	TIME (HR/DAY)	EFFICIENCY (KG/HR)
Hunting		
Forest	0.96	0.98
Highway	2.40	0.69
Fishing		
Forest	0.60	0.37
Highway	0.81	0.18

Source: Saffirio and Hames 1983

Table 3.3
Hunting and Fishing Time and Efficiency in 10 Amazonian Societies

GROUP[a]	TIME (HR/DAY)	EFFICIENCY (KG/HR)	SOURCE
0. Machiguenga	2.18	0.16	Baksh 1985
1. Waorani	1.34	2.43	Yost and Kelley 1983
2. Cocamilla	1.69	2.02	Stocks 1983
3. Ye'kwana	1.81	1.60	Hames 1979, 1980
4. Shipibo	1.20	1.23	Bergman 1980
5. Yanomama[b]	1.56	0.75	Saffirio and Hames 1983
6. Makuna	2.48	0.72	Arnhem 1976
7. Yanomama[c]	3.21	0.56	Saffirio and Hames
8. Mamainde	1.91	0.52	Aspelin 1975
9. Yanomamö	3.00	0.49	Hames 1979
10. Wayana	2.54	0.47	LaPointe 1970
11. Bari	4.08	0.31	Beckerman 1983

[a] Number indicates a group's location
[b] Forest village
[c] Highway village

sources (or have had them depleted for them) hunt and fish more intensively. Protein resources are sufficiently valuable that hunters will allocate more time to their acquisition should their rates of efficiency fall as a result of game depletion.

A weaker test of the time-allocation hypotheses can be made through a comparison of hunting and fishing efficiency and time allocation across a number of Amazonian societies. Again assuming that hunting and fishing efficiency is a fair index of game and fish depletion, we expect a positive correlation between time and efficiency if conservation is occurring and a negative correlation if short-term maximization is occurring. I was able to gather data on these variables for twelve societies (Table 3.3). The correlation between efficiency and time is negative (Pearson's r = -0.67) and significant (p < 0.01).[2]

Still more indirect evidence concerning conservation among Amazonian populations may be made by comparing hunting efficiency as determined by technological differences and time allocation. Comparisons of shotgun and bow hunting among the Ye'kwana and Yanomamö (Hames 1979) and among the Waorani (Yost and Kelley 1983) indicate that low hunting efficiency leads to more intensive hunting in each case. Although these are not tests of the conservation hypothesis they do again support the hypothesis that protein is valuable enough that hunters exert considerable effort should efficiency decline.

Patch Hypotheses

Data on patch hunting behavior is more difficult to obtain since few ethnographers have been interested in documenting hunting variability through space. Again, Vickers's longitudinal study of the Siona-Secoya is valuable (Vickers 1980; Hames and Vickers 1983). As noted above, Vickers (1980) shows that hunting efficiency for the Siona-Secoya dropped by nearly half after four years of inhabiting the same village site, indicating considerable game depletion. This was particularly true in hunting zones near the village. A comparison of the frequency of hunts to near, intermediate, and distant zones reveals that near-zone hunting decreased in frequency from the first and second to the fourth years whereas distant zone hunting increased in frequency over the same period (Table 3.4). Thus, it appears that the conservation and the efficiency hypotheses are both supported.

Data on the intensity (hours over a 216-day sampling period) and the efficiency of hunting (kg/hr) in hunting zones exploited by the Yanomamö are shown in Tables 3.5 and 3.6, respectively. For both groups the correlation (Pearson's r) between hunting-zone intensity and hunting-zone efficiency is positive and significant. Previously (Hames 1980), I argued that a number of time constraints (and gasoline constraints for the Ye'kwana) did not permit Ye'kwana or Yanomamö hunters to exploit richer and more distant hunting zones more intensively. In any event, data on the Siona-Secoya and Ye'kwana

Table 3.4
Siona-Secoya Settlement Age and Hunting Frequency by Zone

SETTLEMENT AGE	NEAR ZONES $\overline{X} = 5$ km	INTERMEDIATE ZONES $\overline{X} = 11$ km	DISTANT ZONES $\overline{X} = 25$ km
First and second years	61	26	12
Fourth year	48	17	35

Source: Hames and Vickers 1983; Vickers 1980

Table 3.5
Yanomamö Hunting Time and Efficiency by Zone

ZONE[a]	TIME (HR)	EFFICIENCY(KG/HR)
Iguapo	1137	0.79
Manguera	458	0.75
Cua	49	0.00
Makanahama	720	0.47
O'doiyenadu	854	0.42
Shanama'na	72	0.53
Wohokuha	53	0.28
Audaha emadi	105	0.30

Source: Hames 1980: 46, Table 1
Note: Correlation coefficient (Pearson's r) $r = 0.64$, $p < 0.025$
[a] Ranked by decreasing distance from village

Table 3.6
Ye'kwana Hunting Time by Zone

ZONE[a]	TIME (HR)	EFFICIENCY (KG/HR)
Metacuni	302	5.43
Watamu	408	3.39
Sedukurawa	225	2.34
Manguera	39	0.00
Cua	399	0.96
Makanahama	140	1.01
O'doiyenadu	326	0.80
Shanama'na	330	2.07
Wohokuha	99	0.96
Audaha emadi	168	0.12

Source: Hames 1980: 46, Table 1
Note: Correlation coefficient (Pearson's r) $r = 0.50$, $p < 0.05$
[a] Ranked by decreasing distance from village

and Yanomamö tend to support both the conservation and efficiency hypotheses. It is obvious that another test, which distinguishes between the two hypotheses, must be devised.

Taboo and Diet-breadth Hypotheses

Almost by definition a taboo animal is conserved. Nevertheless, the important issue is whether the taboo is designed to conserve threatened species or whether conservation is a "fortuitous effect" (Williams 1966) of some other process (such as sumptuary laws). McDonald (1977) contends that native Amazonian game taboos function as a "primitive environmental protection agency" designed to conserve endangered species. Others insist that Amazonian game taboos are better understood for their social and ideological significance—from differentiating cultural groups and social statuses to serving as structural mediators of symbols in native ideological systems (see Kensinger and Kracke 1981).[3] Ross (1978), in contrast, implies that large Amazonian game species (tapir, deer, and capybara) with low natural densities and low rates of reproduction are those most frequently in danger of local extinction. As large game become depleted they are tabooed by small, nonexpanding Amazonian village populations to maximize hunting efficiency and/or minimize variance in protein intake. Throughout his paper Ross considers the costs (for example, pursuit of game into swamps and rivers and the price of trade shotguns and shells) and risks (for instance, pursuit of game near enemy villages) of taking taboo game animals, variables that are also critical in diet-breadth models. He states, "From the point of view of long-term adaptation, behavior of this kind, which promotes a sustained yield rather than encouraging maximal resource use, has far greater selective advantage." He adds, "A consequence—not necessarily a function—of some taboos might be to restore certain seriously reduced animal populations" (1978: 28). It is clear that Ross's model has much in common with an optimal foraging model of game taboos (Kiltie 1980; but see Ross 1980: 544).[4]

The existence of game taboos on animals that, because of their large size, *appear* to be very profitable animals to hunt presents a formidable challenge to optimal-foraging theory and especially to the optimal-diet-breadth model. A basic prediction of the model is that, as high-ranked game become depleted, lower-ranked game that were formerly outside the diet breadth will enter into the new diet breadth, but high-ranked game will continue to be taken when they are encountered on a hunt regardless of how rare or depleted they become. From a diet-breadth perspective, one could simply assert that taboo animals are animals with low rates of return and outside of the optimal diet breadth, that is, they are tabooed because hunting them would lower hunting efficiency. Although I believe that this is essentially the correct answer to some game taboos (other animals, such as jaguars, may be tabooed because they are dangerous to hunt, and others, such as carrion eaters, because handling them may cause infection), it is a rather unconvincing statement because it does not, for example, clearly account for the taboo on tapir and capybara for the Achuara and the lack of such taboos for the Yenewana.

In the diet-breadth model the rank of a game species is determined by its net rate of return upon encounter. The rank of a species can vary as a result of hunting technology and what is sometimes called behavioral depression (Charnov *et al.* 1976). The blowgun provides a good example of how technology can affect the rank of a game species. Many observers (Butt-Colson 1973; Ross 1978; cf. Yost and Kelley 1983) have noted that the blowgun is usually incapable of killing, or killing easily, animals greater than about 20 kg. For blowgun hunters most big game such as capybara, tapir, and peccary would fall out of their diet breadth. A taboo against hunting them would allow them to hunt more efficiently by deterring them from pursuing game species which would lower their hunting efficiency.

Behavioral depression refers to changes in habits of game species that have been subjected to heavy predation. Charnov *et al.* (1976) point out that these animals can become nocturnal, more wary of predators, or move into habitats which make search, pursuit, and approach more difficult. Through behavioral depression it is entirely possible that an initially high-ranked game species can have its rank lowered sufficiently to move it out of the optimal diet breadth. This would occur when "handling time" (stalking, pursuit, and retrieval) has increased such that it is more efficient to ignore game formerly taken upon encounter and spend time searching for other game. However, if depleted game has taken refuge in certain habitats (patches), or has been wiped out everywhere else but in a particular habitat, an optimal-patch-choice model (Charnov 1976; Charnov and Orians 1973) could be employed to predict game taboos. In this case, handling time of a game animal could remain the same (or even decrease) yet the animal would be taboo.

A simple diet-breadth model assumes that game animals are randomly distributed in a homogeneous environment. On the other hand, a patch-choice model assumes a heterogeneous environment in which certain patches (habitats) have different rates of return for a hunter because they have, for example, different mixes of game species. Therefore, in a "patchy" environment, hunters will search for game only in patches that have high rates of return. They will ignore patches with low rates of return (that is, a patch or patch type will be ignored if its marginal rate of return is less than that of the currently used set of patches). In a sense, a hunter will have an optimal-patch-breadth—a restricted set of areas in which he will search for game. An effective way of signaling to a hunter that a particular patch is not worth the effort to hunt would be to place a taboo on an animal that is restricted or has become restricted to that patch. (One would not want to taboo entry into the patch because it could, for example, contain important plant resources.) It is perhaps significant that Ross notes that tapir and capybara are found in riverine environments where they escape from pursuing hunters by swimming and diving and that deer are found in swamps where pursuit is difficult; all three animals are taboo. In swamps, huntable terrestrial biomass is low; moreover, low floristic diversity may mirror low aboreal game diversity and density. Furthermore, from this perspective it is easy to see how technology could influence diet breadth through patch choice. Blowgun hunters are forced to hunt small game (less than 20 kg) that may be abundant in patches where large, commonly tabooed game is absent or rare.

Conservation or Efficiency?

The quantitative evidence shown above evaluated on indicators of conservation is equivocal. Nevertheless, it casts doubt on the existence of conservation in lowland Amazonia. This is particularly true of the time-allocation studies: the data consistently indicate a negative correlation between hunting efficiency (as an index of game depletion) and time allocation. This relationship was found within a single village through time (Siona-Secoya), between two villages of the same culture over space (Yanomama highway and forest villages), and cross-culturally in Amazonia. It appears that native Amazonians regard protein as sufficiently valuable to intensify their efforts for it in the face of depletion. Furthermore, it indicates that they are pursuing a strategy of short-term maximization and that they are not, as yet, constrained by the possible long-term consequences of their actions.

It should be noted that the time-allocation hypotheses and patch-allocation hypotheses are not truly independent. One could argue that increases in overall time allocation are the result of trying to exploit more distant hunting zones while suffering lower rates of return in the process—because of added travel time—in order to conserve resources in near zones. This interpretation can be ruled out only by showing that distant zones have higher rates of return. Data on hunting efficiency and distance of hunting zone from the village for Ye'kwana and Yanomama hunters (Hames 1980: 46, 50) demonstrate positive correlations in each case (Ye'kwana $r = 0.92$, $p < 0.001$; Yanomamö $r = 0.72$; $p < 0.01$). Therefore, hunters increase their efficiency by hunting in distant zones.

Another problem in seeking indicators of conservation or efficiency was seen in the patch-time hypotheses in which conservation and efficiency predictions were identical. This problem has been noted and explored in detail by human socioecologists (Durham 1981: 226-228; Winterhalder 1981: 97; Smith 1983). One way to distinguish between these two predictions is to hypothesize that conservationists passing through game-depleted patches will not take game found there. Instead, they will go on until they reach undepleted patches. In contrast, efficiency maximizers will take everything they encounter. This test is possible because all Amazonian horticulturalists are central-place foragers and must move through near hunting zones to reach distant ones. My experience suggests that the efficiency hypothesis will hold: Ye'kwana and Yanomamö hunters will take game in near zones if they encounter it, even when they are passing through these zones in order to reach more distant hunting zones.

Although the role of wild-protein procurement in Amazonian native adaptations is a contentious issue (Gross 1975; Ross 1978; Beckerman 1979; Chagnon and Hames 1979; Vickers 1980; Johnson 1982), I think it is fair to say that most human ecologists believe that gaining adequate dietary protein at a reasonable cost is a problem faced by all Amazonian populations. The issue is how people respond to decreases in consumption and/or increases in acquisition costs of high-protein foods. I have shown that native Amazonians intensify their hunting efforts and/or extend their hunting ranges. Identification of these responses does not rule out others, such as village relocation or fission, nor does it rule out complex interactions among variables. Nonetheless, attempts to relate informant-generated data on historic patterns of settlement relocation to problems of protein acquisition fail to yield significant results, even though informants cite other environmental problems (Gross 1983; Hames 1983).

Optimal-foraging theorists (Pyke *et al.* 1976; Krebs 1978) recognize that conservation can evolve as a means to maximize hunting efficiency. Simple optimal-foraging models assume that hunters are unconstrained by time limitations such that the goal of hunting is to maximize the net rate of return over the short term while foraging. However, if one assumes that the goal of hunting is to maximize efficiency over a period of, say, several years, hunters may accept lower rates of return (by ignoring rare, high-return game, for example) over the short term to maximize hunting efficiency over the long term. From this time perspective it should be clear that conservation and efficiency are not mutually exclusive and that the former is a means to the latter. The conditions under which conservation or long-term maximization could evolve are explored below.

Conditions for the Development of Conservation

The important question is not whether Amazonian horticulturalists are conservationists but rather, what are the conditions under which conservation is likely to evolve? That is, we need a theory of conservation. Toward this end I would like to suggest three conditions that would have to exist in order for conservation to develop.

First, the local population would have to be territorial, that is, able to defend their resources against outsiders who might subvert their conservation plans. According to Dyson-Hudson and Smith's (1978) application of Brown's (1964) model of economic defendability, territoriality is most likely to evolve in groups that occupy areas characterized by relatively dense and predictable resources. In Amazonia, riverine areas—oxbow lakes in particular— may have fish resources that meet this condition. However, territoriality would not be necessary for conservation if a group had a home range that did not overlap with a neighbor's home range.

Second, local populations must have mechanisms for dealing with their own members who might decide to break conservation conventions. The spatial organization of village domestic structures characteristic of most of Amazonia (especially long houses, or *malocas*) would make cheating without detection very difficult, as would the perceived seriousness of supernatural sanction (for example, the Ye'kwana believe that killing an anaconda will cause the entire village to be destroyed by a flood) or even social sanctions meted out to those who break conservation rules.

Third, conservation implies that unregulated hunting and fishing or population growth places so much pressure on a group's resource base that increases in work effort and/or decline in the consumption of limiting resources will ultimately result in a crash of the group's population. However, the probability that conservation will develop to prevent a population crash depends on at least two factors: (1) availability of unoccupied areas for resettlement (that is, expansion), and (2) alternative sources of or substitutes for limiting resources (that is, economic or technological innovation). In Amazonia there is considerable evidence to indicate that population density in native areas is considerably less today than it was in pre-Columbian times (Denevan 1976). For example, the Yanomamö have a population density of approximately 0.20 persons/km (Hames 1983: 425), and many villages have the option to expand into unoccupied areas. This fact alone may account for the lack of evidence of game or fish conservation among Amazonian populations. As for economic innovation, Roosevelt (1980) has suggested that depletion of fish and game resources by a number of prehistoric riverine Amazonian populations led to increased reliance on the protein in maize as a substitute for animal protein.

It is risky to assume that native ideology about potential fish and game resources can offer any proof that strategies of conservation or short-term efficiency are being followed. Ideological factors are best regarded as approximate causes of behavior that dispose individuals to behave predictably under particular circumstances. In turn, determining whether the behavior generated by ideology is an adaptation designed to solve some problem set forth by the environment requires a different level of analysis—one that measures human impact on the environment and how that interaction ultimately affects human survival and reproduction. If researchers wish to convincingly demonstrate conservation they must enter the field with a theory of conservation from which hypotheses can be deduced and then tested with quantitative data.

Notes

I would like to thank the following colleagues for reading this chapter and offering thoughtful suggestions: Kristen Hawkes, Bonnie McCay, Alan Osborn, Eric Smith, Tony Stocks, Dennis Werner, and Bruce Winterhalder. It is based on a paper presented at the 82nd Annual Meeting of the American Anthropological Association, Chicago, Illinois, November 16-20, 1983.

1. Both hunting and fishing provide animal protein for the diet. Since they are alternative means of solving the same adaptive problem they are analyzed separately (Tables 3.2 and 3.3) and jointly (Table 3.4), depending on available published data.

2. All critical values of Pearson's r used are one-tailed.

3. It should be noted that, following Ross (1978), I am dealing with blanket or general taboos that extend to all group members regardless of status or condition, rather than specific ones on consumption of game animals determined by an individual's status (male, female, old, young) or condition (pregnancy, illness) found in complex array in lowland Amazonia. It should also be noted that not everyone in the Kensinger and Kracke (1981) volume regard social and ideological analyses of food taboo systems as necessarily competing with ecological explanations (e.g. Taylor 1981); for that matter, neither does Ross (1978: 1). See Johnson (1982) for discussion of levels of analysis in Amazonian research.

4. Kiltie (1980), in a comment on Ross's work, suggests that a model of diet-breadth contraction and expansion could account for Amazonian taboos. In reply, Ross (1980: 544) discounts the diet-breadth model by suggesting that increased search time needed to encounter large, depleted game would drop large game from the

optimal diet breadth. Ross errs. In an optimal-diet-breadth model, search time is not charged to individual prey species because it is assumed that all game within a patch are searched for simultaneously. Thus search costs are equalized for all species. Ross also distinguishes between "maximal resource use" and "optimization" without clearly defining either concept.

4

Behavioral Ecology of Conservation in Traditional Societies

Bobbi S. Low

"I am as free as nature first made man,
Ere the base laws of servitude began,
When wild in woods the noble savage ran."
–John Dryden (1672)

Today, as we face increasingly complex environmental problems of ever-enlarging scale, we are faced by a dilemma: our ideas about what we should do to solve these problems are based on largely unexamined conventional wisdoms about our conservation ethics in the past and today. One "wisdom" is the perception that people in preindustrial ("traditional") societies, being more directly and immediately dependent on the ecology of the natural systems around them, were more conserving and respectful of those resources than we. A related belief is that people in traditional societies were likelier to be willing to sacrifice personal benefit for the good of the group when conditions demanded it. Thus, some current environmental strategies urge us to become more like this conception of traditional societies: ecologically aware, and environmentally altruistic. Here I want to examine the assumptions made by such approaches.

The concept of "noble savage," in the sense of morally superior human, uncorrupted by civilization, was strong throughout the sixteenth to nineteenth centuries. Dryden (in *The Conquest of Granada,* 1672) seems to have been the first to use the term. Rousseau, of course, used the concept effectively to anathematize civilization—and in fact, even today, there is a strong streak of romantic distaste for human company and human machinations in various environmental movements (see Budiansky 1995). The concept has been expanded from moral to ecological arenas, most notably by Redford (1991a). A typical environmental argument runs that we have, in important ways, "lost

Reprinted by permission of "*Human Nature*" 7:4, pp. 353-379, Copyright © 1996 Walter de Gruyter, Inc. Published by Aldine de Gruyter, Hawthorne, NY.

touch" with ecological constraints as we have developed technological insulation against ecological scarcity and fluctuations—and thus that we may have drifted away from ideals of ecological concern and from cooperativeness. For example (Oelschlaeger 1991), some environmental historians argue that hunter-gatherers lived in a "balanced and harmonious" state, altering nothing, in contrast to ideals of ecological subjugation common today.

A peculiar conjunction of historical events may have exacerbated possible misinterpretations like Oelschlanger's. The nineteenth century was the heyday of ethnographic explorations, as well as the height of the Romantic approach and the growth of Manifest Destiny in the United States. There is much imputation, but few data, about the pre-contact practices of traditional peoples. A mythology developed, probably fueled by the Romantic view, of a pristine America barely peopled, and populated by societies virtually without ecological impact (see Denevan 1992). American primitivist writers like Cooper and Thoreau probably fueled this image. Threatened by constraint, even extinction, it may have paid indigenous peoples to acquiesce in an overstatement of the globally conserving long-term nature of their policies.

Today we retain remnants of this history. For example, the words of Chief Seattle regarding an impending ecological crisis, and reminding us that "the earth does not belong to man, man belongs to the earth" are widely quoted—yet those words were written by a scriptwriter in the early 1970s, and there is no empirical evidence suggesting that Chief Seattle had any such attitude (e.g., see Budiansky 1995:32-34). And, as anyone who remembers George Bush's desire to be remembered as the Environmental President can attest, stated attitudes and actual practice may remain far apart (e.g., Tuan 1968, 1970).

Romantic misconceptions might not matter, except that the conventional wisdoms arising from them generate normative prescriptions: that in addition to more information about the impact of our actions on ecological balances, we need to become more reverent, to move closer to the ideal we hold of traditional peoples' patterns (e.g., Bodley 1990; Bunyard 1989a; Reichel-Dolmatoff 1976). At the extreme, the Romantic view surfaces as the Gaia concept of Lovelock and Margulis (1974) and Lovelock (1979, 1988), which argues that the earth is a homeostatic, self-regulating unitary entity especially suited for human life (see Williams 1992 for a cogent critique).

Current strategies of environmental and conservation education reflect our faith in such ideals (see Budiansky 1995), but typically without tests of whether any of the assumptions are true. Some approaches muddle the pragmatic and the normative: if only we could recapture the reverence and cooperativeness of traditional societies, and expand it, we could solve our problems. Finally, the Noble Savage paradigm postulates individual willingness to sacrifice for the group; it assumes that individuals routinely take on costs for the good of the group. Are such arguments really true? In fact we have few data.

The few empirical examinations that test the Noble Savage hypothesis (e.g., Alvard 1993; 1994a; Alvard and Kaplan 1991; Hames 1987b, 1989, 1991) suggest an alternative evolutionary hypothesis consistent with what we know about resource use in other species: individuals strive to get sufficient resources efficiently, in competition with other (also striving) individuals. This resource efficiency hypothesis arises from optimal foraging theory (e.g., Smith 1983). "Excess" resources, over and above what's needed for health, are sought if they are reproductively useful (generally true of male mammals; see Hawkes 1993; Low 1990a, 1992, 1993b).

To examine these hypotheses, we really need to know not only whether a society expressed an ethos of restraint, but whether that ethos was effective—whether societies with an expressed ethos of restraint were in fact more conserving of their resources than other societies. We can also ask whether those societies that mute the teaching of competitive striving to their children suffer less environmental degradation. Detailed empirical data come principally from a few societies such as the Ye'kwana, the Piro, and the Yanomamö; my purpose here is to see whether existing cross-cultural data support either hypothesis. Thus I will step back from complex situations and examine the following questions in traditional societies: (1) Is there evidence of environmental degradation in traditional societies? Does degradation show any pattern with expressed need or ecological conditions? (2) Do people in traditional societies express a conservation ethic? Under what ecological conditions? Does this ethic bear any relationship to degradation? and (3) Restraint in consumption can imply confluence of interest and trust among individuals within the group. What conditions foster trust and cooperation? Do societies which teach children to strive suffer more environmental degradation than those which teach trust and generosity?

Materials and Methods

The sample comprised the 186 societies of the Standard Cross-Cultural Sample (Murdock and White 1969), which is stratified for geographic distribution and language group, and for which ethnographies by qualified ethnographers resident with the society for a substantial period of time are available. Data from this source, because they were gathered without reference to the current hypotheses, can be crude and there is a risk of missing information, but it is the only broad sample available, and if we can ask questions in appropriate ways, even crude data may be useful.

[This is a shortened version of the original and here some text was omitted.]

Results and Discussion

Resource Use and Environmental Degradation in Traditional Societies

Ethnographers commented that environmental degradation was severe in 39/122, or 32% of the societies. In fifteen of these cases, the ethnographer described other recent shifts, either increased population density ($n = 8$) or enhanced technology ($n = 7$). Of the six societies with sacred prohibitions, degradation was reported in three. Degradation was positively correlated with recent population increases ($n = 122$, rho = 0.39, $p \leq 0.0001$) and recent technological improvement ($n = 122$, rho = 0.32, $p = 0.0005$). There was no association between the existence of sacred prohibitions and either use pattern ($n = 122$, d.f. = 2, $\chi^2 = 1.9$, $p = 0.38$) or environmental degradation ($n = 122$, d.f. = 1, $\chi^2 = 0.9$, $p = 0.33$); religious prohibitions do not seem to afford general environmental protection.

Just over 5% of the societies (7/122) expressed some reason to leave some resource untouched; there were so few cases that I could detect no pattern. In 73/122 (59.8%) societies, people were not reported by the ethnographer as making any comment. People in 42/122 (34.4%) societies expressed a need to take all one could of what was available. The more a society relied on gathering ($n = 121$, rho = 0.19, $p = 0.04$) or hunting ($n = 121$, rho = 0.19, $p = 0.04$), the more likely they were to express the need to get all they could from the environment. Interestingly, there was no pattern of people's expressed need/conservation ethic with plant productivity ($n = 118$, rho = -0.017, $p = 0.85$). People expressed a need to use everything in societies described by ethnographers as having frequent resource shortages ($n = 122$, d.f. = 4, $\chi^2 = 15.6$, $p = 0.004$); 21 societies in which resource scarcity was reported also "used everything" (13 expected). Similarly, people's expressed need was more frequent in the face of recent population increases ($n = 122$, d.f. = 2, $\chi^2 = 10.7$, $p = 0.005$) and recent environmental degradation ($n = 122$, d.f. = 6, $\chi^2 = 51.9$, $p < 0.00001$).

Ethnographers for twenty of the societies (16.4%) commented that people made use of all parts of resources, including a number of items the ethnographers themselves found repugnant. People in 15 of the 122 societies failed to use all material, although this frequently appeared to be an incidental side effect of the technology (e.g., cliff jumps for buffalo, which produced more meat than could be processed). Ethnographers made no comment beyond describing the process for the remaining 87/122 (71.3%) societies. There was no pattern to degree of use and the importance of any major subsistence type (0.1 [hunting] $< p > 0.9$ [fishing]). Completeness of utilization was not related to the existence of societal or religious prohibitions ($n = 122$, d.f. = 2, $\chi^2 = 1.9$, $p = 0.39$). Not surprisingly, plant productivity was negatively correlated with completeness of use ($n = 118$, rho = 0.21, $p = 0.02$).

Importance of major subsistence types showed no pattern with ethnographers' perceptions of resource scarcity (0.2 [gathering] $< p > 0.8$ [agriculture]), although gathering and hunting are somewhat more important in areas of higher plant productivity (Low, 1989c). Ethnographers' perceptions of resource scarcity were marginally associated with recent population increases ($n = 122$, d.f. $= 2$, $\chi^2 = 4.0$, $p = 0.13$) and changes in technology ($n = 122$, d.f. $= 6$, $\chi^2 = 4.0$, $p = 0.13$).

Conservation Ethic and Degradation

Conservation ethic/expressed need co-varies with utilization pattern ($n = 122$, $\chi^2 = 14.2$, $p = 0.007$); the significance of this association arises simply because people in societies which express a need to use everything are described by the ethnographer as having relatively complete utilization. Conservation ethic ($n = 122$, $\chi^2 = 7.5$, $p = 0.11$) and utilization pattern ($n = 122$, $\chi^2 = 8.5$, $p = 0.08$) show a marginal pattern with the storage of food. However, utilization and conservation ethic/expressed need do not co-vary in many other regards. While utilization increases as plant productivity decreases (above), expression of a conservation ethic does not ($n = 118$, rho $= -0.02$, $p = 0.85$). Similarly, complete use is strongly associated with occurrence of famine ($n = 75$, $\chi^2 = 10.1$, $p = 0.03$), but expressed need/conservation ethic shows no pattern with famine ($n = 75$, $\chi^2 = 1.2$, $p = 0.88$).

Expressions of desire to conserve are rare in this sample, and perhaps simply because they are rare there is no strong pattern to their occurrence. Expressions of need to take everything are far more common. In fact, the statistical patterns that are found arise because expressed need is related to resource scarcity (above). Using logistic regression, 50% of the variation in utilization pattern is explained by patterns in resource scarcity and plant productivity (-2logLikelihood = 162.79, d.f. $= 2$, $p < 0.00001$); 46% of the pattern in expressed need/conservation ethic is explained by patterns of resource scarcity (-2logLikelihood = 190.63, d.f. $= 1$, $p = 0.0002$) (plant productivity adds nothing to this model).

In some cases, it appears that "conservation" constraints and taboos are restricted to nonessential species. Religious protection can also be quite selective. For example, among the Semang, Wazir-Jahan (1981:98) noted that while many religious taboos about resource utilization exist, generally "it appears that plants and trees with a utilitarian value are mythically depicted as destructive rather than beneficial to humans . . . subjected to a curse . . . which enables humans to exploit them for food and other kinds of economic uses." And some reported practices are perverse from a purely conservation perspective, as the prescription among the Montagnai that one should kill all beaver one encounters, whether useful or not, for any surviving individual will tell its kin that you are hunting, and future hunts will be more difficult. Religious beliefs may well have the potential to be an important proximate cause of conservation—but the effectiveness of such prohibitions cannot be assumed (Tuan 1968); they seem likely to be more effective when individual or familial benefits accrue to conserving behavior. That is, protective religious reverence for nature is likely to be more effective as ecological protection when kinship structures are strong and families benefit from conservative management.

These results are consistent with optimization models of resource acquisition, and "economic" approaches to resource use (e.g., Alvard 1993, 1994a; Alvard and Kaplan 1991; Hames 1987a, 1987b, 1989), as well as with the archaeological ant paleontological records. Traditional people are capable of overhunting (Anderson 1984; Bodley 1990; Clay 1988; Denevan 1992). Human introductions of crops or domestic stock or inadvertent plant or animal "weeds" can also cause ecological damage. From the Quaternary, documented extinctions have occurred as a direct result of human activity (Martin 1984). On the Hawaiian Islands, 54% of endemic birds went extinct due to activities of early Polynesians (Olson and James 1984); the same pattern, better- or worse-documented, can be made for most other continents or islands (Martin 1984). Diamond (1984) analyzed factors associated with resource destruction and human activity, including weaponry, stock, pigs, dogs, swift predators like cats, and agriculture. Usually they were deliberately brought by humans. In addition, rats and mice often accompanied humans inadvertently. Of the four prehistoric extinction waves definitely attributable to humans, three "resemble what literate Europeans have been doing on numerous oceanic islands" (Diamond 1984:852).

In Pleistocene North America, humans caused the extinction of large birds, and large and small mammals. Late Pleistocene extinctions in South America and Australia followed much the same pattern (Anderson 1984; Cassels 1984; Crosby 1986; Trotter and McCulloch 1984). Part of the pattern is attributable to deliberate human exploitation like hunting, but introduced stock, predators, and plants, and unwanted commensals like rats were also important. These problems have probably existed throughout human history. The abandonment of Mayan cities appears to have been related to agricultural failures (Deevey et al. 1979; Turner 1982). Ancient Sri Lankans, by forest clearing on mountainous regions, created flooding and reservoir siltation (Lowdermilk 1953).

Sometimes efficiency in getting a resource was not matched by efficiency in handling or storage. Plains Indians who hunted bison were highly selective in their use of meat, hides, and other by-products of the hunt (Haines 1970; Speth 1983), seeking particularly fat and fatty meat, and leaving heavy, less nutritious parts at the kill. When cliff jumps were used, they produced far more bison meat for less effort than competing technologies, but were hardly "conserving." Huge amounts of meat rotted at the base of cliffs; hunters took only the choicest meat.

These examples combine efficient technology and sometimes (islands) limited geographic area. Even in large areas, when efficient technology and profit are combined, impact is likely to be severe; neither alone is sufficient. In North America, for example, Great Lakes Indian societies had sufficient technology to have significant impact on beaver populations; nonetheless, these populations remained relatively stable until the Hudson's Bay Company entered the area and introduced a market economy, changing the value of beaver pelts. A male beaver pelt, in Albany Fort in 1773, was worth a brass kettle, or twenty steel fishhooks, or two pounds of Brazilian tobacco (Newman 1989:60). In this case, the technology was sufficient for some time without resulting overexploitation; what was lacking was immediate advantage to a hunter for continued intensive hunting. Similarly, when efficient steel axes were introduced in New Guinea, with an existing market (Salisbury 1962), serious ecolog-ical degradation followed. Nevertheless, among the Ye'kwana (Hames 1979), game could neither be stored nor traded in a market economy; its availability did not lead to feast and famine conditions. In these conditions, enhanced technology increased hunting efficiency, but not exploitation. Many examples of postcontact degradation appear, like these examples, to relate simply to changes in technological efficiency and profit. In sum, the cross-cultural and historical data suggest that, while people in traditional societies may be excellent ecological managers, at least in the short term, when conditions are harsh, they respond not by conserving, but by using more of what's available.

[This is a shortened version of the orginal and here some text was omitted.]

Resource Coalitions in Traditional Societies

Bleak forecasts of overexploitation due to individually rational, group-destructive behaviors are common (e.g., Hardin 1968), and this is part of the appeal of the Noble Ecological Savage paradigm. A number of large-scale environmental problems today have the property of some types of commons: it is individually profitable not to do what is in the common good—so, as Hardin noted, we should expect defection to be routine. Both economists and biologists recognize self-interest focused on resources as important in human behavior. Certainly the data analyzed and reviewed here are consistent with self-interest, and suggest that the principal factors leading to an appearance of "restraint" are simply ineffective technology, low population density, and low profit (Figure 4.1).

Many definitions of conservation imply cooperation, and loss of individual gain for the good of some larger entity. But cooperation, particularly when it imposes individual costs, is not widespread, nor is it equally likely under all conditions. People in small, tight-knit (highly interrelated) communities in which monogamy and maternal-kin biases are present, and warfare is infrequent, more strongly teach their children to cooperate. These data are consis-

Figure 4.1

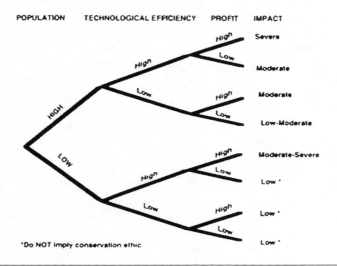

Environmental impact results from the contributions of population density, degree of available technology and the impact of that technology, and the degree of profit to pursuing resource use. There are situations in which environmental impact is low simply because, for example, the population density is low and the technology insufficient to accomplish a degree of utilization sufficient to have serious environmental impact. Such situations (*) do not imply deliberate conservation. In contrast, in conditions of high population density, highly developed technology, and high profit to exploitative enterprise, were we to find low environmental impact, we could confidently infer a strong conservation ethic.

tent with Ostrom's (1990) analysis of successful self-regulating common-property resource management systems as typically small, stable-membership groups in which individuals have both convergent long-term interests and relatively equal voice. In traditional societies individuals are still more likely to cooperate with their kin and long-term reciprocators than with others.

These results remind us again that a central question, in both traditional and modern resource use, is: Whose benefit drives the success of any strategy? Wynne-Edwards (1962) and others hoped to find the answer to limited population fluctuations of some species in terms of group benefit, and recent group-selection resuscitation attempts (Wilson and Sober 1994) blur functional categories, making no distinction between, for example, inclusive fitness maximization/kin selection and true genetic altruism. The immediately obvious problem in systems depending on genetic altruism is that "cheaters" in any population win; any system depending on individuals to perform genetic sacrifices for the benefit of strangers collapses quickly (e.g., Lewontin 1970), as the altruists are out-competed and out-reproduced by reproductively selfish individuals. The overwhelming determination, since George Williams's (1966) classic work, is that genes, carried about by individuals and shared by relatives, are what matters (Dawkins 1982, 1989; Hamilton 1964; for an overview, see Daly and Wilson 1983 and Krebs and Davies 1991). Hawkes and Charnov (1988) have elegantly summarized the arguments in anthropology.

Importantly, the group may sometimes *appear* to benefit as a result of the cumulative selection on individuals (e.g., Williams 1966: a herd of deer may be fleet, but only as an incidental result of selection on each individual deer for escaping predators—that is, we see a herd of fleet deer, rather than a fleet herd of deer). Even apparent reciprocity may be highly self-interested, and mutual benefit difficult to infer (Connor 1986). This can lead to misattribution of cause—the same logical flaws as the "population regulation" arguments of some anthropologists (e.g., Howell 1986; Viazzo 1990). Recent work (e.g., Alvard 1993, 1994a; Hames 1987b, 1991) has produced strong evidence that individual cost-benefit analyses fit the data better.

Yet long-term human resource-use cooperation does exist (though usually to get resources, not to conserve them for others), and cooperation can be a powerful force. Resource coalitions exist among men, among women, and between the sexes. Recent work (e.g., Keohane 1984; Martin 1993; Ostrom 1990; Putnam 1993; Taylor and Singleton 1993; see overview by Keohane, Ostrom, and McGinnis 1993) begins to delineate the conditions under which we expect more, rather than less, cooperative behavior in resource acquisition, and the conditions under which we predict longer-term (less discounting, less extractive) strategies of resource use. The general patterns of resource cooperation in preindustrial societies include ability to exclude outsiders, high genetic relatedness and long-term reciprocal interactions (facilitated by living in small, non-stratified groups), relative isolation, and no large markets. Thus, resource cooperation with "restraint" is more common when economic payoff for cheating is nonexistent or low, and social payoffs for restraint are high. Ostrom (1990) has described a very few larger-scale commons; in these cases, the local group still has considerable control, and its decisions are recognized as valid by larger governmental entities.

The problems are still complex, and they exist at a variety of levels, from small communities to international.

[This is a shortened version of the original and here some text was omitted.]

Conclusions: The Contexts of Conservative Attitudes

These data are clearly both very general and somewhat preliminary. Yet several things stand out. First, consistent with Hames's (1987b, 1989, 1991) and Alvard's (1993, 1994a) findings for specific societies, people in traditional societies do not, at least to their ethnographers, express a widely held conservation ethic. To the contrary, in most societies, people express individual need rather than some sort of communal long-term planning. Second, while there are clear ecological correlates to actual patterns of use, there is little pattern to the expression of conservation ethic—neither peoples in extreme environments, nor people with plenty, are more apt to suggest conservation, although perceived resource scarcity does seem to lead to expressed need to utilize everything. Third, expression of sacred prohibition may not help conserve a resource, and sometimes it may unexpectedly accelerate degradation (Tuan 1968). Fourth, traditional societies can indeed cause environmental degradation, usually when either population or technological changes are rapid. Finally, coalitions of several sorts surround resource and land utilization, and their impact can differ.

In contrast to the Noble Savage view, evolutionary theory leads one to predict (e.g., Alvard 1993, 1994a; Alvard and Kaplan 1991; Hames 1991; Low and Heinen 1993) that several patterns should be apparent in resource use by traditional societies. First, when use of a resource profits an individual or his/her family and the technology is sufficient to accomplish the exploitation, use will be unrestrained; second, when this condition exists in situations of high population density, real ecological damage can result. There is some suggestion that land tenure rules may influence the degree of restraint (inherited lineage land, privately owned, may be more conserved than common property), but codes for degree of ownership/common use are difficult to obtain.

In most hunting and gathering societies, the reproductive benefits to striving (e.g., being an excellent hunter) mean that while cultural elaboration (e.g., religious beliefs) may surround hunting, men who are good hunters can get more wives, and men's attitudes about hunting (and various achievements) center on how hunting skill helps their social and reproductive success (e.g., Chagnon 1988). Men use both time and resources to foster male-male bonding and negotiations that constitute "politics." These appear to have reproductive payoff for individuals, both in children's survivorship and in access to mates (e.g. Chagnon 1988; Hames 1979; Hawkes 1993; Kaplan and Hill 1985a; Low 1990a). This appears to be true even in societies in which hunting rewards are shared (Hawkes 1993): men hunt in a manner that is not optimal from the point of view of caloric return, but may be from a reproductive viewpoint, as they pursue a "show-off" strategy, going for the big kills. Thus, consumption is profitable for both survival and reproduction, and individual interests seem to be paramount.

The crude data here support such a view. People appear to seek resources more intensively when resources are scarce; for example, Alvard (1994a) found that among the Piro, hunters took the prey that returned higher calories for effort, without conservation considerations. The rare expressions of sacred prohibitions do not have, in this sample, any impact on use; neither do they protect environments from degradation. Even when resources are exclusively managed by small groups of relatives and neighbors likely to interact for long periods of time, overexploitation is possible, especially when there is profit to continued exploitation (e.g., when external markets exist).

Traditional peoples often do express concerns about the long-term "health" of the ecosystems on which they depend. They can have deep and unique knowledge of sometimes subtle ecological relationships because such knowledge has had survival value (e.g., Atran 1993). And they sometimes can manage resources sustainably, even when new highly efficient technologies make over-exploitation easy, particularly when the interests of kin or local community are vested in sustainable use (e.g., White 1988). But these are possible outcomes, not driving forces. So far most empirical studies suggest that people take resources from the environment to feed themselves and their families (e.g., Smith 1983), to garner favorable attention (especially for men in sexual selection; e.g., Hawkes 1993), and to reduce variance through reciprocity (typically biased by kinship relations, with closer kin receiving more aid e.g., Hames 1987a). Often, traditional societies view resource acquisition directly as a form of mastery over the environment (Stearman 1994). Empirical studies suggest that people take as much as they need (or sometimes as much as they can); there is typically no reason to consider potential impacts of maximum use if technology changes (e.g., Johnson 1989). This, I suspect, is the source of the pattern in the cross cultural data here that degradation frequently follows technological change. In the Romantic view, this is an evil of society; the data suggest that maximal extraction is simply a characteristic of humans as well as other living things.

The common assertion that utilization is deliberately modified in traditional societies to achieve long-term conservation seems to arise from the observation that many preindustrial societies have less impact on their surroundings than industrial societies—but this is a misattribution of cause, a misinterpretation of the incidental outcome of combinations of low population density, low utility, and/or inefficient extraction technology. In fact, *ceteris paribus*, the degree of impact is simply a function of the population density, the relative impact of the technology used, and the profit arising from utilization of the resource. As Figure 4.1 shows, several combinations of population density, profit, and technological efficiency will yield low impact, but do not imply conservation. To infer that low impact must imply conservation ethic is simply not logical; we can infer deliberate conservation only when impact is low even though there is high profit to be made, the technology is sufficient, and/or the population density is moderate to high.

Conservative resource use is likely to be favored when there is rapid and clear feedback regarding the impact on family and individual welfare (when overexploitation carries clear individual and familial costs). Neither of these conditions are necessarily related to an expressed ethos of conservation or reverence. Resource degradation is likely when population density is high, extraction technology efficient, and profit great (Figure 4.1: Low 1995; Low and Heinen 1993). Because this combination of high population density, prof-itability, and efficient technologies is typically recent, it is not surprising that widespread concern about a conservation ethic is more or less a twen-tieth—century phenomenon (e.g., Callicott 1989; Hargrove 1988; Spoehr 1956; Strong 1988). The bottom line for inference, then, is that environmental impact may be low, even if personal striving is intense—if technology is insufficient to accomplish resource extraction, or if extraction yields the individual no profit (see Alvard 1993, 1994a; Hames 1991). "Noble Savage" arguments that impute conservation ethic from low levels of degradation are simply wrong.

Traditional societies can cause destruction or extinction, and may do so in a variety of ways. They may do so directly, as in hunting a species to ex-tinction. Extinction or environmental degradation most often follow deliber-ate exploitation when technology is sufficient and individual payoffs exist (Figure 4.1), especially if feedback on human impact is slow. Throughout most of our evolutionary history, these conditions were seldom met, although humans have apparently always modified their environment (see above). Deliberate exploitation arises from individual striving, while introductions result more often from a lack of information (sometimes not only about the ecological relationships, but even the presence of the introduced species). Throughout most of human evolutionary history, involving both accidental and deliberate ecological destruction, technology was typically not widespread, and humans had principally local and limited environmental impact (but see Lowdermilk 1953 for an example of widespread impact).

The analytic issues of individual versus group benefits and long- versus short-term context are crucial for understanding resource problems, from traditional societies to industrial nations. It may seem like a long jump from traditional societies to current industrial societies, but many of our conflicts of interest, and difficulties in getting long-term cooperation among strangers, may make sense when set in this context. Problems like acid rain, ozone depletion, and global warming are classic individual/group-long/short-term problems. NIMBY ("Not In My Back Yard") is a "commons" problem (Oye and Maxwell 1995), again involving the difficulties of individual versus group costs and benefits. Consistent with the data presented here, I argue that there is no evidence that human nature has somehow changed, as we have become estranged from our ecological "roots"; rather, we have added new, and novel, layers of resource conflicts because we exist in a novel evolutionary environment, sometimes living among very large groups of strangers and often moving frequently.

In a way, it is encouraging that the data suggest that we have always oper-ated by the same rules; perhaps new, expanded versions of the same rules can still promote cooperation (Low 1995; Low and Heinen 1993). The most easily solved problems appear to be those in which the costs and benefits accrue to the same individuals—probably usually true in traditional societies, and untrue for many current large-scale problems. To avoid or solve such problems, information about the outcomes of different behaviors may be all that is required (e.g., recycle and thereby avoid higher taxes resulting from costs of landfills). Exhortation may also be useful in problems with these characteristics: when "doing the right thing" has social payoffs, especially if it is not otherwise costly, people may be relatively willing (Low and Heinen 1993). Strategies of "Small Wins" (picking manageable pieces of the problem to ensure positive reinforcement; Weick 1984) can sometimes be useful.

Today's large-scale problems often involve novel conditions. The individuals suffering from acid rain effects, for example, are often far removed, and in different political jurisdictions, from the producers of its precursors. The producers would like to avoid the costs of scrubbers (the costs of not having a scrubber are, in economic terms, externalized—something that has been advantageous throughout our history); those who pay the costs of acidification would like to send those costs home to the producers. The most commonly invoked strategies in western industrial nations may be those of economic incentives (taxes, tax relief) and government regulations; for either of these to be successful, effective within-society, coalitions are required (Heinen and Low 1992; Low and Heinen 1993)—and not always easy to accomplish.

These are complex and troubling issues, and I argue that simplistic "Noble Savage" exhortations are neither correct nor useful. Perhaps the most cogent view has been offered by Marcos Terena, President of the Union of Indigenous Nations (UNI), in Belem, Belize, in 1990 at an international seminar on ecological problems in Amazonia. His words (for which I am indebted to Professor Emilio Moran) were:

> Why do you white people expect us Indians to agree on how to use our forests? You don't agree among yourselves about how to protect your environment. Neither do we. We are people just like you. Some of us view nature with a great sense of stewardship whereas others must perforce destroy some of it to obtain what they need to eat and pay for expensive medical treatment and legal counsel.

Solutions, then, must begin with understanding why and how we evolved to use resources, and how individual costs and benefits influence our resource exploitation patterns. If self-interest is unlikely to disappear, perhaps we can, through understanding how and why it evolved, learn to use it in solving environmental problems (e.g., Low 1995; Low and Heinen 1993; Oye and Maxwell 1995; Ridley and Low 1993).

5

Evolutionary Ecology and Resource Conservation

Michael S. Alvard

Recent field work has shown that, contrary to commonly held beliefs, subsistence hunters do not conserve prey resources. Evolutionary ecologists have approached this problem by using foraging theory to show that subsistence hunters prefer short-term returns over the potential long-term returns generated by resource conservation. An important reason for this outcome is that the resources exploited by subsistence hunters are often open-access, which means that collective-action problems can ensue. Ownership is critical for conservation to pay, but even if resources are privately owned high opportunity costs can minimize the long-term benefits of restraint. Because the benefits of conservation accrue in the future, the benefits must be discounted. This is because future benefits may not be realized for a variety of reasons. Recent efforts to understand evolved human time preference suggest an evolved discount rate between 2% and 6% annually, depending on many factors. If the growth rate of a potentially conserved prey population is less than the discount rate, the long-term benefits of conservation will fall short of the short-term benefits of exploitation.

Why Resource Conservation?

Many important aspects of human nature revolve around common problems associated with acquiring, defending, and distributing resources. It is increasingly evident that foraging constraints,[1] as well as competition, and cooperation with conspecifics,[2-4] selected for increased intelligence in primates. The cognitive tools for responding to risk and reward likely evolved in a context of resource acquisition and distribution during our evolutionary past, in the so-called environment of evolutionary adaptedness.[5-7] The cognitive skills required for cooperative hunting, such as cheater and cooperator detectors,[8] and especially the accounting required for fitness-enhancing harvesting, distribution, and consumption (e.g., time discounting[9]), are reflected in behaviors we exhibit today. Our evolved psychological disposition to respond economically is likely derived from past ecological relationships with the beasts our hominid ancestors preyed on and the organisms that preyed on hominids. It is not hard to imagine that many human social and political institutions have developed as solutions to complex economic production and distribution problems.[10] Thus, it not surprising that human evolutionary ecology borrows from economic theory[9,11-15] and, increasingly, vice versa.[16-20]

It is in this context that the topic of resource conservation has been the focus of theoretical development and hypothesis-testing by evolutionary ecologists. As I will show, the initial evolutionary approach to resource conservation has contrasted predictions generated from optimal foraging theory with predictions based on assumptions of conservation. This work has primarily examined subsistence hunters and their relationship to prey resources. The general conclusion is that prey conservation is rare in these contexts. In spite of this, it has been noted that while truly altruistic conservation is unexpected, evolutionary theory does not rule out conservation *per se.*[21] Both the short-term and the long-term benefits of behaviors performed in the present must be considered to predict the pattern of resource acquisition. Conservation is expected to develop in ecological situations in which the long-term benefits outweigh the short-term costs.[22]

The timing of resource use is an implicit decision all organisms must make. Organisms can use resources in the present or defer use to some later time. The adaptive choice depends on many parameters. It should be noted that both immediate and delayed consumption are quite common in nature. Rodents and birds hoard resources to ensure survival through seasons of scarcity.[23] Some carnivores kill large prey they can not consume at one time and then bury or cache the carcass for later consumption.[24,25] Whether these activities constitute "conservation" is not important at the moment. The point is that some organisms defer resource consumption. Reproductive restraint is also quite apparent in nature; it is what Wynne-Edwards[26] tried to explain with his version of group selection. Lack[27] showed how reduced clutch size could, in the subsequent generation, maximize reproductive success through selection on individuals. Modern post-industrial populations postpone reproduction for years past maturity.[14] In fact, much of life-history theory is designed to understand restraint and the timing of expenditures. Growth, for example, can be thought of as trading off present gains for future ones, if reproduction is delayed until a time when the organism is larger and its reproductive value greater.[28] Conceptually, resource conservation fits squarely within this set of adaptive solutions.

Conservation by Traditional People:
An Example of a More General Issue

It has long been suggested that game conservation is characteristic of many traditional people living in subsistence economies.[29-42] There is a widespread belief that western capitalist ways upset a harmonious balance of humans in nature. Sahlins' *Stone Age Economics*[43] is a classic example of this perspective. These unsubstantiated claims have helped stimulate empirical and sophisticated research on the problem of resource conservation among pre-industrial people. The growing consensus from this empirical work is that resource conservation in traditional economies is not as common as was previously thought. The evidence is particularly strong when the resources involved are hunted prey. Less attention has been paid to plant resources, domesticated resources, or non-renewable resources such as lithic materials. Models that assume that individuals behave to maximize their short-term interests and those of their families have been particularly successful at understanding the behavior of people in subsistence economies.[21, 44-55]

One source of the original error is the fact that many behaviors are *apparently* conservative because they are associated with sustainable harvests or the harvests that are biased in ways consistent with genuine conservation. A sustainable harvest is not evidence of conservation. It is an error to conclude that native groups are conservationists simply because they are not over-exploiting their resources.[21,44,53,56] This error has contributed much to a misunderstanding of hunting and gathering economies. Small and mobile groups can use resources in a sustainable manner, but are not necessarily conservationists. Apparent conservation in such a context has been termed *epiphenomenal conservation*.[57] If a growing population is not yet resource-limited, conservation is not an adaptive strategy for individuals and is therefore predicted to be rare. Consuming only what one needs *is not necessarily* conservation.

For example, during the breeding season, male Stellar's sea lions are much easier to kill than females are.[58] This results in a harvest that is biased towards adult males and one that is more sustainable than a harvest that is random with respect to sex or one that is biased towards females. Observing a male-biased harvest might lead to the erroneous conclusion that this bias represents conservation when it does not.[59] Archeological data show that prehistoric arctic whale hunters biased their kills towards juvenile individuals.[60,61] Krupnik[62] has suggested that this pattern is explained as "a utilitarian orientation towards an easier and more accessible prey." The same pattern can be observed with non-human predators.[63,64]

Another source of confusion is that many claims of resource conservation have been made without empirical data. Dietary prohibitions—food taboos—are a good example of traits that have often been attributed to conservation intent, usually with no empirical support.[65,66] Aunger[67-68] has quantitatively examined the nutritional impacts of food taboos in the Ituri forest of eastern Zaire. Although both the Lese horticulturalists and the Efe Pygmies have a large number of food taboos, Aunger found that these taboos had a minimal impact on diet. Less than 2% of calories were restricted for the Lese and less than 1% for the Efe because individuals frequently invoked "exception rules" to allow themselves to eat otherwise taboo food. Because the Efe and Lese kill and eat the animals in spite of the taboos, it is difficult to argue that the taboos are maintained because of their conservation effects.

An Operational Definition of Conservation

Many of the problems I have discussed can be ameliorated with an operational definition of conservation and empirical hypotheses testing. Conservation is when individuals reduce their level of resource use below what would be fitness maximizing in the short term in exchange for long-term, sustainable benefits in the future.[21,69] By short-term, I mean over the course of a number of foraging bouts; by long-term, I mean over the course of years or generations.

Using the short-term cost criterion alone, behavior that has unintended conservation-like consequences but no short-term costs can be rejected as conservation. Thus, a male biased harvest of Stellar sea lions can be rejected as conservation if it is determined that this strategy maximizes the hunter's short-term harvest rate. While necessary, a short-term cost is not a sufficient criterion for conservation, however, Smith[70] easily imagines a food taboo that is costly and also reduces hunting pressure, yet is unrelated to conservation. He points out that in order to identify conservation, it is necessary to demonstrate intent on the part of the actor or design via natural selection. The long-term sustainable benefits of the behavior must be the reason the trait is maintained. Thus, to be deemed conservation, a behavior must exist because of its conservation effects; in other words, the behavior must be an adaptation.[45,21] (See Smith and Winterhalder's[71] and Elster's[72] insightful discussions of functional and intentional explanation.)

The problems of identifying adaptation have been duly noted by many workers, notably Williams[73] (see Reznick and Travis[74] for a recent review). Design is often inferred from outcomes, or, as stated by Smith and Winterhalder,[71] "...the highly regular relationship between an organism's way of life, its environment and the kind of characteristics that will actually improve its fitness" (p. 27). Errors in identifying design can be made if different designs result in the same outcome, as can be the case for true versus epiphenomenal conservation. The answer in this case is for the researcher to look for the "...highly regular relationship" that is expected of design. Operationally, this requires careful field testing. For the case of prey conservation, the argument for design or intentional conservation is strengthened if tests show that people consistently engage in a number of costly behaviors that all mitigate game depletion.[69]

Optimal Foraging and Resource Conservation

Foraging theory has provided useful models for predicting subsistence deci-sions by human hunters by assuming that hunters maximize foraging return rates.[15,21,75-78] Prey choice and resource patch departure times are the decisions most often modeled.[79] Foraging theory can also provide a model for decision mak-ing by nonconserving foragers. I and other evolutionary ecologists have compared the models to actual forager behavior as a test for conservation.[21,44,54,80-83] This research design does not rule out the possibility that resource conservation may be an adaptive strategy in some contexts. Rather, it merely tests for the presence or absence of a conservation strategy in particular cases by contrasting models that use alternative time frames for assessing outcomes. The prey-choice model describes foragers that maximize short-term benefits (net harvest per foraging bout), while the resource conservation models describe foragers that maximize long-term benefits.

An example is Smith's[81] interpretation of Cree hunting-zone rotation. Tradi-tionally, the Cree are boreal forest hunters living in Canada. Feit[84] argued that Cree mobility was designed to avoid prey depletion. He proposed that the Cree rotated hunting zones every few years to avoid depleting prey populations. Moore[85] made a similar claim for the Naskapi, arguing that scapulamancy by shamans was a way of directing hunters randomly across the landscape to avoid resource depletion in any one area. Smith noted that foraging theory provides an alternative explanation for the observed behaviors. Foraging theory's mar-ginal-value theorem predicts that rate-maximizing hunters will not completely exhaust a patch because the return rates from depleted patches eventfully drop below what could be had from leaving and traveling to an undepleted patch.[11] When hunters direct harvests away from depleting areas in this way, it appears to be conservation, yet these behaviors are more parsimoniously explained as harvest rate maximizing. Because the hunters pay no short-term costs when they leave a depleting patch, they are not conserving. This example demonstrates the usefulness of both operational definitions like that developed for conservation and theory-based analytic approaches for disentangling the sorts of confusion I have described.

One critical issue is the length of time over which the models assume forag-ing return rates to be maximized. There is some ambiguity here. Stephens and Krebs[79] maintained that the models generated from the Holling's[86] disc equation assume long-term rate maximization as the proxy currency to fitness. They and other workers[87] rejected the proposal by Templeton and Lawlor[88] that per-encounter rate maximization is more appropriate. Stephens and Krebs also rejected the other extreme when they noted that infinite time-rate maximizers would value future harvests as much as present harvests. Such foragers are predicted to delay harvesting for considerable lengths of time to obtain a sufficiently large future gain. This is an unrealistic assumption of a

0% discount rate.[89] Stephens and Krebs are equivocal, however, about where the prey-choice model falls between these two extremes. It is apparent that the models assume a time interval somewhat less than infinity, but greater than the time span of one encounter. Stephens and Krebs suggested testing various time frames to see which fits the facts of a given case. It is reasonable to assume that the time frame for which return rates are maximized will vary in adaptive ways depending on the circumstances of the organism.[87]

To understand why a 0% or a infinite discount rate is unrealistic in most foraging contexts, time preference must be understood. Economists have long known that humans display time preference,[90] or the tendency to favor certain schedules of resource consumption over others. Much evidence shows that people tend to prefer present consumption over future consumption.[91,92] The rate at which future benefits are discounted and measured in terms of present value is the discount rate. Other organisms also tend to discount the value of future rewards. Experimental research shows that foraging birds, for example, will prefer immediately available prey over delayed prey of the same value and, in some circumstances, even larger delayed prey.[89]

Time preferences are thought to exist for two reasons. The first is some probability that future benefits will not be realized; the second is the loss of compounding gain associated with delay. For example, for a variety of reasons a peccary will not be as useful a year from now as it is in the present. A future peccary is discounted because future benefits may not be realized: The peccary may be killed and eaten by someone else or the forager may die. However, even if the future peccary is 100% assured its future value may still be discounted because present use of the peccary might result in higher long-term benefits if the gains compound over time. Cash resources can be put into a bank with compounding gains equal to the interest rate. When the resource is a peccary, it can be fed to one's offspring or used to obtain additional mating opportunities, and in these ways be converted into descendants at a certain reproductive rate.

In relation to this issue, Hill[93] notes that the prey-choice foraging model assumes that no matter the nature of the harvest today, it will not have an impact on the profitability of the prey types tomorrow. Essentially, the prey models assume no depletion effects. Avoiding depletion is an important reason why we might expect the restraint predicted by true long-term rate maximizers. Because such costs are excluded by the models, the rate-maximizing time frame they assume is better described as relatively short-term[9] (over the course of one or a series of foraging bouts). This time frame is short compared to conservation strategies whereby present-day foraging can have an impact on return rates years or generations in the future. It should be noted that the Marginal Value Theorem (patch-choice model) assumes depletion effects within patches, but only during single visits to each patch.[11]

According to the prey-choice model, whether a particular prey type among a suite of types should be pursued depends on its profitability and the opportunity costs of not pursuing it.[79] Two characteristics determine a prey type's profitability. The first is the expected net gain from pursuit after encounter; the second is the expected time spent handling the prey if it is pursued (handling includes pursuit and processing costs). Hence, profitability is the expected units of currency harvested per unit of handling time. Opportunity costs accrue to a forager from missed chances to do better. Thus, in additional to its profitability, the decision to pursue a given prey type should also depend on the encounter rates of more profitable prey.

The conservation hypothesis predicts prey choice to be based on prey reproductive parameters such as sex, age, density, and reproductive rate. Variability in these factors among harvested prey can have strong effects on hunting impact and sustainability. When ecological circumstances are favorable to research design, the predictions of each hypothesis are mutually exclusive and there exists an opportunity to test. Short-term rate maximization predicts that, all else being equal, larger and easier-to-harvest packages will be preferred over smaller, more difficult types. All prey that fall within the optimal diet should be taken upon encounter, but larger and easier-to-catch prey should still be preferred during simultaneous encounters. Resource conservation predicts intraspecific selectivity similar to a husbanding or management strategy. To maximize long-term return rates, hunters should prefer harvests biased towards males and low reproductive value, and both older and younger individuals.

Stiner,[94] in a comparative study of prey choice by nonhuman predators and human paleolithic and holocene hunters, concluded that human hunters are the only predators that regularly produce mortality distributions biased towards prime-aged prey, a pattern that is not consistent with the conservation hypothesis. Data I have collected on intraspecific prey choice by Indonesian Wana and Peruvian Piro subsistence hunter-horticulturalists show a similar pattern. Both groups hunt or trap for the majority of their meat.[21,69,95,96] Neither Piro nor Wana hunters avoid killing prime, reproductive-aged prey in favor of the young or old (Figure 5.1). Hunters from both groups kill primarily prime-aged adult primates, and take peccary and deer in proportion to a living age structure. The tapir, capybara, and anoa kills include more immature individuals.

It is hypothesized that more immature members of the large-bodied species are killed because of intraspecific variability in profitability due to differences in body size and handling times for different-aged individuals of the same species. In spite of being more common, the immature and infants of the smaller-bodied prey species are hypothesized to fall out of the optimal diet, at least for shotgun hunters. Analysis of the Piro data does show a significant relationship between adult body size and the proportion of immature individuals killed for each species.[69] Other factors could also be important because, as noted earlier, in addition to body size, pursuit decisions are predicted

Figure 5.1

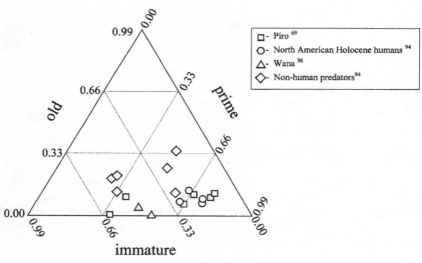

This figure plots three-category kill profiles of prey harvests from sources including Piro shotgun hunters,[69] Wana trappers and spear hunters,[96] and prehistoric North American hunters,[94] as well as nonhuman predators:[94] the Indian tiger, African lion, African wild dog, spotted hyena, and wolf. Using empirical data on animal populations, Stiner[94] developed this method to examine the age-selective prey choice of nonhuman and hominid predators. Points falling toward the top of the graph represent old-adult-biased prey selection, points to the lower right represent prime-adult-biased selection, and points toward the lower left represent a bias toward immature prey. Human hunters tend toward a prime-dominated profile relative to the nonhuman predators. Piro hunters' selection of immature prey may be related to prey body size. Immature individuals are harvested in greater proportion among larger bodied species, suggesting that the young of small-bodied species fall out of the optimal diet.

to depend on both handling times and encounter rates with higher-ranked prey. The immature bias found in the Wana pig kill could be the result of trapping technology that takes young and inexperienced individuals.[96] Neither the Piro or Wana hunters avoided females to bias their kills toward males. The data show that the sex ratios of harvested prey did not significantly differ from those of the wild populations.[69]

The strongest evidence for a lack of conservation is resource depletion resulting from subsistence activity. Data from the Piro and Wana show that both groups have locally depleted some of the more vulnerable large primate species. For example, the Wana have nearly depleted their area of macaques (*Macaca tonkeana*). Figure 5.2 shows encounter rates for macaques and pigs (*Sus celebensis*) during transects in two unhunted and one hunted Wana area.

While there is no difference between encounter rates in the two unhunted sites, the macaques are significantly less common in the hunted areas, to the point of near depletion. Similar results show that the Piro have seriously depleted the large primates in the area around their village.[95] Other studies on contemporary,[98-100] as well as prehistoric societies[101,102] also show that subsistence hunting can have serious negative effects on prey populations. Interestingly, the data show that the Wana hunt pigs and the Piro hunt peccaries sustainably. Studies that have shown evidence for over-exploitation by subsistence hunting populations usually do so only for a limited number of the total prey species hunted. That is, some species are hunted sustainably while others are not.[100, 103,104] This emphasizes the point that hunters who lack behaviors designed to truly conserve will not necessarily overhunt.[83]

The general conclusion of researchers who have applied foraging models to human subsistence hunters is that hunters pursue prey types in their optimal diet regardless of the prey's vulnerability, reproductive value, or state of local depletion, and often hunt species to the point of local depletion. In other words, the data indicate that hunters maximize their short-term return rates.

Figure 5.2

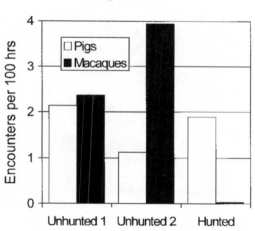

Comparisons between encounter rates during census transects in one area hunted by Wana and two unhunted areas in Morowali Nature Reserve, Sulawesi, Indonesia.[96] The data show that pigs were equally common in all three areas. There was also no significant difference in the size of pig groups encountered at the three sites. The lack of any significant decline in encounter rates in the hunted area suggests that the Wana pig harvest is sustainable. In contrast, the macaques show considerable local depletion in the hunted area. Specifically, macaques were encountered 10 times more often at the two unhunted sites than they were in the hunted area. There was no significant difference in the size of macaque groups encountered at the three sites. The differential impact may be related to differences in the reproductive rates of these two species. Primates are notoriously vulnerable to over-hunting because of their slow reproductive rates. Habitat disturbance from gardening activities may also have differential impacts, independent of hunting, on these two species.

Although conservation is theoretically possible and has been observed in some contexts, it has not been observed in the simple subsistence economies like those of the Wana and the Piro, where hunting is important. Numerous other studies concur. Smith[77] found "little support for the idea of indigenous game management among the Inuit" (p. 256). Data on the Ache show a "complete absence of any practices obviously designed to check overharvesting of resources"[82] (p. 236). For the Cree, "conservation may be the incidental effect of efficient foraging in a heterogeneous environment"[105] (p. 97). Data from a large sample of Amazonians show that, "Hunters and fishers who have depleted game resources hunt and fish more intensively"[44] (p. 96).

Why is Conservation Uncommon in Simple Subsistence Economies?

Rogers[22] points out that conservation will evolve when its benefits outweigh its costs. While this is not a particular revealing conclusion, it does place the question in an economic context. If we can understand why prey conservation is rare where the costs outweigh the benefits, we can glean insight into the sorts of contexts in which the benefits outweigh the costs. I will discuss three related reasons why conservation is uncommon in societies like the Piro and the Wana. The first and perhaps the most important reason is that most of the resources used in simple subsistence economies are open-access and, for this reason, are vulnerable to the tragedy of the commons. Second, even assuming private ownership, resources may be relatively abundant (supply exceeds demand). This decreases resource value and the benefits of both conservation and private ownership. Finally, even if resources are private and valued, if the opportunity costs of conservation are high, conserving may not pay in the long term.

Ownership

The standard argument from economic and political theory tells us that conservation is unlikely when resources exploited are open-access, as is often the case in traditional subsistence economies.[106] This results in the classic "collective action problem," in which individuals are unlikely to provide a public good by altruistically sacrificing their own harvest if others who have not necessarily sacrificed anything are free to share the benefits.[107,108] This dilemma is known as the tragedy of the commons.[106] The world's marine fisheries are a prime example. Marine resources are the last major global nondomesticated or husbanded food source, and the last of the open-access resource stocks. The ocean is so vast and most marine resources so mobile that defense is exceedingly costly. National ownership in the United States, for example, is extended by the Magnuson Act of 1976 to only a 200 mile "exclusive economic zone."[109] And even with that Act, defense is only tenable within a prescribed spatial territory, while the marine resources move in and out of it with ease. As a result, the world's fisheries are notoriously difficult to regulate and often are overexploited.[110] The Salmon War during the summer of 1997, in which Canadian fisherman blockaded an American ferry at the docks of Fort Rupert, BC, demonstrates the conflicts that often result over such resources.[111]

The standard solution is private ownership.[106] Owners are motivated to conserve because they reap the benefits of their sacrifice. Ownership helps prevent competitors from taking a free ride on the work of others and can help solve the collective-action problem. What is not often discussed is that ownership does not come cheaply. Owners must pay costs to defend a resource against outsiders. Ownership, in this sense, implies resource defense and territoriality. Cashdan[112] defined territoriality as "maintenance of an area within which the resident controls or restricts use of one or more environmental resources" and, interestingly, even views it as "a type of resource management that depends on controlling and limiting access to resources" (47-48).

Only if defense costs are sufficiently low or the value of the resource sufficiently high will conservation be likely. If territorial defense costs are high and private ownership difficult, conservation is less likely because of collective-action problems. Many factors affect the cost and benefits of defense. Classic cost-benefit theory of territoriality[113,114] suggests that the spatial distribution and predictability of resources are important for determining the pay-off that can be obtained by monopolizing them. Clumped resources are less costly to defend because of economies of scale.[112,114,115] In addition, if resources are scarce relative to demand, the benefits of defending even widely dispersed resources can be high.

In certain cases, sufficient territoriality combined with self-imposed restraint can result in genuine conservation. A good example of how territoriality can provide a context favorable to genuine conservation comes from the work of Acheson[116,117] and Wilson[118] on traditional Maine lobstermen. Maine lobstermen have norms of ownership and resource use that are enforced through threats of violence and damage to equipment. In this way, the commons become privatized.

Acheson described two types of territorial arrangements, perimeter defended and nucleated areas, both of which are territories in the sense that access to them is limited. Nucleated areas are associated with harbors and have boundaries that become fuzzy as distance from the harbor increases. Rights to work such areas are relatively easy to obtain. Also, there is a high ratio of fishermen to area. Perimeter-defended areas are controlled by lobstermen united by ties of agnatic kinship who vigorously defend their boundaries. If cheaters are detected, retribution is swift. Acheson referred to such arrangements as "fiefs." Territoriality and associated policing have apparently increased the benefits of conservation. Groups that own and defend perimeter-defended areas practice restraint that results in long-term benefits. Restraint includes trap limits and closed seasons. As a result, lobstermen in perimeter-defended areas find a higher mean number of lobsters in their traps than do those in nucleated areas. In addition, the lobsters are larger and higher priced, resulting in greater profits.

While ownership and resource defense are necessary to help avoid the commons syndrome, they are not sufficient. Like nonhuman animals, many human societies are territorial to some extent but do not practice conservation. Lions guard territory and work to exclude outsiders, but group members do not conserve the resources so guarded. In the same way, territoriality in the form of fishing-ground tenure does not necessarily mean conservation, although this is how it has been perceived.[119,120] Marine tenure in Oceania has been interpreted as a way in which individuals maintain control over scarce resources and exclude others. This view proposes that such territoriality is motivated by self-interest: It does not represent a means to conserve resources but rather to intensify exploitation. Polunin[121] pointed out that fishing rights are usually given by the "owner" to anyone who asks, and argued that the function is not to conserve, but rather to enhance the reputations of and increase obligations to the owners. In other words, territoriality can simply result in excluding others from the fruits of over-exploitation enjoyed by those strong enough to control the resource. Thus, while ownership does limit access to resources and is necessary for conservation to be effective, the presence of territoriality is not a sufficient test for conservation.

Resource Value and Abundance

For a resource to be economically defendable and for conservation to provide benefits, the resource must have sufficient value. Some level of scarcity adds to the marginal value of a resource, assuming it is one for which there is demand.[15,122,123] Marginal value is an economic term that refers to the value obtained when an actor consumes an additional unit of resource. If a resource is abundant and not limiting, an additional unit of resource has less value to a consumer than if the resource was scarce. If resources have less value, the returns from the defense required to make them private goods decline and conservation is less likely.

In the Piro case, shotgun hunters harvest no more per consumer per day than do the nearby, bow-hunting Machiguenga. Both groups harvest around 240 grams of meat (gross) per consumer per day.[124] The Piro could easily harvest more prey with their efficient shotgun technology, but they do not. This suggests relatively abundant resources, diminishing returns to additional harvesting for the Piro, and, perhaps, one reason they do not conserve prey resources.

The corollary is that the less abundant a resource is, the greater its value, the greater the return for its defense and ownership, and the greater the likelihood that conservation will pay. Of course, the scarcer the resource, the more risk those without it are willing to take to acquire it, and the higher the defense costs to keep control. This is in many ways analogous to marginal analysis of tolerated theft sharing.[15]

Opportunity Costs

Even if a resource is a private good and economically defendable, conservation is unlikely if the opportunity costs of conserving it are high. Opportunity costs accrue from not doing something. The opportunity costs of conserving accrue from not exploiting the resource immediately. The opportunity costs of killing an animal accrue from being unable to eat its descendants. To paraphrase Alan Rogers,[22] "The costs of conservation are paid in today's dollars, but the benefits accrue with tomorrow's dollars—dollar which are worth less than today's." Because the benefits of conservation are deferred, they must be discounted accordingly. The less the future is discounted, the more conservation will pay off.

This issue can made clear by looking at the work of Colin Clark,[125] a bio-economist critical of maximum sustainable yield models. These models calculate the maximum number of animals that can be removed from a population on a regular, sustainable basis without driving the population extinct. Clark criticized the models because they assume a zero discount rate. According to that assumption, the resources harvested at some distant point in the future—10, 20, or 30 years—are worth as much as resources harvested today. Clark concluded that "for a maximum sustainable yield strategy to be favored, the reproductive rate of the resource population must be higher than the discount rate—the rate of return from the best current alternative investment" (p. 47). Clark's commonly cited example shows that economically rational whalers will harvest whales to the point of near depletion. This will occur if whale reproductive rates are lower than the interest rate. Whale populations grow at a rate of approximately 3% a year; a higher harvest rate will eventually lead to their extinction. If whalers can invest the proceeds from the harvest at 4%, there is no economic incentive to conserve the stock.

This implies that the lower the reproductive rate of the prey, the less interest the harvester should have in conserving its population, even if it is owned (see Figures 5.3 and 5.4). Clark hypothesized that if the discount rate is higher than the reproductive rate of the prey species, the optimal choice is to harvest the resource as rapidly as possible and invest the capital in a current investment with a higher payoff. If the hunter can put those kilograms of meat to good use now—that is, if the meat has greater present value to the hunter than the future discounted value—rapid harvesting is the optimal choice even if it means prey depletion. If the discount rate is higher than the growth rate of the populations, this is the economically efficient choice.

The Piro, however, do not conserve peccaries (nor spider monkeys for that matter) in spite of their high reproductive rate. Why? First, peccaries are an open-access resource that is costly-to-defend. Restraint on the part of any one hunter provides a public good enjoyed by everyone. Second, peccaries are abundant relative to demand. Data do not indicate local peccary depletion at

Figure 5.3

Conserve Collared Peccaries or Spider Monkeys?

Collared Peccary
Average adult weight = 22kg.
r_{max} = 0.84
K = 3,714 for 314 km2
Population at MSY = 1,857
 MSY = 780

Spider Monkey
Average adult weight = 9kg.
r_{max} = 0.08
K = 4,191 for 314 km2
Population at MSY = 2,095
MSY = 84

Collared peccaries and black spider monkeys are two of the prey types hunted by the Piro. This figure presents the annual maximum sustainable yield (MSY) for both species from an area 314 km². This is an area with a radius of 10 km, the distance a hunter can walk and expect to return to camp by dark. MSY depends on the maximum intrinsic rate of increase (r_{max}) and the carrying capacity where MSY = r_{max}K/4.[126] K and r_{max} have been estimated for these species by Robinson and Redford.[127,128] The population at maximum sustainable yield is ½ K, or 1,800 for peccaries and 2,000 for monkeys, but the maximum sustainable yields are significantly different: 780 peccaries and 84 spider monkeys per year.

Diamante; encounter rates during hunts are not different from those at a nearby less hunted site.[95] As mentioned earlier, Piro hunters have the ability to kill much more than they do because they have shotguns, but harvest no more meat per person per day than their bow hunting Machiguenga neighbors.[124] Finally, the Piro may have a short-term time preference (a high discount rate) with respect to peccaries and other prey resources.

Clark's argument can be stated in evolutionary terms. The decision to harvest involves a tradeoff between the value of current and future harvests.[125] Piro hunters do not have credit markets in which to invest their peccary profits, but they do invest in their own reproduction in the evolutionary sense. They can do this by investing the meat they harvest either directly into offspring or into mating opportunities.[129] From the perspective of the Piro hunters, the strategy set can be conceived as consisting of two options.[22] The first is to harvest as much resource in the present as will maximize short-term fertility and survivorship. The second is to pay a short-term fertility and survivorship cost and harvest the resource at a lower, yet sustainable level, in order to reap the benefits of higher reproductive value of future descendants.

In this context, Rogers[22] developed a model to examine how conservation might evolve when inheritance is uncertain. He imagined two strategies. The

Figure 5.4

Conserve Collared Peccaries or Spider Monkeys?

Collared Peccary

Annual MSY = 17,160 kg
Lump sum   = 81,708 kg
Conserve if discount rate < r/4 = 21%

Spider Monkey

Annual MSY = 756 kg
Lump sum   = 37,719 kg
Conserve if discount rate < r/4 = 2%

The strategy of maximum sustainable yield (MSY) returns ~17,000 kg a year of peccaries and only 756 kg a year of spider monkeys. An extreme exploit strategy, or harvesting as quickly as possible, returns an annual total of ~82,000 kg of peccaries and ~38,000 kg of monkeys (assuming the entire population could be killed in a year). For the exploit strategy to be favored for peccaries, the lump sum must be invested in a current alternative investment that has an annual return greater man 21% or $r_{max}/4$ (17,000 kg is 21 % of 82,000 kg). For spider monkeys, exploit pays if the current alternative returns only 2% or more. This exercise shows that peccaries are much more likely to be conserved than are spider monkeys because the opportunity cost of conserving peccaries is much lower man it is for spider monkeys. At the maximum sustainable yield rates, it would require 50 years to recoup the lump sum for spider monkeys but only 5 years for peccaries. As discussed in the text, the opportunity cost of conservation is only one consideration. If peccaries remain an open-access resource, conservation is unlikely in spite of the lower opportunity costs of conserving them.

Conserve strategy (C) pays a short-term cost in reduced fertility, but increases the fertility of those who inherit the territory. The Prodigal (P) strategy does not conserve and hence does not pay the cost of conservation but, of course, bequeaths poor territories. In Rogers' model, individuals need territories in order to breed. The counter-intuitive conclusion Rogers obtains is that conservation is not favored as inheritance approaches certainty. Imagine that Ego conserves the territory and passes it on with certainty, in good condition, to his son. Because the territory he has inherited is in fine condition, the son produces more grandchildren for Ego. At this point in the model, conservation is betrayed by its own success. Because the great grandchildren need more territories than exist, many cannot reproduce. If inheritance is uncertain, conservation is also not favored because conserved resources could end up in the hands of nondescendants. Rogers concluded that this is a difficult hurdle to surmount, made more difficult if one considers diploid organisms (his original model assumed haploid organisms).

One way to mitigate (but not completely avoid) this effect is to imagine the territory as a biological resource such as a herd of ungulates. If the reproductive rate of the herd is as high or higher than the discount rate, it may pay to conserve, especially if inheritance is certain. If tended properly, a herd of ungulates can be managed so that when the time comes to read the will, each offspring receives as much as the parent originally held. As long as the reproductive rate of the herd keeps pace with that of the conservers, there is always a "territory" to inherit. Each great-grandchild can receive a herd as large as that originally held by its great grandfather. One might ask, however, where all the animals will graze? Eventually, the pasture will deplete, the growth rate of the herds will drop below the discount rate, and conservation will no longer be the best strategy. How long a system of conservation like this can be maintained before collapsing into exploitation remains a theoretical and empirical question.

An Evolutionary Discount Rate?

Rogers[9] has calculated an evolutionary discount rate by replacing utility, the standard currency used by economists, with biological fitness. He concludes that the evolved time preference or discount rate (i), should depend critically on three factors: (p), the rate of population growth; (r), the average relatedness between the individual and his or her offspring; and (T), the generation length. They are related in the following equation:

$$i = (\text{-ln } r)/T + p$$

Future reproduction is discounted by the rate of population growth because delayed reproduction contributes a smaller proportion of the larger future population than does current reproduction if the population is growing. In populations that are declining, future reproduction is worth less. This is equivalent to the discounting factor applied by Fisher[130] to calculate the reproductive value of an individual of age x.[131] In addition, if delayed benefits accrue to descendants, they are additionally discounted because the longer benefits are deferred, the less related descendants become. This follows directly from Hamilton's Rule.[132] If the average coefficient of relatedness between parent and offspring is ½ and one generation is about 30 years, selection should favor rates that average 2.4% per year plus the rate of population growth (see Fig. 5.5). The rate will also vary with age; young adults should discount at a higher rate up to 6%.

Rogers assumes that prehistorically, the average rate of population growth must have been zero and that it therefore did not contribute to an evolved time preference. He concludes with an average discount rate of around 2.4%. Recent demographic work with hunter-gathers suggests that a higher rate may be warranted. Hill and Hurtado[133] found that the Ache, a hunting and gathering

Figure 5.5

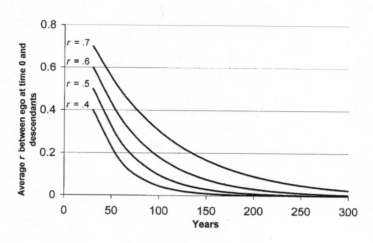

This figure shows how descendants become less and less related with each generation. The X axis is years; the Y axis is the coefficient of relatedness between a conserver in the present and his direct descendents. Recall that the conserver's benefits come in the form of resources invested in future offspring and other descendants. After 240 years, or 9 generations (generation time in this case is 30 years), the coefficient of relatedness between the descendants receiving the conserved animals of the original conserver is $r = 0.002$. At some point in the future, direct descendants become as related to Ego as Ego is related to the average person in the present population. This effect can be considered a cost of sex because it would not accrue to an asexual organism (the descendants of such organisms are clones and 100% related). The average degree of relatedness between individuals in the Piro village of Diamante is $r = 0.027$.[134] Chagnon[135] calculated the average degree of relatedness for 12 Yanomamo villages. Results ranged from $r = 0.058$ to $r = 0.12$, with an average for the 12 villages of $r = 0.086$. In the case of the Piro, after 156 years, direct descendants would, on average, be as related to Ego as any two individuals chosen randomly from the village in the present (assuming $r = 0.5$ between patent and offspring). The four lines on the graph show the effect of four different average degrees of relatedness between parent and offspring. Inbreeding and paternity confidence (for males) can have the significant effect of making descendants more or less related to ancestors, thereby changing the discount rate and the cost to resource conservation.

group in Paraguay, have had a growth rate of more than 2.5% for at least the last 100 years. Hill and Hurtado argue that hunters and gatherers may be characterized by periods of rapid growth interspersed with population crashes. Long-term average population growth remains near zero, but individuals are exposed to a selective environment in which most years are characterized by some measure of growth. If that were the case prehistorically, selection would favor a time preference higher than 2.4%. In the case of the Ache that rate would have been as high as 5%.

Knowing this rate is critical for determining which organisms humans might favor conserving. Compared to chimpanzees, humans have a higher fertility rate for our body size.[131] There is also strong evidence that over the last 2 million years of evolution we have been a species that invades empty niches. *Homo erectus* is characterized by a significant increase in range out of Africa.[136] Archaic *H. sapiens* migrated into the cooler habitats of Eurasia,[137,138] after which modern *H. sapiens* expanded into the high latitudes[139] and, only recently, into the New World.[140] Contrary to the general assumption,[141] these lines of evidence suggest that humans may be more characteristic of an r-selected species than a K-selected one, at least relative to our hominoid precursors. This has profound implications for the benefits that could accrue through any type of restraint, reproductive or subsistence. Compared to Holocene conditions, the environment of evolutionary adaptedness was characterized by environments in which resources were open-access and abundant, and where conspecific competitors were scarce or easily avoided through continued dispersal. In other words, the environment was one in which subsistence strategies that emphasized short-term interests prevailed over strategies that emphasized long-term ones. It is not surprising that our evolved psychology makes present conservation difficult and rare. Few ecological environments in our evolutionary past would have provided a selection regime that favored conservative behavior.

Future Research

Although prey conservation by subsistence hunters is rare now, and unlikely to have been common in the environment of evolutionary adaptedness, there are tantalizing suggestions of conservation in other subsistence contexts. Moreover, there are no theoretic obstacles to the expectation of conservation in these contexts. The research is moving on from simply debunking myth to sophisticated analyses aimed at understanding the contexts that do and do not favor resource restraint. Circumstances have changed considerably in the last 10,000 years. Since the beginning of the Neolithic, for example, resource value has increased as relative abundance declined because of a larger human consumer population. The implications of such a change are extremely interesting when viewed in terms of long- versus short-term subsistence strategies. I have recently hypothesized that a saturated environment could have favored a strategy of resource defense, privatization of open-access wild resources and, perhaps, nascent conservation in the form of animal and plant husbandry.[142,143]

Related issues will provide other fertile areas for future research. For example, Hardin[106] suggested that one solution to the tragedy of the commons is "mutual coercion, mutually agreed upon" (p. 1247). Inspired by this idea, evolutionary ecologists have directed their theoretical attention to the role of negative sanctions or punishment.[144] A variety of game theoretical work suggests that norms of restraint and punishment can result in fragile but cooperative evolutionarily stable strategies that can lead to conservation.[20,145-147] Mutually enforced norms of ownership within a social group can ensure the property rights required for conservation. A number of empirical observations support the role of norms for solving a variety of collective-action problems. Orchard keepers in Washington State have norms to ensure that the quantity of bee-hives kept by neighbors is in proportion to the size of their orchards. Bees are critical for successful fruit tree reproduction, but do not know property lines. Hive owners contribute to the public good. Cheaters do not maintain their own hives, yet benefit from the hives of others.[148] Ellickson[146] describes a complex system of norms among ranchers in Shasta County, California. Sanctions in most of these systems range from gossip and ostracism to physical violence. One Shasta rancher had his bull castrated by an angry neighbor whose heifers were being inseminated by the stray and unattended bull. In traditional societies, witchcraft accusations have been suggested to serve as a sanction for noncooperators.[149]

The ownership required for conservation might be described as a norm between a number of cooperators. Individuals who subscribe to the norms share a common understanding of the circumstances that determine who can and who can not use available resources. Economists call these property rights and define "them as entitlements defining the owner's rights, privileges and limitations for resource use" (p. 41).[122] An example is a house title registered with Erie County in the State of New York. The norm is that private houses cannot be entered and resources removed without the consent of the owner. Cheaters are those who do not behave in a manner consistent with the norms. If the damages from the sanctions are sufficiently high relative to both the gains from defecting and the costs of enforcement, then polymorphic equilibria that include cooperation can be achieved.[20] Recent work suggests that the certainty of punishment may be as, or even more important than the severity of the punishment for ensuring norm compliance.[150] Winterhalder[151] notes that by-product mutualism combined with norm enforcement through punishment could stabilize reciprocity. The work of Boyd and Richerson[145] shows that punishers who induce cooperation (conservation, for example) in others for personal benefits can additionally provide a public good as a by-product. As an example, Ruttan and Borgerhoff Mulder[152] describe East African pastoralists studied by Lane.[153] Seasonal access to grazing land is restricted via coercion by wealthy herders who have the greatest interest in seeing the "commons" preserved. This suggests that social complexity and wealth asymmetries may be associated with conservation.

Summary

Field work to date clearly shows that in contexts characterized by subsistence hunting there is a lack of support for what has been termed the ecologically noble savage hypothesis.[21] Data from the Piro of Peru, the Wana of Indonesia, and other subsistence hunting groups show that hunters generally maximize their short-term returns during foraging bouts in spite of negative impacts on prey populations. Apparent conservation in these cases is best described as epiphenomenal. These conclusions have been obtained by combining the analytic tools of economics with evolutionary biology in the form of foraging theory. An operational definition of conservation and empirical field testing are crucial to such analyses.

Although conservation is rare among subsistence hunters, economics and evolutionary biology have converged on a various ideas that are essential to an understanding of other contexts in which humans and other organisms might conserve resources. Supply-and-demand economics says that relative scarcity increases resource value. Higher resource value increases the benefits of territoriality, which is required to make public goods private ones. Private goods are more likely to be conserved than public goods. This suggests that ownership and territoriality may be linked to resource conservation.

The definition of conservation empathizes the payment of a short-term cost in exchange for a long-term gain. Discounting is an idea critical for understanding decision making that involves behaviors where the costs and benefits occur at different times, as is the case for resource conservation. Recent work to identify an evolutionary discount rate is critical for determining which organisms humans might favor conserving. Evolutionary informed discounting has the promise of tying together a wide variety of now disparate topics within the fields of biology, economics, and anthropology.

Acknowledgments

I thank J. Fleagle, S. Gursky, M. Borgerhoff Mulder, R. Hames, H. Kaplan, L. Ruttan, E.A. Smith, and B. Winterhalder for their comments. The Leakey Foundation and the Charles Linbergh Foundation funded research with the Piro. The Nature Conservancy, the Wenner-Gren Foundation for Anthropological Research, the National Geographic Society, and the Faculty of Social Science at the State University of New York at Buffalo supported research with the Wana.

References

1.   King, B. (1994).
2.   Whiten, A., Byrne, R. (1997).
3.   Byrne, R., Whiten, A. (1988).
4.   Byrne, R. (1997).
5.   Kurland, J., Beckerman, S. (1985).

6.   Tooby, J., Cosmides, L. (1992).
7.   Foley, R. (1995).
8.   Cosmides, L., Tooby, J. (1992).
9.   Rogers, A. (1994).
10. Hirshleifer, J. (1991).
11. Charnov, E. (1976).
12. Maynard-Smith, J. (1982).
13. Hill, K. (1988).
14. Kaplan, H. (1996).
15. Winterhalder, B. (1996).
16. Becker, G. (1976).
17. Simon, H. (1993).
18. Hodgson, G. (ed.) (1995).
19. Bergstrom, T. (1995).
20. Sethi, R., Somanathan, E. (1996).
21. Alvard, M. (1993).
22. Rogers, A. (1991).
23. Vander Wall, S. (1990).
24. Houston, D. (1979).
25. Weeks, H. (1993).
26. Wynne-Edwards, V. (1962).
27. Lack, D. (1947).
28. Charnov, E., Berrigan, D. (1993).
29. Speck, F. (1939).
30. Gorsline, J., House, L. (1974).
31. Dasmann, R. (1976).
32. Reichel-Dolmatoff, G. (1976).
33. Nelson, R. (1982).
34. Hughes, J. (1983).
35. Posey, D. (1985).
36. Todd, J. (1986).
37. Clay, J. (1988).
38. Bunyard, P. (1989 b).
39. Oelschlaeger, M. (1991).
40. Pearce, F. (1992).
41. Alcorn, J. (1993).
42. Budiansky, S. (1995).
43. Sahlins, M. (1972).
44. Hames, R. (1987 b).
45. Hames, R. (1991).
46. Kay, C. (1990).
47. Kay, C. (1994).
48. Kay, C. (1995).
49. Redford, K. (1991).
50. Diamond, J. (1992).
51. Simms, S. (1992).
52. Low, B., Heinen, J. (1993).
53. Alvard, M. (1994a).
54. Low, B. (1996).
55. Ruttan, L. (1998).
56. Vickers, W. (1994).
57. Hunn, E. (1982).

58. Hildebrandt, W., Jones, T. (1992).
59. Lyman, R. (1995).
60. McCartney, A., Savelle, J. (1985).
61. Savelle, J., McCartney, A. (1991).
62. Krupnik, I. (1993).
63. Temple, S. (1987).
64. Fitzgibbon, C. (1990).
65. Ross, E. (1978).
66. McDonald, D. (1977).
67. Aunger, R. (1992).
68. Aunger, R. (1994).
69. Alvard, M. (1995 a).
70. Smith, E. (1995).
71. Smith, E., Winterhalder, B. (1992).
72. Elster, J. (1983).
73. Williams, G. (1966).
74. Reznick, D., Travis, J. (1996).
75. Hill, K., Hawkes, K. (1983).
76. Beckerman, S. (1983).
77. Smith, E. (1991).
78. Smith, E. (1992).
79. Stephens, D., Krebs, J. (1986).
80. Hames, R., Vickers, W. (1982).
81. Smith, E. (1983).
82. Kaplan, H., Hill, K. (1985).
83. Winterhalder, B., Lu F. (1997).
84. Feit, H. (1973).
85. Moore, O. (1957).
86. Holling, C. (1959).
87. Turelli, M., Gillespie, J., Schoener, T. (1982).
88. Templeton, A., Lawlor, L. (1981).
89. Kagel, J., Green, L., Caroco, T. (1986).
90. Fisher, I. (1930).
91. Lang, K., Ruud, P. (1986).
92. Moore, M., Viscusi, W. (1990).
93. Hill, K. (1995).
94. Stiner, M. (1991).
95. Alvard, M., Robinson, J., Redford, K., Kaplan, H. (1997).
96. Alvard, M. (2000).
97. Bodmer, R. (1995).
98. Peres, C. (1999).
99. Peres, C. (n. d.).
100. Fitzgibbon, C., Mogaka, H., Fanshawe, J. (1995).
101. Dewar, R. (1984).
102. Anderson, A. (1989
103. Vickers, W. (1988).
104. Vickers, W. (1991).
105. Winterhalder, B. (1981).
106. Hardin, G. (1968).
107. Olson, M. (1965).
108. Hawkes, K. (1992).
109. Diamond, N., Bartholomew, K., McCarthy, S., Farmer, N. (1987).

110. Gordon, H. (1954).
111. *The Economist* (1997).
112. Cashdan, E. (1983).
113. Brown, J. (1964).
114. Dyson-Hudson, R., Smith, E. (1978).
115. Cashdan, E. (1992).
116. Acheson, J. (1987).
117. Acheson, J. (1988).
118. Wilson, J. (1977).
119. Johannes, R. (1978).
120. Johannes, R. (1982).
121. Polunin, N. (1984).
122. Tietenberg, T. (1996).
123. Beckerman, S., Valentine, P. (1996).
124. Alvard, M. (1995b).
125. Clark, C.W. (1990).
126. Caughley, G. (1977).
127. Robinson, J., Redford, K. (1986 a).
128. Robinson, J., Redford, K. (1986 b).
129. Trivers, R. (1972).
130. Fisher, R. (1958).
131. Hill, K. (1993).
132. Hamilton, W. (1964a).
133. Hill, K., Hurtado, M. (1996).
134. Alvard, M., Kaplan, H. (1990).
135. Chagnon, N. (1988).
136. Rightmire, P. (1993).
137. Roebroeks, W., Conrad, N., van Kolfschoten, T. (1992).
138. Lahr, M., Foley, R. (1992).
139. Klein, R. (1995).
140. Meltzer, D. (1993).
141. Rogers, A. (1992).
142. Alvard, M. (1996).
143. Alvard, M. (1997).
144. Clutton-Brock, T., Parker, G. (1995).
145. Boyd, R., Richerson, P. (1992).
146. Ellickson, R. (1991).
147. Wilson, D.S. (1998).
148. Cheung, S. (1973).
149. Jones, D. (1996).
150. Klepper, S., Nagin, D. (1989).
151. Winterhalder, B. (1997).
152. Ruttan, L., Borgerhoff-Mulder, M. (n. d.).
153. Lane, C. (1991).

6

The Tragedy of the Unmanaged Commons

Garrett Hardin †

In science, as in other human endeavours, progress is sometimes delayed by a "double-take," to borrow a term from the comic theatre. The message is first met with silence, only later to be followed by a painful dawn of understanding. Recall the silence that greeted Mendel's theory of heredity in 1866. Not until 1900 did three scientists experience a double take and alert the world to genetics. Delay of the "take:" 34 years.

Human ecology furnishes another example. In 1833 William Forster Lloyd published his Oxford lectures.[1] Little notice was taken of this work until 1968, when I expanded the theory in my essay, "The tragedy of the commons."[2] Contributing to the long neglect, no doubt, was a 43-word summary in a massive review published in 1953 by the United Nations.[3] This book, the work of an anonymous committee, had the thrust of Lloyd's argument exactly wrong.[4] Duration of the "double take:" 135 years.

The intellectual climate of the times no doubt contributed to the delay. *Laissez-faire* was the dominant attitude after Adam Smith published *The Wealth of Nations* in 1776. The prevailing spirit was wholly optimistic and non-interventionist. Let each man pursue his own interest, economists said, and the interests of all will be best served in the long run.

Not necessarily, said Lloyd. Let a number of herdsmen turn their cattle loose in a pasture that is jointly owned and soon the common will be ruined. Why? Because the pasture has a limited "carrying capacity" (to use a modern term), and each herdsman gets the full benefit of adding to his herd, while the disbenefits arising from over-exploitation of the resource (e.g. soil erosion) are shared by all the herdsmen. Fractional losses are not enough to deter aggressive cattle owners, so all the exploiters suffer in an *unmanaged common.*[5]

Alternatives to the unmanaged commons can be classified under two headings. In privatism, the resource is subdivided into many private properties.

Reprinted from *Trends in Ecology and Evolution*, Vol. 9, No. 5, Garrett Hardin "The Tragedy of the Unmanaged Commons," p. 199, May 1994, with permission from Elsevier.

Each owner is responsible for the management of his plot: those who manage well, prosper; those who manage poorly, suffer. In socialism, the resource is "common property," but the property-owners ("the people") appoint a manager to control its exploitation. Theoretically, an incompetent manager can be fired. In practice, when "the people" is a nation of many millions, it is all too easy for empowered managers (bureaucrats) to survive by hiding their mistakes.

Both privatism and socialism can either succeed or fail. But, except in the smallest of communities, *commonism* cannot succeed. An unmanaged common fails because it rewards individual exploiters for making the wrong decisions —wrong for the group as a whole, and wrong for themselves, in the long run. Freedom in the commons does *not* produce a stable prosperity. This is Lloyd's revolutionary point. Popular prophets, intoxicated by *laissez-faire,* simply could not hear Lloyd.

Apparent exceptions to the theory need to be accounted for. First, a trivial case. When a resource is present in abundance, an unmanaged common may actually be the most efficient. The general rule, "freedom of the seas", led for centuries to the economical exploitation of oceanic fisheries.

Second: "scale effects" must be kept in mind. People of the Hutterite faith in northwestern United States and adjacent Canada live by the Christian ideal that (ironically) Karl Marx expressed best: "From each according to his abilities, to each according to his needs." Farms are owned in common: and everybody is supposed to pitch in and do his share of the work, while taking out no more than a fair share of the products. Conscientious Hutterites make a nominally unmanaged commons work—but only so long as the operational unit is less than 150 people. As the number approaches this, more and more commune members shirk their tasks. (Perhaps we should say a community below 150 really is managed—managed by conscience.)

If such devout and hard-working people cannot make an unmanaged common work, there is no reason to think that anyone can. And it's a long way from 150 to the millions that make up a modern nation. Scale effect rules out the unmanaged commons as an important political possibility in the modern world. Modern nations are a changeable hodge-podge of socialism and privatism.

Some ecologists have failed to see subtle signs of management in traditional societies. For instance, the survival of the Turkana people in Africa under a system of common ownership of grazing was recently cited as an instance of the success of commonism.[6] Yet the same account noted that access to resources was effectively controlled by the elders of the tribe. Such a managed commons presents no problem to the theory of unmanaged commons.

The survival of today's industrialized nations is now threatened by a different sort of commonization. Decades of well-intentioned propaganda in favor of a "world without borders" have stripped sophisticated moderns of psychological defences against truly entropic forces. "To each according to his needs" implies that *needs create rights*. Such rights can be fulfilled on a global scale only if national borders are effectively liquidated. The resulting poverty will accelerate the destruction of environmental wealth. Gresham's Law of economics— "bad money drives out good"—has, under a global system of *laissez-faire,* its cognate in the environmental sphere: "Low environmental standards drive out high." Poverty displaces wealth—globally.

"To each according to his needs" is an immensely seductive phrase to religious people, but in a world without national population controls it is a sure recipe for disaster. Those who are really concerned with the environment— concerned with the well-being of posterity—must give the carrying capacity of the environment precedence over discontinuous human needs, however much these needs may tug at our heartstrings. Of every impulse to globalize wealth the ecologist must ask his ultimate question, "And then what?" What happens after globalized wealth degenerates into globalized poverty? What happens then to the environment for which posterity will hold us responsible?

References

1. Lloyd, W.F. (1833).
2. Hardin, G. (1968).
3. United Nations (1953).
4. Hardin, G. (1993).
5. Hardin, G. (1991).
6. Monbiot, G. (1994).

7

Closing the Commons: Cooperation
for Gain or Restraint?

Lore M. Ruttan[1]

Research concerning the value of communal resource management is limited in two respects. First, while many studies present evidence that communal management is common among traditional societies, a strong theoretical basis is lacking to explain why individuals participate in monitoring and sanctioning efforts. Second, few studies have actually demonstrated resource conservation. There are several ecological and economic reasons for thinking that groups may find it harder to design appropriate conservation measures than to prevent free-riding. However, if groups can surmount these problems, communal management may have advantages over privatization or government control. These arguments are illustrated using results from a pilot study of the communal management of mother-of-pearl shell *(Trochus niloticus)* in the Kei Islands of Eastern Indonesia. It is found that villagers successfully cooperate to defend access to, and regulate their own harvest of trochus. In doing so, they are able to prevent free-riding, and to provide themselves with a long-term source of cash income. However, it is here argued that their aim is "gain rather than restraint."[2]

Introduction

Hardin's (1968) article entitled "The tragedy of the commons" is one of the most widely cited, and mis-cited, articles in the social sciences. This is partly due to his own misuse of the term "commons" by which he meant an open access, common pool resource.[3] Fortunately, Hardin's error has inspired many researchers to document the geographical and historical prevalence of institutions for communally managing common property, especially those designed to prevent free-riding (see chapters in Berkes, 1989a; McCay and Acheson, 1987; Ostrom, 1990; Ruddle and Akimichi, 1984; Ruddle and Johannes, 1989). Yet, we still lack a strong theoretical basis for understanding cooperation, especially the second-order problem of cooperation. It is not sufficient to

Reprinted by permission of *Human Ecology* 26:1, Lore M. Ruttan "Closing the Commons: Co-operation for Gain or Restraint?" pp. 43-66, Copyright © 1998 by Kluwer Academic / Plenum Publishers N.Y.

demonstrate that cooperation persists when free-riders are punished. We must also explain why community members cooperate to sanction offenders when they themselves could free-ride on their own duties as enforcement agents. In addition, further empirical support is needed to demonstrate that communal management systems actually do conserve resources.

In the ensuing sections, I begin by drawing on evolutionary theory in developing a basis for understanding such cooperation. This is followed by a discussion as to whether we should expect traditional, communal management systems to be sustainable. In the second portion of the paper, the results of a pilot study on the communal management of mother-of-pearl shell (*Trochus niloticus*) on Kei Besar Island in eastern Indonesia are presented. I argue that the village described has found a rational and equitable means of organizing to prevent free-riding. However, it would appear that the primary intent behind territory defense and self-regulated use is to maintain exclusive access in a cost-effective manner rather than to conserve the resource *per se*. At present, conservation may be the incidental, if not intentional, product of their efforts, yet changing economic conditions could alter this balance.

Can Communities Cooperate to Close the Commons?

The paradigm of the prisoner's dilemma has been used to explain the overuse of many renewable and nonrenewable resources, but especially those that are open access, common property. The dilemma presented by this game's payoff structure is that pairs of cooperators receive higher payoffs per capita than do pairs of noncooperators, but an individual could receive even higher payoffs by cheating when the other cooperates. Without any mechanism to prevent such free-riding, the perverse outcome is that no one cooperates. Classic evolutionary theory makes essentially the same argument, but in a more general form. Groups of cooperators are at a selective disadvantage since they can be invaded by selfish individuals who free-ride on their efforts. Hence, over the long term, altruism is expected to decline in frequency (Williams, 1966). And yet people do manage to cooperate in very diverse ways. Communal resource management is just one example that has been particularly well documented by social scientists.

Much has been made of this contradiction between theory and empirical research, but the discrepancy is more apparent than real. Modern evolutionary biology provides several theories to explain the evolution of cooperative behavior. Principal among these theories are reciprocal altruism (Trivers, 1971), asymmetrical altruism, kin selection (Hamilton, 1964a; Maynard Smith, 1964), and cultural group selection (Boyd and Richerson, 1985; Soltis *et al.,* 1995).[4]

By far, the greatest amount of attention has been given to reciprocal altruism and the work of many social scientists makes sense in terms of this framework (Ostrom, 1990; White and Runge, 1995). The problem here is to understand how an individual ensures that others reciprocate their cooperative efforts since selection should act against those who cooperate indiscriminately. Most models of reciprocity have been formatted as an iterated prisoner's dilemma, usually between two players (Axelrod and Hamilton, 1981). Although it turns out that "tit-for-tat" is not actually an evolutionarily stable strategy, some behavioral strategies such as "Pavlov" or "suspicious tit-for-tat" can withstand invasion by a wide variety of other behavioral strategies (Boyd and Lorberbaum 1987; Nowak and Sigmund, 1993). However, stability is dependent on there being a high probability that the same individuals will interact repeatedly. In addition, individuals must be able to monitor the behavior of others, and they must have a means of punishing cheaters. These same conditions are also critical in games of reciprocity with groups larger than two. Relatively larger group sizes are possible when group members also adopt "moralistic" and "retributive" strategies. That is, when they punish nonpunishers as well as defectors, and when punishments are selectively directed at defecting individuals rather than at the entire group, as would be the case if punishers simply withheld cooperation (Boyd and Richerson, 1992).

It should be noted that one of the most common criticisms of this approach is that typical models of an iterated prisoner's dilemma do not permit communication during a play (Dugatkin *et al.*, 1992). Clearly, this is unrealistic. At the same time, humans are very adept at deceit and thus honest communication is itself a public good that is susceptible to cheating, unless there are widespread norms of trust. However, the existence of such norms requires a further level of explanation.

The second form of cooperation is asymmetrical altruism and is based on Olson's theory of the privileged group (1971, p. 50). Most models of cooperation have assumed that all actors receive the same benefits if they pursue the same strategies, but perhaps some forms of cooperation have asymmetrical payoffs that individuals accept voluntarily. For example, if one or a few individuals are willing to support the costs of enforcement in return for an extra share of the benefits, then stable cooperation may ensue since this privileged group has a greater incentive to maintain production (Olson, 1971; Vehrencamp, 1983). Of course, it will still be necessary for subordinates to monitor the behavior of elites and to have ways of punishing their free-riding (i.e., exploitation), withdrawing cooperation being one example. Alternatively, subordinates may have few options but to cooperate. In the latter case, the cost to elites of maintaining cooperation becomes even cheaper since small sanctions may be a powerful deterrent to those with few other opportunities (Hechter *et al.*, 1990). Little work has been done to model this form of cooperation but it is anticipated that this may permit somewhat larger group sizes.

The third potential explanation for cooperation is kin selection. By helping kin in proportion to one's degree of relatedness to them, one's own "inclusive fitness" can be increased (Hamilton, 1964a).[5] Roughly speaking, in evolutionary terms aiding close kin is nearly as good as helping oneself. Hence, kin may be less inclined to free-ride on the efforts of close relatives and simultaneously be more tolerant of those that do. Although this has been an unpopular approach to human behavior in the past, in point of fact, we frequently do observe resource use organized around kinship. This is not only the case for those societies in remote regions of the world but is also common in many developed nations (Palmer, 1991; Wilen, personal communication). However, the benefits of kin selection rapidly decrease as the size of the group in question (and average genetic distance) increases. Thus, as is the case with reciprocity, only very small groups can be expected to cooperate purely on the basis of kin selection.

Cultural group selection is a fourth explanation for the evolution of cooperation. While genetic group selection is believed to act much too slowly to have a very large effect on rates of evolution (Williams, 1966; but see Wilson, 1983), the conditions permitting cultural group selection are thought to be much less strict (Boyd and Richerson, 1985). In particular, if conformity is an important part of socialization, then, immigrant "selfish" individuals would adopt cooperative attitudes along with other locally adaptive skills. It might be imagined then, that social norms which improve the persistence of communities are the product of cultural evolution (regardless of whether the norms are antagonistic to, or in support of, individual interests). Such norms may include rules or attitudes toward cooperation and use of resources. For example, they may operate by helping to coordinate people's behavior, thereby decreasing the costs of cooperation, or conversely, by increasing the costs of cheating. This, in turn, may permit larger groups to cooperate. Such a perspective meshes well with much of the more recent work on common property resources that has emphasized the role of ideology and social norms in maintaining cooperation (see chapters in Anderson and Simmons, 1993; also Runge, 1981; White and Runge, 1995), but has neglected to provide much theoretical justification for how such norms might have evolved and persisted in the first place (cf. Sethi and Somanathan, 1996).

Finally, while social norms can reinforce cooperation, they are not written in stone. Human culture evolves rapidly. Without the threat of actual sanctions, it is to be expected that norms of cooperation would deteriorate (Feeny *et al.*, 1996). Likewise, kin selection alone may not be a sufficient explanation of patterns of cooperation. Given a choice, individuals should cooperate with kin who do reciprocate rather than with those that do not (Wilson and Dugatkin, 1991). This suggests that cooperation is most likely to be the product of two or more of these mechanisms acting simultaneously.

Defining Conservation

Before turning to a discussion of whether communities are able to design sustainable management regimes, let us briefly consider how to define and measure conservation. There are principally three means of detecting whether a particular method of resource use conserves. These are firstly, determining by interview that there is an intent to conserve, secondly, measuring a behavioral effect, and finally, observing actual behavior. Each of these methods is problematic (see also Alvard, 1995a and invited commentary therein).

While a "conservation ethic" may be an important feature of a sustainable system, it is insufficient in itself. Intent will not translate into actual conservation if communities are not able to determine the effects of their actions on stock levels. For this reason, it might seem more appropriate to measure the actual effects of a particular management system. However, if human populations are at a low density relative to the resource base there may be an appearance of conservation, and yet the same behavior may not be sustainable at high population densities (Hames, 1990; Alvard, 1993). Some authors call this "epi-phenomenal conservation" to emphasize that we may not understand the causal mechanism underlying the apparent conservation (Hunn, 1982).

For this reason, Hames (1987b) and Alvard (1993) suggest we observe actual behavior rather than effects. In particular, we should examine whether individuals stint on their use of resources.[6] Advantages of this approach are that it allows the researcher to sidestep the issue of whether the behavior is the product of a conscious intent to conserve, or is the result of evolved norms of conservation—food taboos being one example of the latter.[7] It also removes the methodological problem of having to determine whether a population is at low or high density relative to a resource.

In the context of territorial hunter-gatherers who lack complex rules of resource use, it may make some sense to equate conservation simply with the act of forbearing from targeting a particular species.[8] Yet, even here, we may observe acts of stinting that may also be labeled "epiphemonemal conservation." For example, a group might forego using resources found in one particular area solely because of the threat of warfare there (M. Borgerhoff-Mulder, personal communication). Nor is stinting a sufficient criterion of conservation in societies with more complicated rules of resource use, such as limits to the numbers of users or on the duration of an open season. These rules are also types of stinting but the precise number of users given permits, or the timing of the open season, may correspond more closely with economic considerations than with biological ones. As a case in point, it is not unusual for communities in Indonesia and Oceania to close a fishery for a few days, or even weeks, in order to allow a large school of fish to congregate in their harbor. The fish can then be harvested more efficiently and in larger numbers (Johannes, 1978; Hooper, 1989; Hviding, 1989; Zann, 1989; Bailey and Zerner, 1992). This form of short-term stinting has the potential to decrease long-term stock levels.

While it is not the aim of this paper to provide a definitive answer as to how to assess whether a particular institution or behavior leads to conservation, it is worthwhile noting the problems associated with each approach. These concerns are more than academic. If we do not understand the full context of specific management systems we cannot hope to develop effective policy recommendations that will be robust to changes in human population density or in economic conditions.

Can Communities Design Sustainable Management Regimes?

If we return to our original question of whether communal management can conserve the resource base, and if we adopt our initial assumptions that resources are limited and that there is a temptation to cheat, then there can be little argument that if groups cannot prevent free-riding there will be a "tragedy of the commons." However, it is not at all the same thing to argue that if groups are able to prevent free-riding, then sustained use will be the result. There are two principle reasons for this.

First, it may be far easier for groups to prevent free-riding than to invent and adopt rules for managing resources in a sustainable manner, even if that is their intent. One may question whether communities have the ability to design appropriate management rules not because people are ignorant (numerous examples in the literature clearly demonstrate that many pre-industrial societies have a very sophisticated knowledge about the resources which they use) (Johannes, 1981; Bulmer, 1982), but rather because it may not always be easy to evaluate the effects of one's actions on resource levels (Carrier, 1987). Although individuals can monitor each other's behavior on a daily basis, it might be years before one could observe the effects of such behavior on stocks and thus determine whether one's choice of regulations was appropriate. This may be especially true of resources with complicated population dynamics, for example, species of arctic mammals whose population sizes fluctuate over long time scales.

Determining a sustainable yield may be even more difficult, or nearly impossible, for species whose population sizes are density independent, that is, where recruitment is independent of population size above some minimum threshold. In these cases, there may be little apparent correspondence between resource levels and human use. Given the more recent view that population sizes of many marine species are density independent or chaotic (Acheson and Wilson, 1996), it is no wonder that some groups of indigenous peoples do not believe that variations in catch are due to real declines in stock size. They frequently believe, instead, that the prey have become wary or have moved away (Carrier, 1982, 1987). However, when novel technologies and market incentives lead people to harvest a much larger portion of the standing stock, then recruitment may be truly diminished (Johannes, 1978) but, local knowledge systems will prevent them from recognizing this.

The second point is that the primary aim of these management systems is "gain rather than restraint," although conservation may be an incidental effect. While this may seem hardly surprising to some readers, there is a body of work which does argue that traditional societies were "in balance with nature." In some cases, the authors mean simply that there was a conservation ethic which stressed not taking more than was needed (Johannes, 1981; Nelson, 1982; Kay, 1985a). Since needs were fewer prior to the intrusion of capitalism, there was in effect more conservation. However, other writings cross a fine line and imply that it was the aim of native peoples to limit their needs and so, to conserve resources (Jermy, 1983). It has even been argued that warfare existed so as to limit human population size and thus conserve resources (Rappaport, 1969; cf. Peoples, 1982). Similarly, food taboos were taken to be signs of the good management practices of hunter-gatherers (Ross, 1978; Harris, 1979). Unfortunately, it is this latter view of the "ecologically noble savage" (Redford, 1991b) that seems to have taken hold of the mind of the nonspecialist.

Along these same lines, much has been made of limited entry as a conservationist measure (Christy, 1969; Johannes, 1978, 1981; Baines, 1982; cf. Carrier, 1987). However, restricting access to a resource does not necessarily change how much is harvested, just *who* gets it (Polunin, 1984). For that matter, the imposition of territorial boundaries may itself be a response to increasing competition that is due to increased harvesting efforts (Dyson-Hudson and Smith, 1978; Hviding, 1989; Zerner, 1990). Similarly, groups of individuals may cooperate to limit entry so as to raise prices (McCay, 1980). Thus, restraint in numbers of users leads principally to greater gains for those privileged to be in the group.

It is true, however, that a smaller group is potentially more capable of imposing harvest limits and so conserving the resource. Yet still we must question whether the specific form of such regulations is dictated by socioeconomic rather than by biological needs (Polunin, 1984; also Bulmer, 1982). Groups that are able to trade surpluses may find it more profitable to harvest a high proportion now and invest the profits in enterprises whose value appreciates more rapidly (Clark, 1973). Or, more likely in today's world, they may find it necessary to harvest increasing amounts of fish simply to service debt on capital investments (Johannes, 1981). In conclusion then, while we are interested in determining the conditions under which traditional resource use is sustainable and conserves biodiversity, this is not necessarily the aim of local communities.

When is Communal Management Advantageous?

Proponents of communal control have generally argued that local people have a greater stake in protecting the resources upon which their livelihood depends and to which they may have customary and prior rights. Therefore, they might be expected to do a better job of managing them. In addition, it is argued that communal management may be more economical for national governments insofar as local users can double as enforcement agents (Johannes, 1978; Berkes *et al.*, 1989; Johannes and MacFarlane, 1989; Bailey and Zerner,

1992). On the other hand, this perspective has tended to ignore differences in political and economic power within a community. One must consider how local users, whose own interests are at stake, can be counted on to act fairly. This is even more problematic today given the manner in which changing economic opportunities, particularly differential access to capital, are rapidly altering the balance of power in many communities (Baines, 1982).

Despite the cautionary tone of the foregoing discussion, there may well be specific cases where traditional management is sustainable, or even more likely, there may be specific types of regulations that are sustainable and equitable. If these problems can be resolved, then in some cases, communal management does have advantages over private or government control. Some types of resources lend themselves to particular management/tenure regimes more easily than do others. Dyson-Hudson and Smith (1978) predict that territories are most likely to be "economically defendable" (Brown, 1964) when resources are at high densities and are highly predictable. However, they neglect to dis-tinguish between individual vs. group held territories, the issue of interest here. Other work on animal foraging behavior suggests that grouping is advanta-geous when a resource is low in abundance and predictability, but is patchily distributed (Clark and Mangel, 1984), or, when a forager's hunting technique is poor enough to make food effectively scarce (Packer and Ruttan, 1988). However, a full ecological explanation of when individuals should both live in groups and hold territories is lacking.

Preliminary intuition would suggest that high predictability combined with low abundance would be one situation favoring group, as opposed to private, tenure. An example is pastures that do not yield enough vegetation to make individual ownership worthwhile but may be valuable to a group because of economies of scale in the costs of defense (Stevenson, 1991, in Taylor, 1992; Eggertson, 1993). Note however, that while ecological features may make it more advantageous to cooperate in defending a territory than to hold such resources individually, individuals will still be faced by temptations to free-ride. Thus, a group must still find ways of organizing themselves to deter such cheating.

Trochus Management on Kei Besar Island, Indonesia

Marine tenure for fin-fish appears to be less common in Indonesia than in Oceania. It has been suggested that the productive, volcanic soils of many Indonesian Islands has led inhabitants to be more concerned with terrestrial food production (Polunin, 1984). However, an absence of fin-fish territories does not necessarily mean a complete absence of marine tenure as more recent research has documented (Zerner, 1990, 1994; Andamari *et al.*, 1991). In many instances, even when people are dependent on the land for food, they still limit access to areas where cash products such as mother-of-pearl shell, pearls, dried sea cucumber and sea weed, can be found. For example, on the eastern shore of Kei Besar Island in Maluku Province, Indonesia (Figure 7.1), there is no fin-fish tenure but there is communal tenure for mother-of-pearl shell, *Trochus niloticus,* or "lola" as it is called in Indonesian. This island has fertile, if somewhat thin, soils and only a narrow band of fringing reef on its eastern shore. In contrast, neighboring Kei Kecil is a low-lying sandy island with extensive fringing reefs to the west. These people are said to defend fin-fish fishing grounds (R. Topatimasang, personal communication).

Trochus is a marine gastropod found on coral reefs at a depth of roughly 1–10 m. The whole shell can be cut into buttons, or it can be ground and added to automobile paints to make them iridescent. Skin-divers harvest a few trochus per dive, finding them hidden in the cracks and crevices of the reef. Although

Figure 7.1

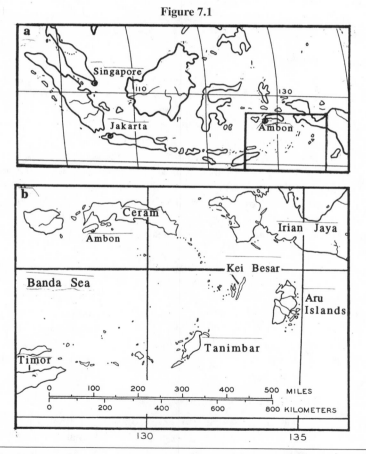

Location of study site: (a) the position of central and southeastern, Maluku Province within Indonesia, (b) an enlargement of (a) inset and location of Kei Besar Island in relation to Ambon, the capital of Maluku Province.

trochus is still abundant along the eastern shore of Kei Besar Island in South-eastern Maluku Province, it has been nearly fished out in some other parts of Maluku Province and most of western Indonesia.

Officially, the state owns all marine resources and it is illegal to harvest trochus. However, in remote areas at least, local communities still maintain effective control.

Traditional management of trochus is particularly strong in the village of Ohoirenan, Kei Besar. For this reason, the author accompanied a team of biologists from the Indonesian Institute of Sciences (LIPI) to this village in the beginning of March 1994, remaining an additional 5 weeks after their departure to conduct a pilot study of trochus management. During this time members of this community were interviewed to learn details of their harvesting practices,[9] and the "buka sasi lola," or open harvest, for trochus was observed in this and a neighboring village.

The Sasi Practice in Maluku Province, Indonesia

Maluku Province is the second most easterly Province in Indonesia. Composed of many archipelagos which include hundreds of islands and nearly as many different indigenous cultures, the area is historically important as the original source of cloves, nutmeg, and mace. Although it is hard to date the origins of the commercial market for marine products, Wallace (1986) described an extensive commercial trade in a variety of marine products, including trochus, that was based in the Aru Islands of Southeastern Maluku Province during the middle of the nineteenth century. During the middle of this century, the commercial importance and value of trochus grew immensely (Bailey and Zerner, 1992). An intensification in use seems to have has contributed to the hardening of territorial boundaries and the increased complexity of marine regulations (Zerner, 1990, 1994).

Throughout most of this region, customary law includes a practice called the "sasi" in Indonesian, which is used to regulate the communal use of terrestrial and marine resources (Bailey and Zerner, 1992; Zerner, 1990, 1994). The word "sasi" literally means prohibition and is used to describe the closing and opening of access to resources, usually by the resource owner or by the village leaders. In practice, imposing a sasi and setting the timing of harvest also includes defining the extent of resources controlled, who has access, and how the resource is gathered.

Managing Trochus in Ohoirenan Village, Kei Besar Island

The village of Ohoirenan (population 906) controls roughly 8 km of shoreline, more than any of its neighbors. In large part this is because it was the first coastal village in this part of Kei Besar Island. Prior to its establishment in 1815, the Evav people lived in the interior of the island. While their principal source of food is still their gardens, they supplement this diet with some fish.

Cash is earned through the sale of copra, coconut oil, and trochus shell, and by the time of this paper's publication, the villages' first coffee plants should be reaching maturity.

The majority of inhabitants of Ohoirenan belong to two castes.[10] The upper of these two castes, the *mel-mel,* is comprised of all the individuals inhabiting the 53 houses closest to the sea and who are all closely related through marriage. The village chief and other principal village functionaries, such as the secretary, are drawn from this caste, which will be referred to hereafter as the "central families." However, members of both castes do hold seats on the village council, a governing body of 18 individuals. It should be noted that real political power is not restricted solely to these 18.

In Ohoirenan, the trochus harvest takes place sometime between February and May. The uncertainty arises because there is a trade-off between the advantage to waiting until April or May when daytime neap tides are lower and diving is easier and the disadvantage of waiting too long and having the East Monsoon begin. All 202 adult males who live in the village of Ohoirenan have the right to dive for trochus. However, not everyone actually participates since the village council announces the "buka sasi lola" only days beforehand and some individuals are invariably absent. Prior to the start of the harvest, village representatives negotiate with traders for a sale price.

During the closed season, known as the "tutup sasi lola," poaching either by villagers or by outsiders is discouraged by the threat of supernatural sanctions such as death or illness. Scientific researchers are subject to these restrictions just like everybody else. When the sasi is opened, these sanctions are lifted temporarily through a series of Protestant, civic, and animistic rituals, and meetings. In 1994, on the day before the season's opening, the village's Protestant pastor led a pre-harvest ceremony. The following morning, the village chief spoke to the divers reminding them of the rules of harvest and their responsibilities to the village. Finally, with the divers waiting in the water, a traditional harvest steward made an appeal to the gods for a good harvest. The conclusion of his invocation in the Evav language was the signal for all to shout in unison and then to commence diving. This last step of the ritual clearly had strong symbolic value to villagers and played an important role in reinforcing group solidarity.

The open season does not mean a free-for-all. In the first place, as mentioned above, only men of the village are allowed to participate. Any outsiders found "stealing" trochus are beaten or required to pay the village an antique, Portuguese brass cannon (the Portuguese controlled this area prior to the Dutch). Few of these cannons are left in circulation and they are worth several fortunes to a villager. In the last 15 years, there have been three such incidents, all by outsiders. Two of the offenders were beaten, and one had to pay a cannon.

Second, the actual harvesting is carefully orchestrated. The total shoreline is divided into four sections, each with its own steward. Only one area at a time is open for harvesting trochus and each area is fished twice over the course of 8 days, weather permitting.[11] On any given day, the harvest begins at a set time in the morning after the steward gives the signal, but ends whenever the divers tire and the tide grows too high, roughly 2 hours later. A handful of divers now have modern masks, instead of goggles made from bamboo and glass, but no scuba equipment is allowed. Upon returning to the village landing, divers must immediately report how many trochus they have gathered to a waiting secretary. An additional restriction is that on any given day, spearfishing is allowed only in the one area where trochus were already harvested that morning. Otherwise, spearfishermen might learn where they can expect to find large concentrations of trochus the next morning, and thus, have an unfair advantage.

Divers are allowed to keep 30% of the profits from the trochus they harvest from three of the four areas. The remaining 70% is pooled and redistributed, supposedly according to age, marital status, and the needs of school-age children (Tables 7.1 and 7.2). However, it is unclear precisely how funds were actually redistributed. At one time, it was said that the village chief and one or two other high ranking officials received extra large shares, while at another time it was said that all the central families received extra large shares. Yet, calculations from the actual figures provided do not support the claim that there was a separate redistribution class for elites. Discrepancies in the figures can be explained if high ranking individuals receive their extra pay by being given more than one share rather than a larger portion.

The fourth harvest area, the "dapur desa" or village kitchen, is managed differently than the other areas. All the money from the sale of all trochus gathered in this area is used to fund village needs such as school equipment, oil for the village generator, repairs, etc. Divers are not allowed to keep any

Table 7.1
Trochus Yield in Kilograms from the Village Kitchen and from the Other Three Areas

Year	Kg from kitchen	Kg from other	Total (kg)[b]	Rupiah/ kilo	Total rupiah[b]
1988	1917	14,635	16,552	5500	91.0 mil
1990	1600	10,000	11,600	8500	98.6 mil
1992	1200	4,800	6,000	6000	36.0 mil

[a] Figures from the final sale in 1994 are not available, but were expected to be similar to 1992. Also, it is not clear whether there was another harvest in 1991.
[b] These figures are author's calculations.

Table 7.2
Rupiah Received by Each Class from Redistribution[a]

Year	Young (n = 20)	Single (n = 57)	Couple (n = 438)	School-age (n = 34)	Total rp. accounted for
1988	50,000	60,000	80,000	40,000	40.8 mil
1990	30,000	80,000	160,000	0	75.2 mil
1992	25,000	35,000	100,000	0	46.3 mil

[a] Note that in 1988 and 1990, not all earned Rupiah are accounted for. Presumably, the unaccounted remainder was divided among high ranking officials. In contrast, in 1992, more money seems to be accounted for than was actually earned. Although the numbers of individuals in each category were said to be the same as in 1988, the most likely explanation for this discrepancy is that those numbers did change. Thus, it is not clear then whether village leaders obtained higher shares in 1992. These figures are author's calculations based on reported share sizes and numbers of individuals.

portion of this money for themselves. All villagers are invited to help with the harvest from the village kitchen, however, it was clear that fewer people participated here than in the other areas.

Finally, in 1994, two new rules were instituted by the council. First, divers were not allowed to bring along young boys or women who could paddle their dugouts while they themselves dove. Second, at the urging of Dr. S. Wauthuyzen of LIPI-Ambon, no trochus with a base width narrower than three fingers (roughly 5–6 cm) were taken. These bring in a lower price and are suspected to be pre-reproductive (S. Wauthuyzen, personal communication).

Sasi for Trochus in a Neighboring Village

The nearby village of Ohoi El has an overall shoreline of roughly 2 km which is divided into three areas. As in Ohoirenan, all profits from the "village kitchen" are used for village purposes. While they also own and manage trochus cooperatively, there are six main differences between their methods and those of Ohoirenan. First, a much larger number of people dive for trochus in Ohoi El despite the fact that their territory is much smaller overall. In 1994, somewhere between 300 and 400 individuals dove there. This included many inhabitants of Ohoirenan who were invited by relatives in Ohoi El. Second, in Ohoi El, no special ceremonies take place at the waterfront to appease the ancestors. Instead, diving was initiated by the firing of a gun. Third, there was little trust of divers. Here, the village chief kept a careful eye on the area boundaries during the harvest and the harvest was terminated each day after exactly 2 hours. Fourth, divers are allowed to keep 100% of the profit from their take outside of the village kitchen. Fifth, each diver bargains with and sells his own harvest to the traders. It might be imagined that Ohoirenan is able to drive a harder bargain since it can promise a large lot. Sixth, there are no size restrictions on the trochus harvested. However, as in Ohoirenan, no scuba gear was used.

Discussion

Trochus is a very valuable cash resource in Ohoirenan and one requiring relatively little effort in production. To give some example of its worth, in 1989, 7.5 tons of trochus shell were harvested from the waters of Kei Besar, an island roughly 50 miles long and 5 miles wide (Abrahamsz, 1991, in Bailey and Zerner, 1992). This amounts to about U.S. $65,000. It is to be expected that this would produce an enormous temptation to either steal trochus or to attempt to monopolize the resource. So how and why does the village of Ohoirenan maintain collective ownership and management of trochus?

The Organization of Cooperation in Ohoirenan

In the introduction, four reasons were provided to explain why self-interested individuals might cooperate. Preliminary research suggests that each of these mechanisms plays some part in reinforcing cooperative resource management in the village of Ohoirenan.

Considering one level of cooperation at a time, at the village level it would seem that the hallmark characteristics of cooperation via generalized reciprocity are present. Access to the resource is limited to a group with known members who also have certainty over future rights and interactions. These individuals are easily able to monitor each others' actions on the fishing ground (it's easy enough to paddle by and see what another is catching and fishermen work at all hours of the day and night), or at home (there is nowhere to hide anything in the village, as this author knows well). In addition, those who do break the rules are selectively sanctioned, although it is not clear whether villagers are "moralistic" and punish individuals who themselves do not punish defectors. At the same time, the village is a large group, too large really to cooperate solely on the basis of reciprocal altruism (at least as it is modeled at present). However, reciprocal altruism may operate to maintain cooperation among the 18-member village council and/or among the heads of the "central families."

In turn, relations between these smaller groups and the village as a whole may be governed by asymmetrical altruism. The village chief and the "central families" have the final say in the punishment of offenders, and in management decisions. They are also said to obtain the largest revenues from the sale of trochus. If the latter is in fact the case, this is precisely what is meant by asymmetrical altruism. Some individuals receive extra benefits in return for paying the costs of enforcement. In this way, a delicate balance is maintained whereby village leaders have an incentive to maintain the system, while other individuals have the ability to resist exploitation insofar as they can choose not to dive for trochus. In some villages in other parts Maluku province, this balance has been disrupted with the result being private ownership of trochus by village elites (Zerner, 1990, 1994; S. Wauthuyzen, personal communication; H. Purnomo, personal communication).[12]

Cooperation between the central families may also be maintained by kin selection. All of the men, roughly 53, who belong to the 12 central families claim a common ancestor 6–8 generations ago. Although this would mean low levels of relatedness by now, examination of genealogies compiled by villagers reveals extensive intermarrying since that time. Observed patterns of sharing and exchange of food among these houses supports this view and deserves further study. For example, fishermen from the central families provide the village head with fish (and other foods) for free, but if none are available then fish are purchased from lower caste families.

It is also obvious that norms and ideology strongly reinforce cooperation. This is clearly the case among the central families, members of whom repeatedly assert "we are all one family," but is also true for the village as a whole. The people of Ohoirenan take great pride in pointing out to visitors how clean, organized, and friendly their village is when compared to others. They also cooperate in a wide variety of civic projects besides the trochus harvest, for example, in paving the streets, maintaining community medicinal gardens, paying the pastor's salary, buying supplies for the two elementary schools, etc. They themselves attribute some of this difference to a unity in religious belief. In Ohoirenan, the one pastor draws on strains of civic, animistic, and Protestant traditions in exhorting her flock to obey the "sasi." In contrast, the village of Ohoi El is both Protestant and Catholic, and all of the other 6 nearest villages are various mixtures of Catholic, Hindu, Muslim, and Protestant. Norms of conduct, specifically norms of cooperation, are often embedded in a system of religious beliefs and effectively transmitted to the community of believers. Villagers from the same religious system can thus be expected to share a high proportion of norms. In addition, church groups also provide a means of cross-cutting traditional class-based organization and of providing a new forum for discussion of village issues. In contrast, mixed religion villages may find it harder to find means of uniting their interests, and may have conflicting sets of norms of cooperation.

In point of fact, the opening of the "sasi" in Ohoi El was strictly a civic affair. While this village still does own, manage, and harvest trochus collectively, and while it uses the profits from the village kitchen for community needs, profits from the other areas are not redistributed and individuals do their own bargaining. There is also less physical evidence of other community projects. Furthermore, a larger portion of the standing stock is harvested since larger numbers of individuals are allowed to dive and there are no size restrictions. This supports the view that this community finds it harder to agree on restrictions that would protect their stock over the long term.

Finally, the fact that Ohoirenan is the oldest village in the area, and presumably the richest due to its long shoreline and consequent large trochus harvests, should not be ignored as an explanation of its well-off appearance and tendency to cooperate widely, not only in redistributing profits but also in engaging in many other types of collective works. However, it is difficult to determine whether their wealth is the cause or the effect of their behavior.

Is the Sasi a System of Conservation?

The inhabitants of Ohoirenan have successfully organized themselves to collectively manage and harvest trochus. They have designed institutions that avoid problems of free-riding and rent dissipation, and do so in what appears to be an equitable manner. However, as discussed in the introduction, this does not in itself mean that these practices were intended to conserve trochus, or that they will have the effect of doing so.

Two lines of evidence support Polunin's (1984) argument that the intent behind Malukan marine "sasi" systems is maximization of profit rather than conservation. First, in the case of Ohoirenan, the open season for trochus is not tied to any particular phase in the animal's life cycle but rather takes place when weather permits the largest possible harvest. Second, by claiming a territory and limiting access to it, Ohoirenan is able to control an area of coast that is much larger per capita than other neighboring villages. Third, it is here argued that the main reason for having a closed season is again, not to regulate trochus populations, but to increase exclusivity of access. In these small communities, anyone poaching trochus and storing it outside of the open season would easily be detected. Hence, intensive monitoring of resources need only take place during the short open season, as was observed.

To what extent conservation is the by-product, or epiphenomenal effect, of these regulations has not been adequately measured in this pilot study. Accounts of harvests since 1988 show declining yields, but whether this is a real trend is unknown. Certainly, villagers are genuinely concerned about the future of this resource. For this reason, they have adopted new measures whose effect may be to protect the stock even if the intent is to improve long-term profits. In particular, they have introduced new restrictions on the size of trochus collected and are now experimenting with harvesting trochus only every other year.

Ecological Reasons for Collective Management

Trochus is a good example of a resource that might be best managed collectively rather than privately. First of all, since the animals are slow-moving and thus not ephemeral, patches can be claimed and defended, i.e., territoriality whether private or communal can occur. Secondly, they are moderate to low in abundance meaning that large areas are usually necessary to make collection profitable. If this single large area were to be subdivided, the length of boundary needing monitoring would be greatly increased. Guarding access to individual holdings could be difficult and time consuming. As a result, defense would be prohibitively costly and unless yield were to increase even faster this might make individual holdings unprofitable. Enhancing production in such a manner seems unlikely for a resource like trochus.[13]

However, privatization or subdivision might be profitable, even with a fixed yield, if there is a reduction in the number of people having rights to the resource. In this case, the increase in profitability may offset the increased costs of defense. In fact, in the one case of privatization described to the author, only a few families now have rights to the resource. However, privatization seems less likely in a remote village such as Ohoirenan where villagers have few other cash opportunities and no market within walking distance. For this reason, villagers of high and low caste alike remain dependent on each other and all villagers have retained their rights to the communities' resources.

At this time, the village still has strong incentives to maintain trochus stocks and to manage for the future since they have few other sources of cash income. Moreover, they are very aware that their status as a rich village depends on good trochus harvests. However, as their economic opportunities diversify and as they become more closely tied into the national and global economies they may face situations where short-term concerns make liquidating trochus stocks seem appealing, as has happened in other parts of Maluku Province. These economic changes could also affect the balance of power within the village, perhaps leading to the alienation of rights to the resource from the majority of inhabitants. Whether either of these outcomes actually occurs will depend both on the foresight of village leadership and the main-tenance of their currently strong beliefs in working together as a village.

Conclusion

It is here argued that the intent of traditional resource management systems is to increase human well-being, not to conserve species *per se*. However, long-term resource conservation may be a by-product of traditional man-age-ment in some specific cases. Communities that do use resources sustainably must both be able to prevent free-riding and be capable of implementing appropriate management strategies. The former is perhaps not as much of a problem as theory suggests, particularly if groups are composed of kin and have a stable membership with shared norms of conduct. However, designing a sustainable management regime may be somewhat more difficult either because it is difficult to predict the effects of human action on some resources, or because immediate economic imperatives outweigh long-term ecological concerns. Thus, true conservation is most likely to have evolved in societies where relatively high population densities are combined with limited opportunities to sell surplus produce. For all these reasons, it is a mistake to assume that a "conservation ethic" is universal among indigenous peoples and to make this a basis of policy decisions.

Despite these precautions, it is also true today that many communities are highly motivated to manage their natural resources sustainably. Furthermore, in some cases, communal management may be preferable to either privatization or state control in terms of economic efficiency and/or resource conservation. Focusing on Hardin's version of the "tragedy of the commons," and on debates as to whether indigenous peoples truly are, or are not, in harmony with nature has distracted many authors from addressing the more interesting and practical problems of characterizing which specific ecological or economic conditions permit the evolution and continuation of successful management regimes. Finally, that local peoples have prior and customary rights remains the strongest moral argument for legally recognizing traditional tenure systems.

Acknowledgements

I wish to thank William G. Davis, Monique Borgerhoff Mulder, Peter Richerson, James Wilen, Carl Ford Runge, and three anonymous reviewers for their helpful comments and insightful criticism of this manuscript. My especial thanks go to my Indonesian research sponsors, the Sub Balai Penelitian Perikanan Laut, Ambon, headed by Dr. Tengku A. R. Hanafiah. Without the very generous help of staff researchers, Agus Henri Purnomo and Retmo Andamari, my stay would not have been possible nor enjoyable. I also wish to express my deep appreciation to researchers from the Indonesian Institute of Sciences (LIPI) in Ambon, Dr. Samuel Wathuyzen and Dr. S. A. Putro Dwiyono, for permitting me to accompany their research team to Ohirenan and for sharing with me their extensive knowledge of trochus biology and management in Maluku Province. Thanks also go to the staff of the LIPI station in Tual, Safer Dody, and Amang Radjab, for facilitating my stay in the Kei Islands. Finally, I give especial thanks to the many villagers of Desa Ohoirenan who shared their knowledge, time, and companionship freely. This research was supported by a University of California Fellowship awarded by the Ecology Graduate Group at UC Davis.

Notes

1.   Division of Environmental Studies, University of California at Davis, Davis, California 95616.
2.   This phrase is drawn from Polunin (1984) and is not meant to preclude the possibility that gain may be the result of restraint.
3.   In the critical section where he discusses the incentives leading a herder to add more animals to a pasture he begins by asking the reader to "picture a pasture open to all" (p. 1244).

4. Berkes (1989b) has also noted that three of these four (reciprocal altruism, kin selection, and cultural group selection) may be useful explanations of human cooperation.

5. Inclusive fitness is usually defined as the number of copies of a gene that an individual passes to its own offspring (its direct fitness), plus the number of gene copies that are passed on by the individual's relatives, the latter being weighted by r, the degree of relatedness between the focal individual and their relative. This weighting corrects for the fact that relatives do not have exactly the same set of genes. For example, your full sibling is expected to have only half their genes in common with you. The actual terms under which you would benefit from aiding a relative are when the costs to yourself are smaller than the product of the benefit to the relative times the constant of relatedness, r (Hamilton, 1964a).

6. More precisely, they should stint on foods that would be included in an "optimal diet" as determined by optimal foraging theory.

7. Also note that it is highly debatable whether food taboos lead to conservation (Hames, 1987b; Kaplan and Hill, 1985b). Among other things, taboos may merely shift the diet to another equally vulnerable species (Johannes, 1978).

8. Although one should not expect that hunter-gatherers lacking territorial boundaries or pursuing large migratory animals would conserve in the first place.

9. The following information was obtained through informal discussions with numerous members of the central families. Of especial assistance were the village chief, Julius Rahallus, the village secretary, Yustus Ubra, as well as Jusuf Rahayanaan, Jacop Ubro, Jon Ubro, Enos Rahallus, and Kempes Rahakbauw.

10. There are also a few members of a third, lower caste composed of individuals who have been punished by social ostracization and live on the outskirts of the village. Inhabitants of Irian Jaya who temporarily reside in the village to earn cash and clothing are also considered to be long to this lowest caste.

11. If the weather is poor, they may skip a day or two before resuming harvesting, but in no case is an area fished more than twice.

12. It is also notable that privatization in these villages followed a crash in their trochus population. This crash, in turn, came after a sharp drop in the price of cloves, the main source of cash income in that area. Thus, villagers followed what would appear to be the only sensible solution to a temporary low in clove prices, that is, harvesting and selling as much trochus as possible in the short term (S. Wathuyzen, personal communication).

13. However, in the past, villagers set out fish weirs made of woven papaya leaves. Biologists believe that trochus spawned on these traps, thus enhancing production, but villagers are no longer interested in making these time-consuming products (S. Wauthuyzen, personal communication).

8

Revisiting the Commons: Local Lessons, Global Challenges

Elinor Ostrom, Joanna Burger, Christopher B. Field,
Richard B. Norgaard, and David Policansky

Thirty years have passed since Garrett Hardin's influential article "The Tragedy of the Commons."[1] At first, many people agreed with Hardin's metaphor that the users of a commons are caught in an inevitable process that leads to the destruction of the very resource on which they depend. The "rational" user of a commons, Hardin argued, makes demands on a resource until the expected benefits of his or her actions equal the expected costs. Because each user ignores costs imposed on others, individual decisions cumulate to a tragic overuse and the potential destruction of an open-access commons. Hardin's proposed solution was "either socialism or the privatism of free enterprise."[2]

The starkness of Hardin's original statement has been used by many scholars and policy-makers to rationalize central government control of all common-pool resources[3] and to paint a disempowering, pessimistic vision of the human prospect.[4] Users are pictured as trapped in a situation they cannot change. Thus, it is argued that solutions must be imposed on users by external authorities. Although tragedies have undoubtedly occurred, it is also obvious that for thousands of years people have self-organized to manage common-pool resources, and users often do devise long-term, sustainable institutions for governing these resources.[5-7] It is time for a reassessment of the generality of the theory that has grown out of Hardin's original paper. Here, we describe the advances in understanding and managing commons problems that have been made since 1968. We also describe research challenges, especially those related to expanding our understanding of global commons problems.

Reprinted with permission from Elinor Ostrom, Joanna Burger, Christopher B. Field, Richard B. Norgaard, David Policansky, 1999. "Revisiting the Commons: Local Lessons, Global Challenges." *Science* 284(9): 278-282. Copyright © 1999 by AAAS.

An important lesson from the empirical studies of sustainable resources is that more solutions exist than Hardin proposed. Both government ownership and privatization are themselves subject to failure in some instances. For example, Sneath shows great differences in grassland degradation under a traditional, self-organized group-property regime versus central government management. A satellite image of northern China, Mongolia, and southern Siberia[8] shows marked degradation in the Russian part of the image, whereas the Mongolian half of the image shows much less degradation. In this instance, Mongolia has allowed pastoralists to continue their traditional group-property institutions, which involve large-scale movements between seasonal pastures, while both Russia and China have imposed state-owned agricultural collectives that involve permanent settlements. More recently, the Chinese solution has involved privatization by dividing the "pasture land into individual allocations for each herding household."[8] About three-quarters of the pasture land in the Russian section of this ecological zone has been degraded and more than one-third of the Chinese section has been degraded, while only one-tenth of the Mongolian section has suffered equivalent loss.[8, 9] Here, socialism and privatization are both associated with more degradation than resulted from a traditional group-property regime.

Most of the theory and practice of successful management involves resources that are effectively managed by small to relatively large groups living within a single country, which involve nested institutions at varying scales. These resources continue to be important as sources of sustained biodiversity and human well-being. Some of the most difficult future problems, however, will involve resources that are difficult to manage at the scale of a village, a large watershed, or even a single country. Some of these resources—for example, fresh water in an international basin or large marine ecosystems—become effectively depletable only in an international context.[10] Management of these resources depends on the cooperation of appropriate international institutions and national, regional, and local institutions. Resources that are intrinsically difficult to measure or that require measurement with advanced technology, such as stocks of ocean fishes or petroleum reserves, are difficult to manage no matter what the scale of the resource. Others, for example global climate, are largely self-healing in response to a broad range of human actions, until these actions exceed some threshold.[11]

Although the number and importance of commons problems at local or regional scales will not decrease, the need for effective approaches to commons problems that are global in scale will certainly increase. Here, we examine this need in the context of an analysis of the nature of common-pool resources and the history of successful and unsuccessful institutions for ensuring fair access and sustained availability to them. Some experience from smaller systems transfers directly to global systems, but global commons introduce a range of new issues, due largely to extreme size and complexity.[12]

The Nature of Common-Pool Resources

To better understand common-pool resource problems, we must separate concepts related to resource systems and those concerning property rights. We use the term common-pool resources (CPRs) to refer to resource systems regardless of the property rights involved. CPRs include natural and human-constructed resources in which (i) exclusion of beneficiaries through physical and institutional means is especially costly, and (ii) exploitation by one user reduces resource availability for others.[13] These two characteristics—difficulty of exclusion and subtractability—create potential CPR dilemmas in which people following their own short-term interests produce outcomes that are not in anyone's long-term interest. When resource users interact without the benefit of effective rules limiting access and defining rights and duties, substantial free-riding in two forms is likely: overuse without concern for the negative effects on others, and a lack of contributed resources for maintaining and improving the CPR itself.

CPRs have traditionally included terrestrial and marine ecosystems that are simultaneously viewed as depletable and renewable. Characteristic of many resources is that use by one reduces the quantity or quality available to others, and that use by others adds negative attributes to a resource. CPRs include earth-system components (such as groundwater basins or the atmosphere) as well as products of civilization (such as irrigation systems or the World Wide Web).

Characteristics of CPRs affect the problems of devising governance regimes. These attributes include the size and carrying capacity of the resource system, the measurability of the resource, the temporal and spatial availability of resource flows, the amount of storage in the system, whether resources move (like water, wildlife, and most fish) or are stationary (like trees and medicinal plants), how fast resources regenerate, and how various harvesting technologies affect patterns of regeneration.[14] It is relatively easy to estimate the number and size of trees in a forest and allocate their use accordingly, but it is much more difficult to assess migratory fish stocks and available irrigation water in a system without storage capacity. Technology can help to inform decisions by improving the identification and monitoring of resources, but it is not a substitute for decision-making. On the other hand, major technological advances in assessing groundwater storage capacity, supply, and associated pollution have allowed more effective management of these resources.[15] Specific resource systems in particular locations often include several types of CPRs and public goods with different spatial and temporal scales, differing degrees of uncertainty, and complex interactions among them.[16]

Institutions for Governing and Managing Common-Pool Resources

Solving CPR problems involves two distinct elements: restricting access and creating incentives (usually by assigning individual rights to, or shares of, the resource) for users to invest in the resource instead of overexploiting it. Both changes are needed. For example, access to the north Pacific halibut fishery was not restricted before the recent introduction of individual transferable quotas and catch limits protected the resource for decades. But the enormous competition to catch a large share of the resource before others did resulted in economic waste, danger to the fishers, and reduced quality of fish to consumers. Limiting access alone can fail if the resource users compete for shares, and the resource can become depleted unless incentives or regulations prevent overexploitation.[17, 18]

Four broad types of property rights have evolved or are designed in relation to CPRs (Table 8.1). When valuable CPRs are left to an open-access regime, degradation and potential destruction are the result. The proposition that resource users cannot themselves change from no property rights (open access) to group or individual property, however, can be strongly rejected on the basis of evidence: Resource users through the ages have done just that.[5–7, 13, 15, 19] Both group-property and individual-property regimes are used to manage resources that grant individuals varying rights to access and use of a resource. The primary difference between group property and individual property is the ease with which individual owners can buy or sell a share of a resource. Government property involves ownership by a national, regional, or local public agency that can forbid or allow use by individuals. Empirical studies show that no single type of property regime works efficiently, fairly, and sustainably in relation to all CPRs. CPR problems continue to exist in many regulated settings.[17] It is possible, however, to identify design principles associated with robust institutions that have successfully governed CPRs for generations.[19]

Table 8.1
Types of property-rights systems used to regulate common-pool resources [7]

Property rights	Characteristics
Open access	Absence of enforced property rights
Group property	Resource rights held by a group of users who can exclude others
Individual property	Resource rights held by individuals (or firms) who can exclude others
Government property	Resource rights held by a government that can regulate or subsidize use

The Evolution of Norms and Design of Rules

The prediction that resource users are led inevitably to destroy CPRs is based on a model that assumes all individuals are selfish, norm-free, and maximizers of short-run results. This model explains why market institutions facilitate an efficient allocation of private goods and services, and it is strongly supported by empirical data from open, competitive markets in industrial societies.[20] However, predictions based on this model are not supported in field research or in laboratory experiments in which individuals face a public good or CPR problem and are able to communicate, sanction one another, or make new rules.[21] Humans adopt a narrow, self-interested perspective in many settings, but can also use reciprocity to overcome social dilemmas.[22] Users of a CPR include (i) those who always behave in a narrow, self-interested way and never cooperate in dilemma situations (free-riders); (ii) those who are unwilling to cooperate with others unless assured that they will not be exploited by free-riders; (iii) those who are willing to initiate reciprocal cooperation in the hopes that others will return their trust; and (iv) perhaps a few genuine altruists who always try to achieve higher returns for a group.

Whether norms to cope with CPR dilemmas evolve without extensive, self-conscious design depends on the relative proportion of these behavioral types in a particular setting. Reciprocal cooperation can be established, sustain itself, and even grow if the proportion of those who always act in a narrow, self-interested manner is initially not too high.[23] When interactions enable those who use reciprocity to gain a reputation for trustworthiness, others will be willing to cooperate with them to overcome CPR dilemmas, which leads to increased gains for themselves and their offspring.[24] Thus, groups of people who can identify one another are more likely than groups of strangers to draw on trust, reciprocity, and reputation to develop norms that limit use. In earlier times, this restricted the size of groups who relied primarily upon evolved and shared norms. Citizen-band radios, tracking devices, the Internet, geographic information systems, and other aspects of modern technology and the news media now enable large groups to monitor one another's behavior and coordinate activities in order to solve CPR problems.

Evolved norms, however, are not always sufficient to prevent overexploitation. Participants or external authorities must deliberately devise (and then monitor and enforce) rules that limit who can use a CPR, specify how much and when that use will be allowed, create and finance formal monitoring arrangements, and establish sanctions for non-conformance. Whether the users themselves are able to overcome the higher level dilemmas they face in bearing the cost of designing, testing, and modifying governance systems depends on the benefits they perceive to result from a change as well as the expected costs of negotiating, monitoring, and enforcing these rules.[25] Perceived benefits are greater when the resource reliably generates valuable products for the users. Users need some autonomy to make and enforce their own rules, and they must highly value the future sustainability of the resource. Perceived costs are higher when the resource is large and complex, users lack a common understanding of resource dynamics, and users have substantially diverse interests.[26]

The farmer-managed irrigation systems of Nepal are examples of well-managed CPRs that rely on strong, locally crafted rules as well as evolved norms.[27] Because the rules and norms that make an irrigation system operate well are not visible to external observers, efforts by well-meaning donors to replace primitive, farmer-constructed systems with newly constructed, government-owned systems have reduced rather than improved performance.[28] Government-owned systems are built with concrete and steel headworks, in contrast to the simple mud, stone, and trees used by the farmers (Figure 8.1). However, the cropping intensity achieved by farmer-managed systems is significantly higher than on government systems (Table 8.2). In a regression model of system performance, controlling for the size of the system, the slope of the terrain, variation in farmer income, and the presence of alternative sources of water, both government ownership and the presence of modern headworks have a negative impact on water delivered to the tail end of a system, hence a negative impact on overall system productivity.[27]

Imposing strong limits on resource use raises the question of which community of users is initially defined as having use rights and who is excluded from access to a CPR. The very process of devising methods of exclusion has substantial distributional consequences.[29] In some instances, those who have long exercised stewardship over a resource can be excluded. A substantial distributional issue will occur, for example, as regulators identify who will receive rights to emit carbon into the atmosphere. Typically, such rights are assigned to those who have exercised a consistent pattern of use over time. Thus, those who need to use the resource later may be excluded entirely or may have to pay a very large entry cost.

The counterpoint to exclusion is too rapid inclusion of users. When any user group grows rapidly, the resource can be stressed. For example, in the last 10 years the annual sales of personal watercraft (PWCs) have risen in the United States from about 50,000 to more than 150,000 a year. This has placed a burden on the use of surface water and created conflicts with homeowners, other boaters, fishermen, and naturalists. The rapid rise of PWCs has created a burden on the use of shorelines, contributed to a disproportionate increase in accidents and injuries, and caused disturbances to aquatic natural resources.[30] Traditional users of the water surface feel threatened by the invasion of their space by a new, faster, and louder boat that reduces the value of surface waters. In many other settings, when new users arrive through migration, they do not share a similar understanding of how a resource works and what rules and norms are shared by others. Members of the initial community feel threatened and may fail to enforce their own self-restraint, or they may even join the race to use up the resource.[31]

Given the substantial differences among CPRs, it is difficult to find effective rules that both match the complex interactions and dynamics of a resource and are perceived by users as legitimate, fair, and effective. At times, disagreements about resource assessment may be strategically used to propose policies that disproportionately benefit some at a cost to others.[4] In highly complex

Figure 8.1

The government-owned Chiregad irrigation system (right panel) was constructed in Nepal to replace five farmer-owned irrigation systems whose physical infrastructures were similar to the Kathar farmer-managed irrigation system (left panel). In planning the Chiregad system, designers focused entirely on constructing modern engineering works and not on learning about the rules and norms that had been used in the five earlier systems. Even though the physical capital is markedly better than that possessed by the earlier systems, the Chiregad system has never been able to provide water consistently to more than two of the former villages. Agricultural productivity is lower now than it was under farmer management.[37] Not only do the farmers invest heavily in the maintenance of the farmer-owned system on the left, they have devised effective rules related to access and the allocation of benefits and costs. They achieve higher productivity than most government-owned systems with modern infrastructure. [Photographs by G. Shivakoti (left) and E. Ostrom (right)]

Table 8.2

Relationship of governance structures and cropping intensities[27] [p. 106]. A crop intensity of 100% means that all land in an irrigation system is put to full use for one season or a partial use over multiple seasons, amounting to the same coverage. Similarly, a crop intensity of 200% is full use of all land for two seasons; 300% is full use for three seasons.

Parameter	Farmer-owned systems	Government-owned systems	F	P
	$(N = 97)$	$(N = 21)$		
Head-end crop intensities	246%	208%	10.51	0.002
Tail-end crop intensities	237%	182%	20.33	0.004

systems, finding optimal rules is extremely challenging, if not impossible. But despite such problems, many users have devised their own rules and have sustained resources over long periods of time. Allowing parallel self-organized governance regimes to engage in extensive trial-and-error learning does not reduce the probability of error for any one resource, but greatly reduces the probability of disastrous errors for all resources in a region.

Lessons from Local and Regional Common-Pool Resources

The empirical and theoretical research stimulated over the past 30 years by Garrett Hardin's article has shown that tragedies of the commons are real, but not inevitable. Solving the dilemmas of sustainable use is neither easy nor error-free even for local resources. But a scholarly consensus is emerging regarding the conditions most likely to stimulate successful self-organized processes for local and regional CPRs.[6, 26, 32] Attributes of resource systems and their users affect the benefits and costs that users perceive. For users to see major benefits, resource conditions must not have deteriorated to such an extent that the resource is useless, nor can the resource be so little used that few advantages result from organizing. Benefits are easier to assess when users have accurate knowledge of external boundaries and internal microenvironments and have reliable and valid indicators of resource conditions. When the flow of resources is relatively predictable, it is also easier to assess how diverse management regimes will affect long-term benefits and costs.

Users who depend on a resource for a major portion of their livelihood, and who have some autonomy to make their own access and harvesting rules, are more likely than others to perceive benefits from their own restrictions, but they need to share an image of how the resource system operates and how their actions affect each other and the resource. Further, users must be interested in the sustainability of the particular resource so that expected joint benefits will outweigh current costs. If users have some initial trust in others to keep promises, low-cost methods of monitoring and sanctioning can be devised. Previous organizational experience and local leadership reduces the users' costs of coming to agreement and finding effective solutions for a particular environment. In all cases, individuals must overcome their tendency to evaluate their own benefits and costs more intensely than the total benefits and costs for a group. Collective-choice rules affect who is involved in deciding about future rules and how preferences will be aggregated. Thus, these rules affect the breadth of interests represented and involved in making institutional changes, and they affect decisions about which policy instruments are adopted.[33]

The Broader Social Setting

Whether people are able to self-organize and manage CPRs also depends on the broader social setting within which they work. National governments can help or hinder local self-organization. "Higher" levels of government can facilitate the assembly of users of a CPR in organizational meetings, provide information that helps identify the problem and possible solutions, and legitimize and help enforce agreements reached by local users. National governments can at times, however, hinder local self-organization by defending rights that lead to overuse or maintaining that the state has ultimate control over resources without actually monitoring and enforcing existing regulations.

Participants are more likely to adopt effective rules in macro-regimes that facilitate their efforts than in regimes that ignore resource problems entirely or that presume that central authorities must make all decisions. If local authority is not formally recognized by larger regimes, it is difficult for users to establish enforceable rules. On the other hand, if rules are imposed by outsiders without consulting local participants, local users may engage in a game of "cops and robbers" with outside authorities. In many countries, two centuries of colonization followed by state-run development policy that affected some CPRs has produced great resistance to externally imposed institutions.

The broader economic setting also affects the level and distribution of gains and costs of organizing the management of CPRs. Expectations of rising resource prices encourage better management, whereas falling, unstable, or uncertain resource prices reduce the incentive to organize and assure future availability.[34] National policy also affects factors such as human migration rates, the flow of capital, technology policy, and hence the range of conditions local institutions must address to work effectively. Finally, local institutions are only rarely able to cope with the ramifications of civil or international war.

Challenges of Global Commons

The lessons from local and regional CPRs are encouraging, yet humanity now faces new challenges to establish global institutions to manage biodiversity, climate change, and other ecosystem services.[35] These new challenges will be especially difficult for at least the following reasons.

Scaling-up problem. Having larger numbers of participants in a CPR increases the difficulty of organizing, agreeing on rules, and enforcing rules. Global environmental resources now involve 6 billion inhabitants of the globe. Organization at national and local levels can help, but it can also get in the way of finding solutions.

Cultural diversity challenge. Along with economic globalization, we are in a period of reculturalization. Increasing cultural diversification offers increased hope that the diversity of ways in which people have organized locally around CPRs will not be quickly lost, and that diverse new ways will continue to evolve at the local level. However, cultural diversity can decrease the likelihood of finding shared interests and understandings. The problem of cultural diversity is exacerbated by "north-south" conflicts stemming from economic differences between industrialized and less-industrialized countries.

Complications of interlinked CPRs. Although the links between grassland and forest management are complex, they are not so complex as those between maintaining biodiversity and ameliorating climate change. As we address global issues, we face greater interactions between global systems. Similarly, with increased specialization, people have become more interdependent. Thus, we all share one another's common interests, but in more complex ways than the users of a forest or grassland. While we have become more complexly interrelated, we have also become more "distant" from each other and our environmental problems. From our increasingly specialized understandings and particular points on the globe, it is difficult to comprehend the significance of global CPRs and how we need to work together to govern these resources successfully. And given these complexities, finding fair solutions is even more challenging.

Accelerating rates of change. Previous generations complained that change occurred faster and faster, and the acceleration continues. Population growth, economic development, capital and labor mobility, and technological change push us past environmental thresholds before we know it. "Learning by doing" is increasingly difficult, as past lessons are less and less applicable to current problems.

Requirement of unanimous agreement as a collective-choice rule. The basic collective-choice rule for global resource management is voluntary assent to negotiated treaties.[36] This allows some national governments to hold out for special privileges before they join others in order to achieve regulation, thus strongly affecting the kinds of resource management policies that can be adopted at this level.

We have only one globe with which to experiment. Historically, people could migrate to other resources if they made a major error in managing a local CPR. Today, we have less leeway for mistakes at the local level, while at the global level there is no place to move.

These new challenges clearly erode the confidence with which we can build from past and current examples of successful management to tackle the CPR problems of the future. Still, the lessons from successful examples of CPR management provide starting points for addressing future challenges. Some of these will be institutional, such as multilevel institutions that build on and complement local and regional institutions to focus on truly global problems. Others will build from improved technology. For example, more accurate long-range weather forecasts could facilitate improvements in irrigation management, or advances in fish tracking could allow more accurate population estimates and harvest management. And broad dissemination of widely believed data could be a major contributor to the trust that is so central to effective CPR management.

In the end, building from the lessons of past successes will require forms of communication, information, and trust that are broad and deep beyond precedent, but not beyond possibility. Protecting institutional diversity related to how diverse peoples cope with CPRs may be as important for our long-run survival as the protection of biological diversity. There is much to learn from successful efforts as well as from failures.

References and Notes

1.   G. Hardin (1968).
2.   G. Hardin (1998).
3.   J. E. M. Arnold (1998); D. Feeny, S. Hanna, A. F. McEvoy (1996); F. Berkes and C. Folke, (1998); A. C. Finlayson and B. J. McCay, (1998); R. Repetto (1986).
4.   D. Ludwig, R. Hilborn, C. Walters (1993).
5.   B. J. McCay and J. M. Acheson (1987); F. Berkes, D. Feeny, B. J. McCay, J. M. Acheson (1989); F. Berkes (1989a); D. W. Bromley et al. (1992); S. Y. Tang (1992); E. Pinkerton (1989); C. Hess (1996).
6.   R. Wade (1994).
7.   D. Feeny, F. Berkes, B. J. McCay, J. M. Acheson (1990).
8.   D. Sneath (1998).
9.   C. Humphrey and D. Sneath (1996).
10.   R. Costanza et al. (1998).
11.   W. S. Broecker (1997).
12.   M. McGinnis and E. Ostrom (1996); R. O. Keohane and E. Ostrom (1995); S. Buck (1998).
13.   E. Ostrom, R. Gardner, J. Walker (1994).
14.   E. Schlager, W. Blomquist, S. Y. Tang (1994)
15.   W. Blomquist (1992).
16.   R. Norgaard (1995); C. Gibson (1999); A. Agrawal (1999).
17.   Organisation for Economic Co-operation and Development (OECD) (1997); National Research Council (1999).
18.   H. S. Gordon (1954); B. J. McCay (1995).
19.   E. Ostrom (1990).
20.   C. R. Plott (1986); (1991).
21.   See S. Bowles, R. Boyd, E. Fehr, H. Gintis (1997); E. Ostrom and J. M. Walker (1997); J. M. Orbell, A. van de Kragt, R. M. Dawes (1988); E. Ostrom (1998). In these experiments, the formal structure of a dilemma is converted into a set of decisions made by subjects who are financially rewarded as a result of their own

and others' decisions. See also J. H. Kagel and A. E. Roth (1995). The model is also not as robust in explaining exchange behavior in traditional societies where evolved norms still strongly affect behavior.

22. L. Cosmides and J. Tooby (1992); L. Cosmides and J. Tooby (1994); E. Hoffman, K. McCabe, V. Smith (1996).
23. R. Axelrod (1984); (1986).
24. M. A. Nowak and K. Sigmund (1992); D. M. Kreps, P. Milgrom, J. Roberts, R. Wilson (1982).
25. H. Demsetz (1967); D. North (1994); C. M. Rose (1994); J. E. Krier (1992); F. Michelman (1990); V. Ostrom (1990).
26. E. Ostrom (2001).
27. W. F. Lam (1998).
28. W. F. Lam (1996).
29. G. D. Libecap (1989).
30. J. Burger (1998); in (26); L. Whiteman (1997).
31. F. G. Speck and W. S. Hadlock (1946); C. Safina (1994).
32. J. M. Baland and J. P. Platteau (1996); M. A. McKean (1992).
33. J. Buchanan and G. Tullock (1962); J. B. Wiener (1999a).
34. C. W. Clark and G. R. Munro (1994).
35. See O. Young (1999); *Global Governance: Drawing Insights from the Environmental Experience* (1997); P. Haas, R. Keohane, M. Levy (1993).
36. J. B. Wiener (1999b).
37. R. M. Hilton (1992).
38. This paper profited from ideas discussed at a symposium on "The Commons Revisited: An Americas Perspective" held in conjunction with the X General Assembly of the Scientific Committee on Problems of the Environment (SCOPE), June 1998. We thank the U.S. Environmental Protection Agency, NSF, and NASA for supporting the U.S. National Committee for SCOPE, where this effort began. We thank F. Berkes, A. Blomqvist, P. Dalecki, D. Dodds, K. Dougherty, D. Feeny, T. Hargis-Young, C. Hess, B. J. McCay, M. McGinnis, M. Polski, E. Schlager, N. Sengupta, J. Unruh, O. Young, and anonymous reviewers for their useful comments. Supported by NSF grant SBR-9521918, the U.N. Food and Agriculture Organization (FAO), the Ford Foundation, and the MacArthur Foundation (E.O.) and by U.S. Department of Energy grant Al DE-FC01-9SEWS5084 to the Consortium for Risk Evaluation with Stakeholder Participation and National Institute of Environmental Health Sciences grant ESO 5022 (J.B.)

9

Grassland Conservation and the Pastoralist Commons

Monique Borgerhoff Mulder and Lore M. Ruttan

Introduction

Conservation commonly refers to the maintenance of biodiversity, in terms of the full set of genetic, species and ecosystem diversity at the natural abundance at which they occur (OTA, 1987). Although this definition is quite widely accepted, it has been more difficult to reach consensus on a satisfactory definition of the human acts that might aptly be called "conservation." For most social scientists, conservation occurs when people *intend* to conserve, and they are particularly interested in determining the multiple social and political factors that lead to the successful practice of conservation. In contrast, evolutionary anthropologists (evolutionary ecologists studying human affairs) are cautious about equating conservation with intent and emphasize rather the potentially altruistic, and hence theoretically problematic, aspects of conservation. Their concern then is to understand the conditions under which conservation is individually advantageous

In this chapter, we take the view that aspects of both positions have merit. Retaining both the assumption of individual self-interest and the methodological rigour that characterize an evolutionary ecological approach, we build a model that incorporates the asymmetries in power and interest that typify most human communities and engage social scientists' attention. More technically, we question two fundamental assumptions within the evolutionary anthropologists' intellectual tool kit. First, that economic efficiency is synonymous with short-term efficiency as predicted by optimal foraging models (and its corollary that economic efficiency is antithetical to conservation).

Monique Borgerhoff Mulder and Lore M. Ruttan "Grassland conservation and the pastoralist commons", pp. 34-50 in Gosling, L. M. and W.J. Sutherland (eds): "Conservation and Behaviour" 2000, Cambridge University Press, © 2000 The Zoological Society of London, reprinted with permission.

Second, that an act cannot be designated as conservationist unless it is the true cause of the conservation outcome (e.g., Hunn, 1982).

The chapter is structured as follows. First, we give some history to the study of conservation behaviour in humans, highlighting the significant contribution from evolutionary ecology. We then address the question of whether and how conservation acts might be maintained in a population, using the classic example of a pastoral grazing reserve; here we develop a game theory model following closely on the work of institutional economists and behavioural ecologists. Finally, we end with some brief conclusions concerning the implications of our model, and some new ways of thinking about these matters.

Does the "Ecologically Noble Savage" Exist?

Living in Harmony with Nature

Ethnographers and sociocultural anthropologists commonly argue that indigenous peoples live in balance with their environment and, more generally, in harmony with nature. They base this view primarily on the observation that small populations with limited technology subsist on plant and animal species without driving these resources to extinction, and without causing long-term degradation of the environment (e.g., Alcorn, 1989; Posey & Balee, 1989; IWIGIA, 1992), a point first made by Birdsell (1958). In conjunction they note that indigenous peoples exhibit deep reverence for nature (e.g., Nelson, 1982; Kay, 1985a; Zann, 1989) and often command an extensive knowledge of their prey species' habits (e.g., Johannes, 1981). Somewhat similar equilibria (be they stable or oscillating) are also observed among non-human predators and their prey. Thirty years ago biologists attributed such systems to the behaviour of predators, speculating that a "prudent predator" (Slobodkin, 1968) spares the prime-aged reproductive prey in order to secure a sustainable harvest for the future. While this view has become outdated in biology, it still holds sway in anthropology. Indeed most anthropologists maintain that prudence is a *goal* of subsistence hunters. Accordingly, indigenous peoples are attributed the reputation of being natural conservationists (e.g., Durning, 1993) or, in Redford's (1991b) more colourful epithet, "ecologically noble savages."

As might be expected, evolutionary ecologists have raised theoretical objections to the somewhat romantic view that indigenous peoples are natural conservationists (e.g., Winterhalder, 1997b; Smith, 1983; Hames, 1987b, 1991; Low, 1996; Beckerman & Valentine, 1996; for a model, see Winterhalder & Lu, 1997). Individuals cannot be expected to limit present harvests of resources for the purpose of conserving them for future use if this behaviour entails a cost. As resource economists noted long ago, if resources are not owned privately then any restraint in their use is altruistic insofar as the benefits from restraint will be shared by many while the costs are borne individually (Gordon, 1954; Scott, 1955). Such altruism can easily be exploited and hence is unlikely to evolve between unrelated individuals (Williams, 1966).

Definitions and Tests of Conservation Acts

With these theoretical objections in mind, human evolutionary ecologists began conducting field examinations of whether or not the "ecologically noble savage" exists, with the focus mainly on foraging populations. Hames (1987b) had the nice idea of testing for conservation behaviour by pitting its predictions against a set of alternatives derived from optimal foraging theory, which he termed "efficiency hypotheses." Here, efficiency refers to returns to the *individual* and not to the group, and therefore differs from economists' use of the term (a usage that side-steps the issue of how the costs of providing efficiency to a group are borne). Efficiency maximization is predicated on two assumptions—that more food enhances individual fertility and survival, and that time spent acquiring food incurs opportunity costs (Kaplan & Hill, 1992). It should be noted, however, that most optimal foraging models (but see Benson & Stephens, 1996) are based on the assumption that exploitation of a resource has no effect on its future abundance. Hence, in testing for efficiency, evolutionary anthropologists predict that foragers should make prey choice decisions that maximize the rate at which resources are taken per unit time spent foraging, *irrespective of the effect this harvesting will have on resource availability in the future.*

While the expectations from optimal foraging theory were already well specified (e.g., Stephens & Krebs, 1986), precise conservation hypotheses were as yet undetermined. Hames (1987b, p 93) captured the essence of the meaning of conservation by emphasizing short-term restraint for long-term benefits, incidentally paralleling political economists' earlier revival of the term "stinting" (Tate, 1967 in Ciriacy-Wantrup & Bishop, 1975). Thus, evolutionary anthropologists define conservation as the costly sacrifice of immediate rewards in return for delayed benefits with respect to preventing, or at least mitigating future resource depletion, species extinction, and habitat degradation (Alvard, 1995a; Smith, 1995b). However, because restraint and long-term environmental effects are both hard to identify in field situations, empirical tests of these two hypotheses have tended [but see Alvard (1995a) for some exceptions] to view conservation acts simply as those that are not predicted by optimal foraging theory.

Hames' (1987b) initiative generated valuable empirical research. Without exception, these studies show that foragers choose the prey that maximizes return rates or efficiency (Hames, 1987b; Alvard, 1993, 1994b, 1995a, 1998a; see also Stearman, 1994; Vickers, 1994), seriously weakening the notion that

populations living in apparent harmony with their environments necessarily practice a conservation ethic. Furthermore, foragers eschew prey choice decisions
that might minimize any impact on the population dynamics of their prey,
for example, they do not avoid females and individuals of peak reproductive
value (Alvard, 1995a). These and other studies have stimulated intriguing
speculations by resource economists and evolutionary anthropologists
(e.g., Clark, 1973; Hames, 1987b, 1991; Rogers, 1991) and many other
social scientists (e.g., Ostrom, 1990) on where and when conservation
might be observed, in terms of explicit trade-offs between short-term costs
and long-term benefits.

Despite these advances in evolutionary anthropologists' ability to identify
and predict conservation acts in humans, there is still little consensus on the
validity of such analyses (see, for example, Stearman's (1995) and others' comments in Alvard, 1995a). Here we focus on one issue—the major simplification
that was made (albeit for analytical tractability) in finessing conservation as
an *alternative* to efficiency. To differentiate conservation from efficiency,
Hames (1987b) and others assumed that conservation strategies are economically less efficient for the individual than are optimal foraging strategies because, by definition, the latter maximize rates of return. Yet, rates of
return are measured over the period of time that the behaviour in question
is employed, usually a period in the order of hours. While this simplification was a practical necessity (with recognized short-comings, e.g., Alvard,
1995a; Hill, 1995b), it ignores the fact that in many areas of human endeavour
economic efficiency is synonymous with sustained long-term production,
a point recognized by Berkes (1987), Bulmer (1982) and others. Indeed a
behavioural strategy can be both environmentally friendly and economically
rational for the individual. This is particularly true in production systems where
producers have more control over the resources they depend on than is the case
among foragers. Under such conditions, conservation acts *can be isomorphic
with economic efficiency.*

In a nutshell, our point is not that evolutionary anthropologists are unaware of
this time-scale problem, nor that they think short-term restraint (conservation) is
antithetical to individual self-interest, only that the significance of longer-term
outcomes has been undervalued in empirical analyses of when, where and why
humans act as conservationists.

Applying Game Theory to the Conservation of the Pastoralists' Commons

If our goal is to extend the search for an "ecologically noble savage" beyond the realm of hunting into the realm of food producers, and to identify
conservation acts in a more realistic economic and political world, evolutionary
anthropology needs to expand its intellectual tool-kit. We look here at the
conservation of pastoralists' commons. We retain the lay notion of con-

servation acts as short-term restraint aimed at maximizing long-term benefits. We do not accept the false corollary that only short-term benefits maximize efficiency. However, because restraint on the part of some individuals opens up the possibility for cheating on the part of others, we pose the problem as a frequency-dependent game. Our approach differs from most other treatments of collective action in that we examine whether restraint can emerge as an evolutionary stable strategy (ESS) when the population is heterogeneous with respect to their wealth and interests. Finally, we speculate on the possibilities for coercion to enter into the game. We use the classic example of a pastoral grazing reserve to show how heterogeneous interests in resource use can generate conservation outcomes. [The pastoral commons was also used by Hardin (1968), albeit inappropriately (see Ciriacy-Wantrup & Bishop, 1975), to illustrate the inevitable nature of the tragedy of the commons.] We believe that our approach retains the strength of evolutionary ecological thinking (development of simple testable models), yet admits greater realism with respect to politics, economics and stratification.

Identifying Conservation among Pastoralists

Grazing Reserves and the Barabaig

Despite highly publicized claims that pastoralist communities are respon-sible for overgrazing and environmental degradation (e.g., Lamprey, 1983), pastoralists are also commonly credited with practising a conservation ethic. Much of the evidence that supports the "overgrazing hypothesis" is equivocal (Sandford, 1983; Ellis & Swift, 1988; Homewood & Rodgers, 1991). Furthermore, there are numerous detailed observations of herdsmen following grazing regimes (e.g., McCabe, 1990; Lane, 1996) observing stocking regulations (e.g., Netting, 1976) and maintaining institutional land-use practices (e.g., Galaty, 1994) that seem to protect grasslands from over-use or an annual scale; (between-year range condition is often more dependent on stochastic environmental fluctuations than organizing pressure). Though an evolutionary ecologist might read these provisional conclusions with some scepticism, there are plenty of data showing that herdsmen do adopt strategies regulating the use of grazing areas in the vicinity of their settlements, and that these can be successful in preserving rangeland quality. The evolutionary ecological puzzle, then, is this: are these strategies individually costly, and if so what mechanisms serve to maintain such conservation strategies in the population?

We take as an example the Barabaig (Figure 9.1), a subsection of the Datoga (Tomikawa, 1979; Borgerhoff Mulder et al., 1989). They keep cattle and small-stock on the semi-arid plains of Hanang District, Tanzania, supplementing their diet with grain, obtained through exchange of livestock or shifting cultivation. Rainfall averages 600 mm/year and periodic droughts occur (Lane, 1996). The

Figure 9.1

An Eyasi Datoga Community assembly in 1989, at which crimes and misdemeanors are adjudicated. Game theory can be used to predict the circumstances under which coercion is used by wealthy livestock owners to prevent less wealthy and powerful families from using pastures reserved for dry season grazing (photograph by Monique Borgerhoff Mulder).

dominant vegetation in the area is acacia and commiphora woodland, interspersed with open grassland. In general, as in most East African pastoral groups, grazing is free to all members of the local population, here the entire Barabaig subsection of the Datoga.

Barabaig grazing patterns (Figure 9.2) have been well described by Lane (1996) in the context of a study exposing the devastating social and ecological consequences of land alienation. Most Barabaig maintain permanent homesteads throughout the year on the Barabaig plains, but soon after the rainy season starts (November-December) young men and women take their family herds to the Basotu plains where high-quality forage is available, and set up temporary camps on these wet-season pastures (WSP). Because there are no permanent water sources here, at the end of the rainy season (May-July) the herds are driven back down the Rift escarpment to the Barabaig plains [see Western (1975) for a similar water-limited grazing pattern in East Africa]. Throughout the next 4-8 months grazing becomes increasingly scarce on the dry season reserve (DSR), such that in normal rainfall years animals lose condition, and in years of drought they are forced to enter more densely wooded areas with associated elevated risks of disease and predation. Among the Barabaig, grazing rotation is dictated by water needs (specifically the permanent water of the wells at lake Balangida Lelu on the DSR), by the desire

Figure 9.2

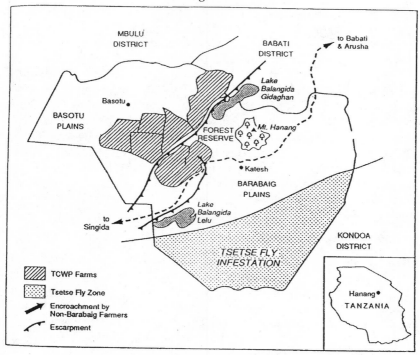

Map of Barabaig grazing movements (adapted from Lane, 1996).

to diversify livestock intake and the opportunity to capitalize on grass species whose productivity varies seasonally. Most critically for our purpose here, herders need to preserve the grasslands adjacent to the permanent homesteads on the Barabaig plains, and elaborate regulations exist to prevent unseasonal use of these pastures. If these are infringed, all households suffer.

Pastoralists almost universally move their herds around in response to the vagaries of climate (and a variety of other constraints), so as to make the best use of grazing and water resources (e.g., Dyson Hudson & Dyson Hudson, 1969). To facilitate this form of foraging, they operate systems of common land tenure. Lane (1996) describes in detail the extensive rights, obligations and prohibitions associated with the common property system of the Barabaig, as well as the customary rules regulating its use, and the tripartite jural structure entailed in rule enforcement. For instance, a herder who takes his livestock into seasonally protected areas will first be brought to a judicial moot (informal court) and asked to desist from the offence. If he refuses he may be fined, and in more serious cases, cursed. Very similar observations come from other pastoral groups, and indeed have stimulated a whole literature attesting to the conservation ethic among pastoralists (mentioned above).

We model the Barabaig grazing system as an example because it is clearly described and simple to conceptualize (even though it is somewhat unusual in its details). Many intriguing aspects of the system described by Lane (1996) are not preserved in the algebra, in view of keeping the model both simple and of broad generality. For example, we do not include the fact that forage on the Basotu plains (WSP) is of somewhat higher nutritional value than that on the Barabaig plains (DSR). We also defer discussion of encroaching Basotu wheat farms until the end of the chapter. We do however allow differences in herd size among rich and poor herders (as a varying ratio), because this is universal among Datoga (Lane, 1990; Borgerhoff Mulder, 1991) and most other pastoral groups (e.g., Fratkin & Smith, 1994). We explore the dimension of heterogeneity in wealth because the present modelling exercise was stimulated by the idea that richer herders (with presumably a greater investment in the pastoral sector) might be more inclined to favour conservationist strategies.

A Game with Asymmetrical Pay-Offs

To keep the analysis simple we imagine there are two types of herders; rich herders, who each own many cattle (m) and poor herders, who each own few (f). The cattle belonging to rich and poor herders are identical insofar as each eats an amount (a) each day. Pay-offs to rich and poor herders are asymmetrical because their value depends on the number of cattle owned by that herder. We model the environmental conditions described above by imagining that the WSP are available for a limited number of days, d, and that the DSR have a fixed area and thus a fixed amount of grass available, A.

The decision we wish to analyse is whether or not a herder should "co-operate," and move his livestock to the WSP during the rainy season, or whether he should "defect" and remain in the DSR all year round. By defecting, a herder avoids paying a cost, C, of moving his livestock and setting up a wet-season camp. However, less forage may then be available for his livestock in the dry season. The currency is measured in arbitrary units of livestock survival and productivity, which are an unspecified function of grass intake, how far the livestock need to travel, and the availability of labour within the household.

The matrix (Figure 9.3) summarize the pay-off to each strategy dependent on the behaviour of the other player. We can see then that if a rich herder co-operates, a poor herder should also co-operate if:

$$(afd) \cdot f/(f+m) > C \qquad (1)$$

A poor herder should co-operate if the cost of going to the WSP is less than the amount they lose during the dry-season, because they themselves "stole" it.

Figure 9.3

	Rich Co-operates	Rich Defects
Poor Co-operates	$\alpha md + A \cdot m/(f+m) - C$ $\alpha fd + A \cdot f/(f+m) - C$	$\alpha md + (A - \alpha md) \cdot m/(f + m)$ $\alpha fd + (A - \alpha md) \cdot f/(f + m) - C$
Poor Defects	$\alpha md + (A - \alpha fd) \cdot m/(f+m) - C$ $\alpha fd + (A - \alpha fd) \cdot f/(f+m)$	$A \cdot m/(f + m)$ $A \cdot f/(f + m)$

Pay-off matrix for a game between two herders, one rich and one poor. Pay-offs to the poor herder are in the lower left corner of each cell while pay-offs to the rich herder are in the upper right.

If a rich herder defects, then a poor herder should co-operate when:

$$(afd - C > (amd) \cdot f/(f+m) \qquad (2)$$

That is, when the amount taken from a distant pasture minus the cost of going there, is greater than what would have been the poor individual's share of what the rich herder is taking from the nearby pasture while the poor individual is away.

If we consider how a rich herder should act when faced with a co-operating or defecting poor herder, we find that the rich individual should base his or her decision on the same criteria. However the asymmetrical pay-offs, resulting from differing numbers of livestock owned by the two parties, lead to there being different cut-off points for the two individuals. A defecting herder with few livestock takes less forage from nearby pastures than does a rich defector but the rich herder would normally get a larger share of that "stolen" forage. Thus, given the same parameters, rich and poor's preferred strategies may not be the same.

We present results for the three variables that show the strongest effects on ESS outcomes, the cost (C) of going to the WSP, the relative wealth differences (or inequality) between wealthy and poor herders [$f/(f + m)$], and the numbers of days the WSP was available (d). Plotting each player's indifference curve we see three ESS spaces (Figure 9.4a). Above both curves the ESS is "Mutual Defection;" both poor and rich should defect. Below both curves it is to the advantage of both parties to co-operate and there is no temptation to defect. The ESS is thus "Mutual Co-operation." In the regions between the curves, the ESS is "Rich Co-operate Poor Defect" (RCPD); the herder with more animals should always co-operate while the herder with fewer should always defect. The pattern remains similar as the wet season lengthens (Figure 9.4b),

Figure 9.4

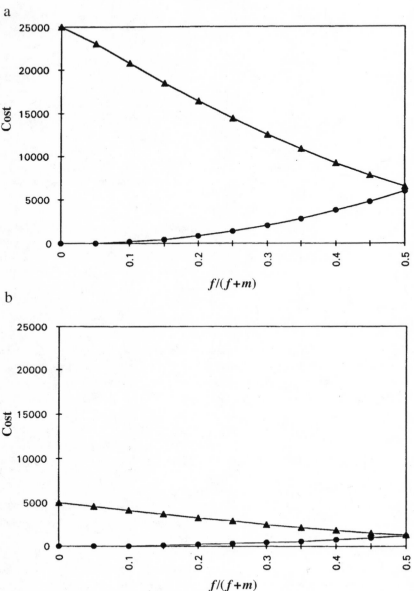

ESS regions for a two-person game. Equations (I) (triangles - rich) and (2) (circles - poor) are plotted as a function of cost of moving to the WSP and the poor's proportion of the total number of livestock owned by rich and poor. Units of forage eaten per cattle per day, a, is set equal to I. In (a) $d = 250$ and in (b) $d = 50$, where d is the number of days the WSP is available.

and when the cost of moving to the WSP is assessed on a per animal rather than per herder basis (not shown).

Three clear patterns emerge, all of which make intuitive sense and suggest that the model captures some important features of grazing patterns among groups like the Barabaig. First, "Co-operation" by either herder is precluded when the cost of using the WSP is very high [perhaps these seasonally available pastures are very remote or perhaps construction and defence of the wet season camps is too labour-intensive, see Sperling & Galaty (1990), Borgerhoff Mulder & Sellen (1994)]; this is particularly the case when the WSP is only available for a short time. Second, "Mutual Co-operation" is most likely when costs are low and wealth inequalities are mild, that is when the pay-offs to any given strategy are quite similar for wealthy and poor herders. Third, "RCPD" (unilateral co-operation on the part of the rich and unilateral defection on the part of the poor) is most common when the costs of using the WSP are moderately high and the inequalities in wealth are sharp, in other words when only some individuals are in a position to benefit from the WSP. Note that this outcome reflects the fact that if a poor herder defects, it still pays a rich herder to leave for the WSP because the value of what a poor herder (with his very few cattle) steals from the DSR in the rainy season does not match what the rich herder (with his many cattle) stands to gain at the WSP.

For the purpose of the present chapter we highlight two general results from this model that are directly relevant to the question of how to identify and account for conservation acts in human communities. First, under many of the conditions we have modelled, respect for dry season reserves emerged as a stable strategy. In other words, conservation occurred for purely selfish reasons, precisely because it contributes to economic efficiency. "Mutual Co-operation" in this context is somewhat similar to "By-product Mutualism" (Mesterton-Gibbons & Dugatkin, 1992; Connor, 1995) insofar as it enhances production of the resources to be shared. Second, because of the asymmetrical pay-offs contingent on the differing numbers of livestock owned by the two players, herders have different thresholds for triggering defection, which leads to the result "Rich Co-operate Poor Defect." Here again there is no "problem" in attaining the conservation outcome, it is simply the product of the rich individual's self interest. Analyses of an *n*-person game indicate that the outcome RCPD still occurs, albeit under a narrower range of conditions (Ruttan & Borgerhoff Mulder, 1999). Note however, that in the RCPD zone a rich herder could get an *even higher net pay-off by* coercing a reluctant poor herder to leave the DSR in the wet season. What are the opportunities for such coercion?

Some Speculations on Coercion and Policing

Among pastoralists, herd owners who benefit most from conservation of the dry season reserves (men with the largest herds) are commonly those who are the most powerful in their local community, as elders. Though wealth is not a prerequisite to elder status, influential and respect elders are usually rich in livestock. Furthermore numerous ethnographic references point to the role of elders in enforcing grazing regulations and policing infractions through committees and moots (e.g., Lane, 1991). As our analysis shows, given the parameter conditions yielding an RCPD outcome, the rich might benefit from punishing herders who utilize DSR in the wet season, at least if the marginal returns from coercion are positive.

Because we have not explored incentives for punishment in an n-person game with more than one possible punisher, we can only speculate on why Barabaig appear to punish non-cooperators. First, the costs of detecting cheaters are small; given the proximity of the DSR to the permanent homesteads of the Barabaig, cattle cannot be grazed there illegally in the wet season without being spotted. Second, the opportunity costs of bringing offenders to trial may be low given that men with one or more wives and grown sons are not directly involved in the day-to-day herding of their stock. In the Datoga, wealthier men do indeed have more people living in their homestead (Sieff, 1995); furthermore, wealthy pastoralists enjoy more leisure (Fratkin & Smith, 1994). It therefore seems plausible that wealthy old men incur few opportunity costs from sitting on committees and councils, attending trials, and adjudicating moots. Third, when offenders are punished a fine is exacted. Among the Datoga and many other groups, disputes can be resolved with fines of cattle, or with obligations to brew honey beer. In either case, the payment is consumed by the elders on the relevant council(s) (Klima, 1965), often immediately after the ruling is reached; this is so even in disputes where women adjudicate (Klima, 1964). Finally, in the Barabaig (Lane, 1996) and several other groups it is reported that some councils operate in total secrecy, so as to protect members from the possible retribution of those that are punished. Each of these considerations serves to reduce the costs associated with punishment and regulation.

Future Directions in Modelling

As with any model, a number of simplifications were made. The most critical of these is that we have not considered what happens when the dry season reserves are exhausted before the rains return, in other words when cattle die and where rangeland is irreversibly degraded. One might suspect that cattle loss would be more costly for poor than for rich individuals, on account of their greater proximity to a minimum threshold for survival; the finding that poor herders seem to keep higher *proportions* of their livestock alive during droughts than do rich herders (e.g., Fratkin & Roth, 1990; Herren, 1990) suggests their greater fear of catastrophe. If the poor are more averse to losing cattle, this could lower threshold for co-operating (in other words make them keener conservationists). Conversely, with respect to rangeland quality, it might be that rich herders discount future pay-offs at a lower rate than poor herders;

this notion arises from the observation that in many populations (e.g., Brad-burd, 1982; Fratkin & Roth, 1990; Sieff, 1995) poor families are much more likely to drop out of pastoral production than are rich families, or at least to diversify production with honey collection, hunting and cultivation. If the rich do value range-land quality more than the poor, this should lower *their* threshold for co-operation (and conservation), and in addition provide greater motivation for them to coerce the poor into co-operating. Finally, the cost of utilizing wet season pastures probably varies according to the wealth of the herder, because poorer families often have insufficient labour to maintain two households (Sperling & Galaty, 1990; Borgerhoff Mulder & Sellen, 1994); this would effectively raise a poor herder's threshold for cooperation. Because the Barabaig evidently punish those who do not conserve, our current work (Ruttan & Borgerhoff Mulder, 1999) is investigating the role of punishment (following Clutton-Brock & Parker, 1995).

Conclusions

General Conclusions and Implications

Using game theory to explore the consequences of asymmetries in power and pay-offs supplies some much-needed political and economic reality to the study of conflicts over resources within and even between communities (see also Ostrom *et al.*, 1994). Failure to place the study of conservation more squarely within a politico-economic context is one of the main criticisms of earlier evolutionary anthropological studies of conservation in indigenous groups. We should, however, acknowledge that many field workers who study common pool resource systems (notably those of pastoralists) reject the view that indi-viduals in traditional communities can be characterized as selfish and liable to defection. Rather they stress the multiple relationships and interdependencies among social actors that reduce an individual's temptation to cheat. We respond by noting that although predicating models on self-interest may oversimplify issues, it has the power to delineate the potential scope and weaknesses of com-mon property systems under different conditions, and hence is a useful tool (see also Ostrom, 1990).

We end by drawing attention to just two conclusions. First, conservation can be an outcome of economic maximization, or efficiency. Second, asymmetries in pay-offs provide the opportunity for both unilateral co-operation and/or coercion of the weak by the powerful. Regarding the first point, the success of our model in generating conservation outcomes throws into relief the restric-tive nature of earlier work that for heuristic reasons opposed conservation and efficiency outcomes. Although it is clear why investigators such as Hames made this practical gambit, now that the search for conserva-tion behaviour is moving beyond the realm of prey choice in foragers, it may be more productive to look for conservation acts *within the framework of efficiency maximization.* One important implication of this approach is that it highlights the fact that not all interactions leading to beneficial outcomes for the group are necessarily structured as a Prisoner's Dilemma Game (or as a Game of Assurance). Another implication is obvious, but more practical: those behavioural acts with conservation outcomes that are also economically efficient will be particularly attractive to potential adopters, and hence of more relevance both to development advisors and community-based conservation activists.

Our second conclusion is that asymmetries in pay-offs contingent on wealth differentials (and other related interests) can generate an array of conservation outcomes. Particularly interesting is the finding that as the degree of asymmetry in pay-offs increases (and the costs of conservation rise) opportunities for unilateral conservation can emerge. From here we argued that if asymmetries in pay-offs map onto asymmetries in power, as they do in many pastoralist communities, dominant individuals are tempted to coerce all individuals into co-operating. The implications of this are the following. First, conservation is most likely to be observed where those with the greatest interest in conservation are those with the most power. Indeed this may be an important reason why pastoralists are often successful in protecting their commons from degradation, at least if these systems are undisturbed by external forces (e.g., Ensminger & Knight, 1997). To take a stark comparison, think of modern industrial societies, where corporate power has far more to gain from environmental abuse than does any single individual—what hope is there in instituting conservationist policies until individuals have equal or greater power than corporate bodies? Second, when field-workers observe conservation outcomes, we cannot assume that the conservation strategy is indeed necessarily beneficial to all individuals (see also Gadgil & Berkes, 1991); some coercion may be going on. Third, in a more applied context, empirical studies need to determine precisely *how* environmentally successful projects obtain their successes; see for example Gibson and Marks' (1995) evaluation of Zambia's ADMADE communal resource management project or Metcalfe's (1994) study of Zimbabwe's CAMPFIRE project, where attempts are made to determine whose interests a project serves.

Our results not only are in line with those of institutional economists and political scientists, but also provide a broader framework within which to accommodate their contrasting results. The specific finding that Mutual Co-operation is most likely to occur when heterogeneity among players is low (in conjunction with low costs of conservation) is consistent with many empirical observations on the management of common pool resources (e.g., Ostrom, 1990; Hanna *et al.,* 1996). At the same time, our finding that opportunities for unilateral conservation are more likely to emerge as the degree of asymmetry in pay-offs increases is consistent with the arguments of other social scientists who propose that a certain amount of heterogeneity facilitates collective action insofar as members of a small "privileged" group find it individually advantageous to support the costs of punishment (Olson, 1967; McKean, 1992; Ruttan, 1998).

Political Issues

Studies purporting to show conservation, or its lack, in traditional communities are often viewed as inflammatory with respect to the contentious political debates surrounding conservation and indigenous affairs (e.g., Alcorn, 1991, 1995; Puri, 1995). However, as Hill (1995b) and many others make quite clear, conservation performance is not a valid reason for divesting indigenous people of their land. Indeed indigenous groups demonstrating no conservation ethic under traditional conditions readily become conservationists as they adjust their behaviour to the novel, political and economic environment (Kaplan & Kopishke, 1992; Vickers, 1994). Furthermore, environmental regulations that are based on traditional custom and sanctioned by community institutions are more likely to be respected than those imposed by external authorities, even where the regulations themselves are very similar (Stevens, 1997). Hence a dispassionate understanding of where, when and how humans engage in conservation is central to the development of strategies designed to develop sustainable patterns of resource use, to protect human rights, and to conserve biodiversity.

The implications of the present study concerning the impact of power differentials on conservation outcomes does not, of course, mean that we sanction the old-fashioned view that the poor stand in the way of conservation. In fact our motivation is quite the opposite. Given the current popularity of communal resource management systems among Non-Governmental Organizations, we suspect that such projects need to be closely scrutinized, to see not only *whether they afford biodiversity protection,* but *how and why they work,* and potentially in *whose* best interests (see also IIED, 1994).

Finally, we note that Barabaig are continually losing access to the Basoto wet season pastures because of agricultural incursions that began in the 1960s and accelerated in the 1980s. Herders are now forced to use the Barabaig Plains at times of the year when these reserves should be resting, and to maintain permanent homesteads in zones that were traditionally prohibited from habitation. Accordingly, elders can no longer enforce traditional sanctions. We suspect that under these unfortunate, and all too common (Fratkin, 1997) circumstances, a game theoretical analysis of herding practices based on individually perceived costs and benefits becomes increasingly valuable, precisely *because* the traditional community norms have been undermined (see also Grabowski, 1988). The current land-use crisis in Hanang is rooted ultimately in economic conflict stimulated by national and international development priorities, and can be remedied only through institutional change at the national and regional level.

Acknowledgements

Lore Ruttan was supported by a MacArthur Fellowship from the Institute of Global Conflict and Co-operation, University of California. We thank Michael Alvard, Peter Coppolillo, Hilly Kaplan, Peter Richerson and Eric Smith for helpful comments and discussion.

10

Conserving Resources For Children

Alan R. Rogers

The ability of the human race to anticipate impending ecological disasters seems to have outstripped its willingness to do anything about them. For example, it is estimated that soil erosion and other forms of degradation are reducing the world's agricultural productivity by the equivalent of some 14 million tons of grain per year (Brown and Young 1990). It is not hard to foresee that this will lead to smaller harvests and to an increase in human suffering. Yet we seem to lack the will to take the steps necessary to reverse this process. Why?

There are at least two reasons. Many resources, of which air is a good example, are owned by no individual or corporation, but are held in common by us all. When an individual restricts his consumption of such a resource, he or she pays 100% of the costs of this action but receives only a small fraction of the benefits, since these are widely shared. No single individual can, by personal restraint, save a resource species from extinction. Restraint only allows competitors to grab a larger share as *they* drive it extinct. Thus, conservation of shared resources may serve the common interest, but it seldom serves the selfish interests of individuals. This problem is variously referred to as the "tragedy of the commons" (Hardin 1971) and the problem of "externalities" (Heller and Starrett 1976).

The second impediment to conservation affects resources held by in-dividuals as well as common holdings. The problem is that conservation pays for today's sacrifices with tomorrow's dollar, but tomorrow's dollar is worth less than today's. For example, conserving the soil of agricultural fields may require long fallow periods. Increasing the period of fallow reduces the productivity of the field this year, but it may increase the productivity to be expected 10 years from now. The fallow period that a farmer decides to use will depend on how much he is willing to pay for this expected future benefit. It will depend, in other words, on how he discounts the future against the present.

Reprinted by permission of *Human Nature* 2:1, pp. 73-82, Copyright © 1991 Walter de Gruyter, Inc. Published by Aldine de Gruyter, Hawthorne, NY.

The schedule by which humans discount future rewards is not well understood. · We are not sure how much less $1000 is worth to us if it is delivered next year rather than today. Economic predictions typically assume that discounting is exponential, at a rate equal to the rate of interest on investments. This assumption implies, for example, that the owners of a stock of economically useful animals will harvest it all at once, exterminating the stock, unless its rate of growth exceeds the interest rate (Clark 1973). The evidence of psychological experiments (Herrnstein 1990), however, suggests a hyperbolic schedule, in which discounting is initially much faster than the exponential model allows, but then slows. This assumption implies that it should be much harder to persuade the owners of resource stocks to conserve them. I have argued elsewhere that, in most past and present human populations, natural selection must have favored a discount schedule quite different from either of these models (Rogers 1994).

Table 10.1
Summary of Notation for Fertilities and Relative Frequencies

Behavior	Territory Quality	Relative Frequency	Fertility
Conserving	High	p_1	$w_1 = 1 - c$
Conserving	Low	p_2	$w_2 = 1 - c - b$
Prodigal	High	p_3	$w_3 = 1$
Prodigal	Low	p_4	$w_4 = 1 - b$

This last result takes account of a parent's effect both on her own future reproduction and also on that of her children. It ignores, however, the uncertainty with which resources are inherited. Resources are not always inherited by offspring; they may pass instead to unrelated individuals. This uncertainty has two effects on the evolution of propensities to conserve resources, of which one is obvious, and the other obscure. It seems obvious that selection is more likely to favor conservation if resources are likely to be inherited by offspring. Otherwise, the fruit of the conserver's labor will often be wasted on the children of non-conservers. It is less obvious, however, that the strength of selection for conservation may decline as the certainty of inheritance is increased. Below, this effect is illustrated using a simple mathematical model.

The Model

We are interested in the effect of the inheritance of wealth on the evolution of conservation. In its general form, this problem is overwhelmingly complex, and we cannot hope to solve it here. The first step in attacking the larger problem is to gain some intuition about how its variables are related. For this purpose, simple models, which one can hope to understand, are more useful than realistic models, which are inevitably complex. Thus, the model discussed below incorporates only enough realism to allow us to assess the effect of varying degrees of certainty in inheritance of resources on both the direction and the rate of evolution.

Consider a population of individuals, each of whom must obtain exclusive access to a "territory" in order to breed. For simplicity, assume for the present that these individuals reproduce asexually (the effect of relaxing this assumption will be considered later). Suppose that individuals of genotype C (for conserving) conserve their territories, whereas those of genotype P (for Prodigal) do not. Conservation reduces the fertility of the conserver, but increases that of the individual who occupies the territory in the subsequent generation. Territories last occupied by conservers and non-conservers are referred to as "high-quality" and "low-quality" territories, respectively. There are thus four phenotypes as shown in Table 10.1. This table also introduces notation for the relative frequency and fertility of each phenotype. Conservation reduces the fertility of conservers by c, the cost of conservation, and reduces that of occupants of poor territories by b, which can be thought of as the benefit of conservation, since it is a cost that the children of conservers avoid.

Assume that territories are passed from parent to offspring with probability h, and that with probability $1 - h$ they are passed to individuals drawn at random from an "offspring pool" comprising the offspring produced by all parents. The frequency, x, of conservers within the offspring pool is a function of the frequencies and fertilities of the four phenotypes, as is shown in the appendix. For the moment, however, let us take x as given and ask how this system will evolve.

There are, in general, only two possibilities: the frequency of conservers may continue oscillating up and down forever, or else it will eventually settle down at some equilibrium value. Our task is to discover what values of $p_1, ..., p_4$, if any, are equilibrium values, and how fast these equilibria are approached. To begin with let us take x as given and ask what relationship must hold between x and $p_1, ..., p_4$ when the system is at equilibrium. The relationship between x and $p_1, ..., p_4$ is elaborated in the appendix (see equation 13).

The frequency of the ith phenotype in the subsequent generation will be denoted by p'_i to distinguish it from p_i, the frequency in the current generation. The frequency of conservers among individuals holding territories is denoted by $z = p_1 + p_2$. Consider now the four recurrence equations,

$$p'_1 = [h + (1 - h)x] z \tag{1}$$
$$p'_2 = (1 - h)x(1 - z) \tag{2}$$
$$p'_3 = (1 - h)(1 - x)z \tag{3}$$
$$p'_4 = [h + (1 - h)(1 - x)](1 - z) \tag{4}$$

which relate phenotype frequencies in one generation to those in the next. To justify the first of these equations, note that the proportion of rich territories in the subsequent generation is the same as the proportion of conservers in the current generation, z. Of these, a fraction, h, will be occupied by conserving offspring who inherited their territories, and a fraction, $(1 - h)x$, by conserving offspring from the offspring pool. Equations 2-4 follow from analogous arguments.

Results

In the appendix it is shown that the change per generation in the frequency, z, of conservers among territory-holding adults is

$$\Delta z = \frac{z(1-z)(1-h)(hb-c)}{\overline{w}} \qquad (5)$$

where $\overline{w} = \Sigma_i p_i w_i$, is the mean fertility. At equilibrium Δz must equal zero. Since $z(1 - z)(1 - h)$ is always greater than or equal to zero, the sign of Δz depends only on $hb - c^*$, then the frequency of conservers increases if

$$hb > c \qquad (6)$$

and decreases otherwise. Selection, in other words, favors conservation when the benefit of conservation, discounted by the probability of that benefit accruing to an offspring, exceeds the cost of conservation. This inequality is similar to "Hamilton's rule" of inclusive fitness theory (Hamilton 1964a), differing

Figure 10.1

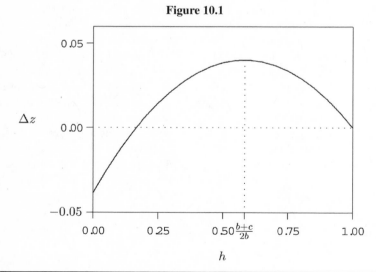

How the strength of selection depends on h, the certainty of inheritance.

only in that the parameter h has here replaced the coefficient of relationship in Hamilton's theory. Inequality 6 is most easily satisfied when h is large, in agreement with our intuition that inheritance of resources should facilitate the evolution of conservation.

We have, however, glossed over something that is less obvious. The *strength* of selection (as measured by $\triangle z$) does not continue to increase as h increases. As Figure 10.1 shows, it rises to a maximum at the point $h_{max} = (b + c)/2b$, and then declines. It is nil when $h = 1$. Thus, selection for conservation is strongest at intermediate values of h, and there is no selection whatsoever when $h = 1$.

To gain an intuitive appreciation of what is happening here, consider the case when $h = 1$, which means that each parent is guaranteed to leave its territory to an offspring. This implies that each parent produces one and only one offspring, since there are no empty territories left over. Since each parent produces a single offspring, there is no variation in reproductive success and therefore no selection. Thus, neither conservation nor anything else can be favored by selection if territories are inherited with complete certainty.

This conclusion sounds strange, but the puzzle disappears when one considers the effect of conservation on the number of a conserver's great-grandchildren. Certainty of inheritance always increases the chances that a conserving parent will succeed in increasing the fertility of its offspring. More grandchildren are thus contributed to the offspring pool. These grandchildren, however, must obtain territories in order to reproduce, and the chances of their doing so decline with h, the probability of inheritance. Thus, large values of h may decrease the expected number of a conserver's great-grandchildren. Conservation is favored most strongly at an intermediate value of h, where these opposing effects just balance. It seems likely that this principle applies far more widely than the artificial model considered here. Genes descended from one ancestor cannot long increase in frequency unless his or her descendants are able to gain access to resources that the ancestor did not control. This requires, of course, that some individuals fail to pass their resources to offspring. Thus, inheritance of resource holdings is a factor that cuts both ways. The obvious advantage it provides for conservation is partially offset by a less obvious disadvantage.

This model also illustrates the fallacy of the often repeated claim that selection always favors those who produce the most surviving offspring. Fisher (1958) counted grandchildren rather than children, in explaining the evolution of the sex ratio, and here we have found it necessary to count great-grandchildren. In other contexts, inheritance of wealth can require an accounting of even more distant descendants (Harpending and Rogers 1990; Rogers 1990b). The strategy that produces the most surviving offspring may or may not be favored by selection.

Confronting Reality

The effect of resource inheritance in this model may be interesting, but the question remains, what does it have to do with reality? Let us consider, therefore, the ways in which real humans differ from the creatures of the model.

To begin with, humans are sexual, having fathers as well as mothers. Consequently, even if every father were guaranteed of leaving his territory to a son, there could still be an open competition between daughters, leading to fairly intense natural selection. This suggests that certainty of inheritance will reduce the strength of selection much less (maybe half as much) in a sexual model. The negative effect of inheritance on conservation, in other words, will be weaker. What about the positive effect? Robert Boyd has pointed out to me that, when the conserver allele is rare, it will occur mainly in heterozygotes, half of whose children will not carry the allele. Thus, conservation has only a 50% chance of benefiting copies of the conserver allele in children who inherit territories. Thus, the positive effect of inheritance on conservation may also be only about half as strong. Without analysis of a sexual model, it is difficult to predict the net effect of these changes. Nonetheless, it seems likely that the negative and the positive effects of inheritance should operate even in sexual models, and as a result, property inheritance should favor conservation less than naive intuition might suggest.

The model above assumes that conservers and prodigals transmit resources to offspring with the same probability. Since the territories inherited by prodigal offspring are always of low quality, these offspring might do better by refusing their inheritance and taking their chances in the offspring pool. I have not studied models of this sort, but it seems clear that this alternative assumption would increase the fitness of the prodigal allele, making it still more difficult for conservation to evolve. Furthermore, this effect should be strong when conservers are common (since a random territory is likely to be of high quality) and weak when they are rare. Therefore, as the number of conservers increases, their fitness advantage should decrease. This might lead to polymorphic equilibria, which do not occur in the model presented above.

Finally, my model is unrealistic in assuming that a territory is no better after 10 generations of conservation (and no worse after 10 of prodigal exploitation) than after 1. In reality, there will often be considerably greater scope for improvement (and degradation) of territories, allowing conservers to increase in frequency even when $h = 1$. But it also seems unlikely that territories can be improved or degraded forever. Eventually, territories should stop changing in quality, and completely certain inheritance would thereafter prevent evolutionary change. The qualitative results of the model presented above should continue to hold.

In summary, realistic models would differ from the one presented here in a variety of ways, but none seem likely to alter the conclusion that certainty of inheritance has a negative, as well as a positive effect on the evolution of conservation. This negative effect may be part of the reason why our species seems so reluctant to make the sacrifices needed to maintain the quality of life on our planet.

Conclusions

Inheritance of property has two countervailing effects on the evolution of conservation. The greater the certainty (h) of inheritance, the greater the chances that conservation will enhance the fertility of one's offspring. However, increasing h also reduces the number of vacant territories, thereby reducing the probability that a grandchild will be able to breed. Thus, large values of h increase the expected number of grandchildren born to conservers but reduce the expected number of great-grandchildren. Consequently, inheritance of resources promotes the evolution of conservation less than intuition might suggest.

I thank Rob Boyd, Eric Charnov, Hillard Kaplan, Elizabeth Cashdan, Steven Gaulin, and Henry Harpending for their comments and suggestions. The work was supported in part by NIH grant MGN 1 R29 GM39593-01.

[This is a shortened version of the original and here some text was omitted.]

Appendix omitted (see: http://www.anthro.utah.edu/~rogers/)

11

Two Truths about Discounting and
Their Environmental Consequences

Norman Henderson and William J. Sutherland

Discounting is the process by which we translate the value of expected future costs and benefits into present-day values. It is a fact of life that, taken as a whole, society discounts future values. Economists note that as borrowers people are willing to pay an exponential interest penalty to acquire resources now; conversely, savers expect to be rewarded for their forbearance with an exponential interest premium on bank deposits and investments. Using this exponential logic, it can appear to make economic sense to harvest rain forests rapidly or to fish the great whales to extinction if the discount rate is higher than the natural asset's growth rate[1]. Similarly, if the high costs of decommissioning a nuclear power station are incurred far into the future, they have negligible impact on our present-day decisions. Ethologists, however, take a hyperbolic approach to discounting. The difference is of considerable importance over long timeframes, with important implications for sustainability.

Under the standard economic perspective,[2,3] the present value, V_p, of a future value, V_f, received after a delay of time, t, is:

$$V_p = V_f/(1 + r)^t \qquad (1)$$

where r is the (exponential) discount rate. The rationale for this exponential relationship is the cost of borrowing capital; bank deposits and liabilities grow exponentially.

However, ethologists, although they agree with economists on the principle of discounting, disagree with them as to the nature of discounting and maintain that:

$$V_p = V_f/(1 + r_h t) \qquad (2)$$

Reprinted with permission from: *Tree,* vol. 11, no. 12, Dec. 1996, pp. 527-528.

where r_h is the (hyperbolic) discount rate. This conclusion is based on studies[4, 5] of human choice, where different sized rewards are linked to differing time delays, and on studies of other animals, such as rats and pigeons, which have also been shown to discount hyperbolically.[4, 5]

When integrated over time from zero to infinity the exponential function produces a finite value, whereas the hyperbolic function produces infinity. In practice, policy makers are rarely concerned with the extreme long-term consequences of resource-use decisions, but do sometimes take the more immediate long term (50 to 200 years) into account. Under the exponential view, at normal rates of discount ($r = 0.04$-0.10), costs and benefits occurring more than 50 to 100 years into the future carry negligible weight when converted to present values. Under hyperbolic discounting, the distant future has more import. The distinction is critical when trying to value projects such as forestry replantings, dam development or species and habitat preservation – all projects whose costs and benefits extend far into the future.

To illustrate this we show data collected by Cropper *et al.*,[6] who asked members of the American public to choose between alternative government spending programmes. The first programme alternative would save 100 (anonymous) lives today, while the second would save a specified larger number of (anonymous) lives in the future. The scenarios tested were set at 5, 10, 25, 50 and 100 years into the future. For example, respondents were asked to choose between saving 100 lives today and 4000 lives 25 years from now. From their data, Cropper *et al.*[6] calculated the implied discount factors (V_p/V_f), shown in Figure 11.1 with the fitted hyperbolic discount factor curve.

Over a period of 10 years (a short-term cost-benefit evaluation period), the distinction between the economist's $r = 0.20$ exponential discount curve and the ethologist's $r_h = 0.21$ hyperbolic discount curve is not necessarily critical for decision making. The two curves approximate each other tolerably well over the length of the evaluation period. But for longer periods, such as 100 years, the discrepancy between the hyperbolic discount curve and any exponential curve necessarily increases.

The diagram also suggests that if the hyperbolic hypothesis is correct, and yet policy makers limit themselves to exponential discount rates, they will tend to recommend lower exponential discount rates as a project's timeframe expands. Over long timeframes, lower exponential discount rates most closely resemble the hyperbolic curve. There is evidence that this indeed occurs, with the US government applying reduced discount rates to water development projects and the UK government doing the same when evaluating forestry projects.[7] Both forestry and water projects have long-term costs and benefits. However, if the hyperbolic hypothesis is appropriate, discretionary exponential discounting policies may still lead to serious overdiscounting in the latter years of a long evaluation period. The problem is most acute for very long-run policy issues such as nuclear waste disposal or the possible over-exploitation of

Figure 11.1

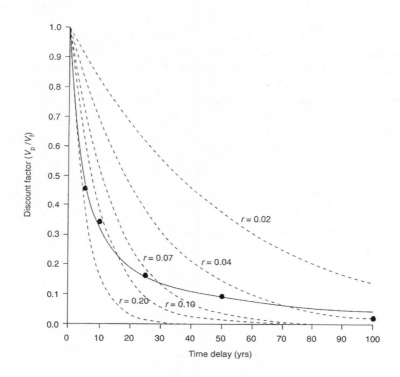

The decrease in the present-day value (V_p) of future (V_f) costs or benefits in relation to the time delay. Solid circles mark Cropper *et al.*'s[6] implied discount factors (V_p/V_f), and the solid line shows the best-fit hyperbolic relationship $r_h = 0.210$, with R^2(adj) = 99.6% when (V_p/V_f) is the single explanatory variable used to predict the implied discount factors at time t (and $V_p/V_f = 1$ for $t = 0$). The dashed lines show exponential discounting at a range of values of r (exponential discount rate). For example, according to the illustrated hyperbolic discount factor curve (solid line), a benefit received 50 years in the future is valued as being equal to one tenth of its original value in terms of benefit received today. Note that the present-day value of costs or benefits received in later years (e.g. year 100) becomes insignificant (tends to zero) when discounted at high- or mid-range exponential rates *(r)*.

a species with a slow reproduction rate. We could be seriously underestimating the costs of decommissioning a nuclear power station or the benefits of sustainable resource use.

The results of standard economic discounting practice for long-term analysis have been recognized by many as being unsatisfactory for some time.[1, 8, 9] Recent concern for long-run values like biodiversity accentuates the discounting problem. While economists debate over which of the many possible market interest rates for returns on capital are the appropriate indicators for the government's social discount rate, they have remained largely unaware of the findings of ethologists and human behaviourists. Like Cropper *et al.*,[6] economists sometimes find their data support the hyperbolic discounting hypothesis, but such evidence is viewed with reticence and is published under headings or titles such as "Anomalies,"[10] "Myopia and Inconsistency"[11] or "Myopic Discounting."[12] For their part, ethologists have yet to explain the discrepancy between the expressed individual preferences they observe and the logic of capital market returns

Policy makers are left to cope with discounting uncertainty. A typical pragmatic response is that of British Prime Minister John Major, who stated that the choice of the discount rate is a matter for negotiation and consultation between individual government sectors and the Treasury.[13] The need to better inform public policy, and the increasing importance to society of resource and environmental sustainability in decision-making, make interdisciplinary work on discounting of pressing importance.

References

1. Lande, R., Engen, S., and Saether, B. (1994).
2. Pearce, D. and Turner, R. (1990).
3. Schmid, A. (1989).
4. Rachlin, H. (1991).
5. Logue, A. (1988).
6. Cropper, M., Aydede, S. and Portney, P. (1992).
7. Henderson, N. and Bateman, I. (1995).
8. Lind, R. (1990).
9. Goodin, R. (1982).
10. Loewenstein, G. and Thaler, R. (1989).
11. Strotz, R. (1956).
12. Winston, G. and Woodbury, R. (1991).
13. Spackman, M. (1991).

12

Sex Differences in Valuations of the Environment?

Margo Wilson, Martin Daly, Stephen Gordon, and Adelle Pratt

Sexual selection theory affords a rationale for predicting that men, especially young men, may be more willing than women to risk harms and to discount the future in the pursuit of short-term gains. These propositions apply to many domains of risky behavior, and it is likely that they apply to decisions involving potential harms to the environment and health hazards as well. Two preliminary studies of university subjects' responses to hypothetical dilemmas that support the predicted sex difference are described. Important understudied questions are, to what extent reckless risk acceptance may be mitigated by material wellbeing, by marriage, and by parenthood.

What do people value in their surroundings? Environmental benefits cover a wide spectrum of pleasures and needs: economic and nutritive, aesthetic and recreational, health and safety. To even enumerate the full range of such benefits and to organize them into some sort of meaningful taxonomy is a formidable challenge, let alone prioritizing them or computing their net worth. Economists have approached the question of the value of environmental goods by simply asking people what they would be willing to pay for them, but this is unsatisfactory (see for example, Kahneman & Knetsch, 1992; Kahneman, Ritov, Jacowitz, & Grant, 1993 regarding the absurdities that ensue from supposing that people can place dollar values on nonmarket environmental goods). The availability of many things that people value, such as drinkable water or inexpensive paper, may be dependent on management practices and environmental protection in ways of which the consumers are scarcely aware. Even those natural resources that we use directly may be taken for granted if they are so reliably abundant and of useable quality that we have never been

Population and Environment: A Journal of Interdisciplinary Studies, Vol. 18, No. 2, November 1996, pp. 143-159 "Sex Differences in Valuations of the Environment?" Margo Wilson, Martin Daly, Stephen Gordon, and Adelle Pratt. Copyright © 1996 Human Sciences Press Inc. with kind permission from Kluwer Academic Publishers.

obliged to think about what they are worth to us. Moreover, many environmental goods such as oxygen are not even detectable by our sensory systems. But in spite of these and other difficulties facing any study of people's preferences and valuations of environmental goods, psychologists, philosophers, economists and other social scientists have not been deterred (e.g., Hechter, Nadel & Michod, 1993).

We propose that the reasons why people have particular preferences and priorities cannot be satisfactorily addressed without consideration of the ways in which our long history of natural and sexual selection pressures shaped the emotional, motivational and information-processing mechanisms underlying the expression of values and preferences. The functional significance of some valuations is obvious: warmth, satiety, and safety are valued states, for example, because of their obvious contribution to survival. But the utility of other preferences and appetites is less obvious, and perhaps less universal. Intangible states such as status, power, and self-esteem are products of psychological mechanisms which have been shaped by selection pressures in the context of social living and reproductive competition. In a long-lived sexually reproducing social species such as *Homo sapiens*, one can anticipate that numerous complex, facultative social and psychological adaptations for social-living, for mate choice, for intrasexual competition, and so forth, have evolved.

In this paper, we briefly outline an evolutionary psychological framework which can be used to derive hypotheses about factors affecting valuations of natural resources. Rather than developing a taxonomy of values or domains of environmental valuations, we present a logic by which one may predict differences in valuations as a function of sex, age, and social and material circumstances. We focus on possible differences in valuations of the environment by women versus men, as well as differences in their reluctance to risk damaging the environment, based on considerations of sexually differentiated adaptations for intrasexual competition.

First of all, is there any evidence for sex differences in environmental valuations? "Eco-feminism" is perhaps the only perspective that is consistently concerned with the issue of gender (Mellor, 1993), but to the best of our knowledge there have been few empirical studies derived from this perspective. One of the pioneering studies of valuations and attitudes toward the environment was conducted by Riley Dunlap and his colleagues (Olsen, Lodwick & Dunlap, 1992) who surveyed some 1600 residents of the state of Washington in 1976. There was a moderate but significant sex difference in expressed concern about the quality of the environment, with women professing to be more concerned than men. Moreover, women were more likely to report being willing to change their habits at home with respect to household products which pollute the environment.

We are presently studying people's perceptions and views of the local wetland ecosystem of the western end of Lake Ontario which has suffered extensive environmental degradation from decades of industrial pollution, inadequate sewage treatment, excessive use of fertilizers and herbicides, and increasing amounts of vehicular combustion emissions (Remedial Action Plan, 1992). The McMaster University campus borders this wetland ecosystem. In a preliminary study, 118 McMaster University students were asked, "How concerned are you about the present state of the environment?" Both men and women expressed strong concern on a 7-point Likert scale where 1 was defined as "not concerned" and 7 as "extremely concerned," but the average was significantly higher for women (6.2) than for men (5.8). This kind of question undoubtedly covers many issues, and does not clarify what specific environmental matters women and men might be primarily, and perhaps differentially, concerned about. People could be concerned about health hazards, resource depletion, reduction in species diversity, and other more specific and local issues. It is also possible that this sort of questioning evokes greater expressed concern from women than from men regardless of the specific issue.

We hypothesize that particularly large sex differences will pertain to valuations that engage psychological adaptations shaped by a history of intrasexual competition. Environmental resources are obviously crucial to the survival and reproductive success of both sexes, but because our sexual selective history has been one of mild effective polygyny (Daly & Wilson, 1983), men may perceive a lesser reduction in the marginal utility of resource exploitation beyond immediate needs, and hence be more willing to deplete them. As discussed below, there is also reason to expect men to discount the future more steeply than women, and to be less deterred by health hazards and other risks. More specifically, we suggest the following seven propositions derived from the logic of sexual selection theory.

1. Men may be expected to be more inclined than women to use material resources to enhance their social status, both because reproductive success has been more tightly linked to status for men than for women (Low & Heinen, 1993) and because the display of resources has been the principal way that men can enhance their attractiveness to women (Buss, 1994).

2. Men maybe expected to discount the value of future resources relative to present resources more than women, both because they are less likely than women to live to see the future and because immediate, even total, resource expenditure can pay off for a man (as mating effort) but is likely to be a disaster for a woman.

3. For similar reasons, men may be expected to disdain health risks more than women.

4. For similar reasons, men may be expected to discount long-term detrimental impacts of polluting or exploiting the environment more than women.

5.   Where material resources are unevenly distributed—where there is much variance in personal income or wealth—there is more intense intrasexual competition, and men may be expected to be relatively more inclined to take a short-term view of the merits of exploiting environmental resources. Of course, women are expected to be affected by inequitable distributions of wealth too, but men are expected to be more affected.

6.   Where life expectancy is relatively low, for whatever reasons, people, especially men, may be expected to be relatively unconcerned about personal health hazards or environmental degradation.

7.   All of the above six propositions are expected to be more germane to young men than older men, and these age-related effects are expected to be modulated, in part, by marital status and parental status. While women may also be affected by age and by marital and parental status, men are expected to be more affected.

Sexual Selection Theory

Darwinian selection favors those phenotypic "designs" that out-reproduce alternate designs within the same population. More specifically, selection favors whatever contributes to out-reproducing others of one's own sex. In a sexual population, all the males are engaged in a "zero-sum game" in which the paternal share of the ancestry of all future generations is divided among them, while the females are engaged in a parallel contest over the maternal share of that ancestry. In a fundamental sense, then, one's principal competitors are same-sex conspecifics.

Darwin (1871) drew a distinction between the process of "natural selection," which favors those improved phenotypic designs that enhance survival and the efficient transformation of resources into growth and reproduction, from "sexual selection," which is a matter of differential access to mates. Sexual selection seemed to Darwin a quite different matter from natural selection because the attributes favored by sexual selection, such as the famous peacock's tail, often seemed positively detrimental in other contexts. Contemporary biologists generally prefer to think of sexual selection as simply a component of natural selection, noting that distinct "selection pressures" may often oppose one another within the realms that Darwin subsumed under natural selection, too, and that selection in all its forms is ultimately a matter of differential contribution to the replicative success of one's genes relative to their alleles.

Sexual selection may further be divided into differential reproduction as a result of differential success in competitive interactions with others of ones own sex and differential reproduction as a result of mate choice by the opposite sex (Andersson, 1994).

Although it is true of both females and males that selection entails a zero-sum competitive contest for genetic posterity, the evolutionary consequences are not necessarily similar in the two sexes. In particular, sexual selection is generally of differential intensity in the two sexes, leading to a variety of sexually differentiated adaptations for intrasexual competition. In most mammals, the variance in male fitness is greater than the variance in female fitness. The male's ceiling on individual reproductive success is higher than the female's, since a male can sire a surviving offspring with very little investment of time or energy whereas each youngster demands considerable female investment (at least gestationally and usually postnatally as well). But it follows that an individual male also has a higher probability than a female of dying without having reproduced at all.

What this sex difference in fitness variance implies is that male mammals are generally subject to more intense sexual selection than females, with the result that the psychological and morphological attributes that have evolved for use in intrasexual competition are usually costlier and more dangerous in males than in females. This claim about sex differences does not hinge on maleness per se, but rather on whichever sex makes the greater investment in offspring (Trivers, 1972). Although that sex is usually the female, and perhaps universally so in mammals, there is a sizable minority of animal species, including some birds, fishes, frogs, and insects, in which males make the greater parental investment and females have the higher reproductive ceiling and greater fitness variance. In such "sex-rolereversed" species, sexual selection has operated more intensely on the females, who are larger and more combative, court choosey males, and die younger (see, e.g., Daly & Wilson, 1983).

The human animal, like other mammals, is one in which males are the most intensely sexually selected sex. The putative record for personal reproduction by a human being is of course held by a male: the Moroccan emperor Moulay Ismail the Bloodthirsty (1672–1727) is credited with 888 children (McWhirter & Greenberg, 1979). Even among despots, Moulay Ismail was a little extreme, but he does illustrate a point of broad applicability: successful men can sire more children than any one woman could bear, consigning other men to childlessness, and they have apparently always done just that (Betzig, 1986). Great disparities of status and power became possible only within the last few thousand years, after the invention of agriculture. But even among relatively egalitarian foraging peoples, who make their livings much as did most of our human ancestors, some high-status men maintain two or three wives while other men are consigned to bachelorhood. In contrast to the men, virtually all women of fertile age are married, and are in a position to reproduce if physiologically able (e.g., Howell, 1979; Hill & Hurtado, 1996). It follows that men compete not merely to attain the highest status, but to avoid the lowest. Indeed, competition is often fiercest near the bottom of society, where a man faced with predictive cues of total failure has nothing to lose by the most dangerous competitive tactics (e.g., Daly & Wilson, 1990; Wilson & Daly, 1985). It is in competitive social contexts that man's competitive psychological systems were designed by selection to be concerned with social comparisons, achievement, and status.

Successful reproduction in ancestral human environments required a long-term commitment on the part of a woman, but not necessarily by a man. Female fitness has been limited mainly by access to material resources and by the time and energy demands of each young, whereas male fitness has been limited, at least in part, by access to fertile females—and not necessarily "access" of any great duration! Therefore, it might be expected that males would find rapid resource accrual, resource display, and immediate resource use somewhat more appealing than would females, and that males would be more inclined to discount the future in their decisions about acquiring and utilizing resources. As Alexander (1979, p. 241) notes, "the entire life history strategy of males is a higher-risk, higher-stakes adventure than that of females."

It is also reasonable to expect that sexually differentiated valuations of natural resources will be especially conspicuous in those life stages in which males have been selected to compete most intensely for reproductive opportunities. The life-stage where claiming and using resources—indeed, exploiting the resources at the expense of the future and hence being disinclined to conserve resources—would have been most rewarded would be that lifestage where such behavior would have had the greatest statistically expected payoff in reproductive success in past environments. And that lifestage for men is young adulthood—once men are themselves husbands and parents then concern for their offspring's wellbeing may result in alterations of their valuations of the environment, especially if the resources would be those of recurring value from one generation to the next, such as land or water rights.

Young Men as Competitors and Risk-Takers

Several lines of evidence about lifespan development support the idea that young men constitute a demographic class specialized by a history of selection for maximal competitive effort and risk-taking. Young men appear to be psychologically specialized to embrace danger and confrontational competition. In various activities, they have been found to be especially motivated by competition and especially undeterred by danger to self (e.g., Bell & Bell, 1993; Gove, 1985; Lyng, 1990; Jonah, 1986). Interestingly, young males are also exceptionally risk-accepting in decisions about investments in stocks and bonds (Blume & Friend, 1978; Mclnish, 1982).

Risk of death by "external" causes such as driving fatalities is maximally sexually differentiated in young adulthood (Wilson & Daly, 1985). Mortality by suicide and homicide is also maximally sexually differentiated in young adulthood (Holinger, 1987; Daly & Wilson, 1988; 1990). The fact that men senesce faster and die younger than women even when they are protected from external sources of mortality suggests that these sex differences in mortality have prevailed long enough and persistently enough that even male physiology has evolved to discount the future more steeply than female physiology. In the case of homicides, young men are not only the principal victims but also the principal perpetrators; indeed, men's likelihood of killing is much more peaked in young adulthood than is the risk of being killed (Daly & Wilson, 1988). All of these things can be interpreted as reflections of an evolved lifespan schedule of risk-proneness.

An alternative to this hypothesis, however, is that age patterns are entirely the result of changes in relevant circumstances which happen to be correlated with age. Mated status, for example, would be expected to inspire a reduction in dangerous risk-taking because access to mates is a principal issue inspiring competition and married men have more to lose than their single counterparts. Marital status is indeed related to the probability of committing a lethal act of competitive violence, but age effects remain conspicuous when married and un-married men are distinguished (Daly & Wilson, 1990). Similarly, men are most likely to be economically disadvantaged in young adulthood, and poverty, too, is a risk factor in homicide, but young adulthood and unemployment are again separable risk factors for homicide (Daly & Wilson, 1990).

Dangerous acts are adaptive choices if the positive fitness consequences are large enough and probable enough to offset the possible negative fitness consequences. Disdain of danger to oneself is especially to be expected where available risk-averse alternatives are likely to produce a fitness of zero: if opting out of dangerous competition maximizes longevity but never permits the accrual of sufficient resources to reproduce, then selection will favor opting in (Rubin & Paul, 1979).

From a psychological point of view, it is interesting to inquire how age- and sex-specific variations in effective risk-proneness are instantiated in perceptual and/or decision processes. Wilson and Herrnstein (1985) argued from diverse evidence that men who engage in predatory violence and other risky criminal activity have different "time horizons" than law-abiding men, weighing the near future relatively heavily against the long term. What these authors failed to note is that facultative adjustment of one's personal time-horizons could be an adap-tive response to predictive information about one's prospects for longevity and eventual success (Gardner, 1993; Rogers, 1991; 1994). Other psychological processes that have the effect of promoting risk-taking can also be envis-aged. One could become more risk-prone as a result of one or more of the following: intensified desire for the fruits of success, intensified fear of the stigma of non-participation, finding the adrenaline rush of danger pleasurable in itself, underestimating objective dangers, overestimating one's competence, or ceasing to care whether one lives or dies. As drivers, for example, young men both underestimate objective risks and overestimate their own skills, in comparison to older drivers (Brown & Groeger, 1988; Finn & Bragg, 1986; Matthews & Moran, 1986; Trimpop, 1994). Apparent disdain for their own lives might be inferred from the fact that men's suicide rates maximally surpass women's in young adulthood (Gardner, 1993; Holinger, 1987). There is also some evidence that the pleasure derived from skilled encounters with danger diminishes with age (Gove, 1985; Lyng, 1990; 1993). In general, sensation-seeking inclinations as measured by preferences for activities which are thrilling and entail some disdain for danger are higher in men than in women, and decline with age (Zuckerman, 1994).

The risk-proneness of young men seems to be remarkably general in domain. Adolescents are relatively unlikely to seek medical assistance or other health enhancing preventive measures (Millstein, 1989). Young men are also relatively willing to take risks with drugs and intoxicants and chances of contracting sexually transmitted diseases (e.g., Irwin, 1993; Millstein, 1993). It therefore seems very likely that they may also underestimate the health hazards of various environmental contaminants. And if they underestimate the risks to self that they are incurring with their activities then it is very likely that they will underestimate the risks that their activities impose on other people, and on other living things.

Inequalities in Intrasexual Competition

From the perspective of sexual selection theory, men more than women might be expected to be especially sensitive to cues of their status relative to their rivals. If intrasexual competition among men has been a major determinant of acquisition of resources (both material and social) which were converted into reproductive opportunities and if there has been a history of high variance in the distribution of resources and reproductive opportunities, then the masculine psyche is likely to be such as to take bigger risks to acquire, display and consume resources, especially when accepting a small payoff has little or no more value than no pay-off, as for example when it leaves a poor man still unmarriageable. This argument treats risk as variance in the magnitude of pay-offs for a given course of action. In life-threatening circumstances people often take the riskier (higher variance) course of action. But people also take great risks when present circumstances are perceived as "dead-ends." For example, history reveals that successful explorers, warriors and adventurers have often been men who had few alternative prospects for attaining material and social success. Such risk-takers acquired fame and fortune by their successes, fame and fortune which would not have otherwise been won. Later-born sons of Portuguese aristocratic families were the conquerors and explorers of the fifteenth-sixteenth centuries; inheritance of the estate and noble status went to first born sons (Boone, 1986). Similar options in life befell more humble folk too. Later-born sons and other men with poor prospects have taken, and still take, the riskier option of emigrating, often with a successful outcome when the destinations were places like North America (Clarke, 1993).

Experimental studies of nonhuman animal foraging decisions have established the ecological validity of a risk preference model based on variance of pay-offs. Rather than simply maximizing the expected (mean) return in some desired commodity, such as food, animals should be—and demonstrably are—sensitive to variance as well (Real and Caraco, 1986). Seed-eating birds generally prefer to forage in low variance microhabitats as compared to ones with the same expected (mean) yield but greater variability, but they switch to preferring the high variance option when their bodyweight or blood sugar is so low as to be predictive of overnight starvation and death unless they can find food at a higher than average rate (Caraco, Martindale & Whittam, 1980). The high variance option of course increases the risk of getting very little or no food, but a merely average yield is really no better—dead is dead—and the starving birds accept the "risk" of finding even less in exchange for at least some chance of finding enough. Such experiments, in which alternative responses yield identical mean return rates but different variances, reveal that several seed-eating birds (Caraco & Lima, 1985; Barkan, 1990), as well as rats (Kagel et al., 1986; Hastjarjo et al., 1990) switch to prefer the high variance option if their caloric intake is sufficiently reduced.

One can imagine numerous human parallels in addition to explorers, adventurers, and warriors. Taking dangerous risks to unlawfully acquire the resources of others might be perceived as a more attractive option when safer lawful means of acquiring material wealth yield a pittance, even if the expected mean return from a life of robbery is no higher and the expected lifespan is shorter. Interestingly, variations in robbery and homicide rates between places are better explained by measures of variance in income than by absolute values of poverty (Hsieh & Pugh, 1993).

People are demonstrably sensitive to variance as well as expected returns. Psychologists and economists, using various hypothetical lottery or decision-making dilemmas, have documented that people's choices among bets of similar expected value are affected by the distribution of rewards and probabilities (Lopes, 1987; 1993), as well as being influenced by whether numerically equivalent outcomes are portrayed as gains or losses (Kahneman & Tversky, 1979). The underlying psychological dimension governing these choices among alternative uncertain outcomes has been considered to be that ranging from risk aversion to risk proneness or risk seeking. In the experimental studies of foraging decisions described above, the below-weight animal preferring the high variance option might be deemed risk seeking. One's relative inclination toward risk aversion or risk proneness presumably reflects perceptions of subjective utilities of the outcomes as well as perceptions of the uncertainties associated with each outcome. From the reasoning of sexual selection theory, people's perceptions of both subjective utilities and uncertainties are likely to reflect their sex, life history variables and cues of relative success in intrasexual competition. Psychological research on variations in risk acceptance has hitherto focused on stable individual differences (e.g., Zuckerman, 1994; Trimpop, 1994), with some consideration of lifespan development (Gove, 1985; Gardner,

1993), but there has as yet been little consideration of panhuman psychological adaptations designed by selection to deal with probability-weighting algorithms for different domains of utilities and for variable distributions of uncertainties.

Trade-offs between Success and Health

The line of reasoning developed in this paper suggests that men might value opportunities for resource accrual, and/or might discount associated costs such as health risks, more than women. One way to test this idea is to present people with hypothetical dilemmas. For example, we asked 197 people to imagine themselves in the following situation in which a career opportunity entailed an attendant health risk.

"You presently live in a small southern Ontario city of 250,000 people, where you were born and where most of your family and friends still reside. For quite some time you have been waiting for some sort of advancement in your position at work, and it has finally come. The company has offered you a promotion that will significantly boost your career, but your employers inform you that it will involve relocating to their newest branch out on the west coast (about 4,000 km away), to a city of 1.5 million people, for at least two years. Most of what you have heard about this large city is appealing, but it is also famous for its smog, and last year 10 percent of the population suffered from severe respiratory illnesses there, compared to a 1 percent rate of comparable cases in your home town."

Do you accept the promotion?

Eighty-five percent of the men and 68% of the women accepted the promotion (Chi-Square 8.9, P < .01). This significant sex difference in the predicted direction is supportive of our general thesis that men will discount health risk in favor of an opportunity to improve one's economic and social status. This result does not imply that men are insensitive to health risks; men might not accept the promotion if the percent of respiratory illness was much higher than 10%. Furthermore, women might not have accepted the promotion even if the health risks were the same in both places. Among the subjects in this study, only 41% of those who were parents said they would accept the promotion, compared to 81% of those without children (Chi-Square = 19.9; P < .001). This difference supports the expectation that parenthood is relevant to valuations and preferences. Parenthood is, of course, correlated with age. Fifty-two subjects in this study were between 21 and 35 years of age and within this age category 50% of the 16 parents and 83% of the 36 nonparents accepted the promotion (Chi-Square = 6.3; P < .02). The sample size was not adequate to address the question of whether parenthood affected men's and women's responses differently.

The results from this preliminary study suggest that this kind of tradeoff dilemma is a promising research method. Various modifications could be added to take into account considerations such as other potential costs and benefits of the promotion and relocation as well as variations in the perceived likelihood of the two outcomes. It is also important to assess the ecological validity of this self-report protocol: how well do self-reports reflect people's actual inclinations and preferences?

Trade-offs between Present Profits with Environmental Degradation Versus Reduced Profits with Longterm Sustainability

In addition to the expectation that young men are more likely to disdain personal health risks for economic and status advantages, we also hypothesize that men are more likely than women to discount future environmental degradation in favor of present profits and material successes. Another set of 104 McMaster University people considered the following dilemma:

"Imagine you are farming a tract of land. Your father, like his father before him, lived off the profits from the farm without taking additional wage work elsewhere. You were fortunate to earn a scholarship to university to study agriculture, and now that you have inherited the farm you are considering changing the techniques of farming to be more scientific and business-like. Prior to inheriting the farm you had a successful career as a broker specializing in agricultural commodities. [After your wife died suddenly, you've decided to leave that job to return to the farm. Your two children are delighted about the prospect of living on the farm.]"

Presently, you are pondering whether to follow one course of action (Plan A) or another (Plan B).

Plan A: Convert the farm entirely to hybrid corn production for livestock feed. Corn is extremely profitable to grow, but it requires heavy chemical fertilization which over time will percolate into the water table with a very high probability that the land will not be usable in 60 years without heavy chemical supplements.

Plan B: Convert the farm entirely to hay for livestock feed. Hay in good years can bring a good market price, but generally hay yields a modest profit. On the other hand, hay production does not diminish the quality of the soil and chemical supplements are not needed.

Which plan did you choose? A or B? _____

Men were significantly more likely to choose the soil-degrading option: 39% of men (N = 36) and 16% of women (N = 68) chose hybrid corn, plan A (Chi-Square = 6.7, P < .01). These men and women ranged in age from 17 to 24. In order to determine whether these "decision-makers" were utilizing sound economic logic we asked four additional questions (Table 12.1). Those subjects who chose the hybrid corn option, whether men or women, were significantly more likely to have agreed that profits can be reinvested in other ventures as a rationale for discounting the costs of damage to the land.

Table 12.1

Average Ratings of Agreement to Questions Intended to Assess the Reasoning Associated with Men and Women's Choices Between Present Profits with Environmental Degradation (hybrid corn) and Choice of Reduced Profits with Longterm Sustainability (hay).

	Subjects' Sex	Hybrid Corn (A)	Hay (B)
Do you agree that because you can always invest the profits from farming in other economic ventures including other farmland that you should weigh profit over damage to the land? *Corn vs. Hay: F = 15.4, P < .0001*	M F	4.6 4.4	2.5 2.9
You would have felt more confident about making this choice between Plan A and Plan B if you knew how much acreage was involved. *Corn vs. Hay: F = 33.5, P < .0001*	M F	4.3 2.5	2.5 2.5
You would have felt more confident about making this choice between Plan A and Plan B if you knew more about the quality of the soil. *Corn vs. Hay: F = 13.3, P < .0001* *Male vs. Female: F = 5.0, P = .03*	M F	4.5 3.0	3.0 3.8
One of the factors you considered was the fact that your children will inherit the family farm. *Corn vs. Hay: F = 11.3, P < .0001*	M F	3.5 3.8	5.0 5.3

7-point rating scale: 1=strongly disagree; 7=strongly agree

This is very likely a common rationale for many short-term decisions with some risk of long-term environmental degradation, especially whenever there is a diversity of economic and relocation opportunities. Men who chose the more profitable but destructive crop appeared to be less confident that they had made the right decision than those who chose hay, indicating that they could have used more information; however, women (who were unlikely to choose corn in any case) showed no such pattern.

One factor which should influence decisions that may have long-term negative effects on the quality of your farm is whether your children are likely to continue farming. This was the rationale for adding the bracketed sentence at the end of the hypothetical scenario for half the subjects:

"After your wife [husband] died suddenly, you've decided to leave that job to return to the farm. Your two children are delighted about the prospect of living on the farm" (the "family" version). A comparison of the family versus no family versions did not result in a significant difference in choice of crop plan. We had anticipated that parental status would be a significant factor in reducing the likelihood of both women and men discounting the possible long-term degradation of the land. But perhaps imagining that one has children cannot evoke the mindset of actual parenthood. In this sample, only four people were married and two actually had children. Even though there was not a significant difference in choice of crop for the family versus no family versions, those men and women who chose the hay option (Plan B) were significantly more likely to have agreed that inheritance of the family farm was relevant to their choice of crop plan (Table 12.1).

Concluding Remarks

We would like to stress that further study of sex differences in valuations of various environmental goods throughout the life course is very likely to pay off handsomely in identifying those factors which might have real impacts on sparing our environment from further degradation. Notwithstanding the fact that men and women seem to be different in discounting risks to self and the environment, men are not impervious to social cues and life prospects. We have not addressed the question of the impacts of variations in the intensity of social competition on environmental issues, but there is good evidence in other domains such as crime that the intensity of social competition can affect men's competitiveness and risk acceptance. Thus it is entirely plausible that social and material equity might increase men's inclinations to value the future and to treat resources less exploitatively so that sustainable development is an achievable goal.

Acknowledgements

The study of environmental attitudes, perceptions, and decision-making has been supported by grants from the Great Lakes University Research Fund (GLURF grant 92-102) and from the Tri-Council Eco-Research Programme of Canada (TriCERP grant 922-93-0005).

13

The Evolution of Magnanimity: When is It Better to Give than to Receive?

James L. Boone

"When goods increase, they increase that eat them: and what good is there to the owners thereof, saving the beholding of them with their eyes?"
—Ecclesiastes 5:11

Human behavioral ecology is concerned with explaining variation in fitness affecting behaviors under varying environmental conditions. In investigating a particular class of behavior, such as foraging, a standard procedure is to weigh the total amount of time or energy expended in a particular strategy (i.e., costs) against the energy acquired (benefit); the predicted, or optimal, behavior is one that maximizes the net energy gain per encounter under a specified set of environmental circumstances, wherein energy is used as a proxy for fitness. In other cases, energy or time expenditures may be weighed against a more direct measure of fitness, such as number of surviving offspring. These approaches have yielded impressive results in terms of explaining a wide range of variation in human subsistence and reproductive strategies (Borgerhoff Mulder 1991b; Smith and Winterhalder 1992b), and the explanatory power of optimization theory seems undeniable.

A not insignificant problem that arises with applying the cost-benefit approach to human behavior is that in many, if not most, ranked and stratified human societies, large amounts of time and energy are expended on behaviors that, on the face of it, seem to have little directly to do with energy acquisition, survival, somatic maintenance, or the production of direct offspring. In many cases, costly activities actually seem to compromise at least some components of fitness. One needs only to consider the astounding wastage embodied in gladiatorial displays and circuses underwritten by Roman elites in memory of their dead ancestors (Hopkins 1983), the sumptuary displays of late nineteenth

Reprinted by permission from *Human Nature*, 9:1, pp. 1-21, Copyright © 1998 Walter de Gruyter Inc. Published by Aldine de Gruyter, Hawthorne, NY.

century Golden Age elites of New York (Wharton 1962), or the elaborate, costly, and often risky recreational activities undertaken by contemporary Americans on their respective "vision quests" to understand that expenditures in this area of human endeavor can be very large indeed. Yet this problem is not limited to the activities of a few wealthy elites. "Conspicuous consumption" (Veblen 1973) permeates the everyday lives of all humans: the decoration of pots, elaborate clothing and housing, and participation in "nonessential" ritual and recreational activity are ubiquitous to the activities of all humans, rich or poor. All of these behaviors involve the investment of time and energy into activities that, while they may be rooted in "functional" tasks, seem to varying degrees to go beyond what is required for the fulfillment of basic survival, maintenance, and reproductive goals. The costs of such behavior seem obvious—but what are the benefits, in evolutionary terms?

As Thorstein Veblen (1973) argued nearly a century ago, much of this kind of display behavior can be understood, at least on a descriptive level, in terms of its function in advertising and reinforcing social status. In this light, I would suggest that a satisfactory Darwinian explanation of conspicuous consumption as a form of adaptive energy expenditure will have two requirements. First, we need to explain why "wasting" time and energy or taking unnecessary risks has the effect of reinforcing social status. Second, we need to explain how and why the often huge expenditures of time and energy associated with status reinforcement are compensated for in terms of fitness benefits. My goal in this article is to present an outline of the form such a twofold explanation of conspicuous consumption might take.

Conspicuous Consumption as Advertising

How can "wasting" time, energy, and resources on conspicuous displays possibly increase fitness? One avenue towards an understanding of this kind of behavior may be found in a body of theory developed by Amotz Zahavi (1975) and formalized by Alan Grafen (1990a, 1990b) based on what Zahavi calls the strategic handicap principle. In Zahavi's and Grafen's view, "wasteful" displays signal or "advertise" some underlying, unobservable phenotypic quality in the sender of the signal, the knowledge of which would be of benefit to others in the population, whom we can refer to as receivers. Knowledge of this unobservable quality on the part of the receivers will in turn cause them to behave in a manner that benefits the sender. Imagine now a population in which there are two or more senders of varying underlying quality. Senders would benefit from having the receivers perceive in them a higher level of quality than they actually possess. On the other hand, receivers will benefit from being able to infer the underlying quality in the sender as accurately as possible. Hence, there is an inherent conflict of interest between senders and receivers over the "honesty" of the advertisement—that is, the degree of correspondence between the level of advertisement and the actual level of the unobserved quality. How then could an "honest" signal of the senders' true underlying quality evolve? This is where the handicap principle comes into play.

Costly signaling operates as a handicap to the sender because it costs something—time, energy, or risk—to undertake. The higher the level of extravagance of the display, the higher the absolute cost of the display. Since all advertisers will benefit from having receivers perceive them at the highest level of quality possible in relation to other senders, both high- and low-quality advertisers could benefit equally from displaying more. But the costly aspect of advertisement will act as a guarantor of the honesty of the display if those of higher underlying quality pay *lower marginal costs per incremental unit of display* than those of lower quality (and not, it should be emphasized, because they get more benefit per incremental unit of display). This would be the case, for example, if the expenditure of time and energy in advertising reduces the amount of energy available for other fitness-related activities such as somatic maintenance, predator avoidance, or parental care. The more energy reserves a sender already has, the less each incremental unit of display will take away from reserves that must be put into other components of fitness. Under these conditions, senders of higher quality maximize their *net* benefit to signaling at a higher level of display than those of lower quality, and vice versa (Figure 13.1). Hence, receivers can infer from the level of display that senders who produce costlier displays must be of higher phenotypic quality than those who display less.

Figure 13.1

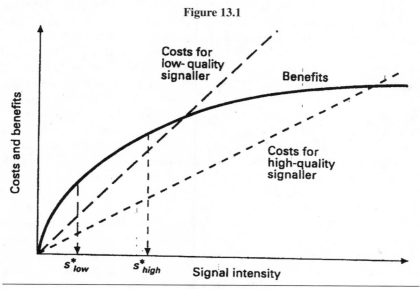

Graph showing that a signaler of underlying quality maximizes its net benefit to signaling at a higher level of signal intensity than a signaler of lower quality when the signal is functioning as a handicap (from Johnstone 1997:168; courtesy of Blackwell Scientific, London, © 1997).

The main illustrative case in point used by Zahavi and Grafen is that of costly plumage displays observed among males in some bird species and the effect these displays have on female mate choice. The costs and benefits of mate choice on the part of females and of displays on the part of males may well vary in character from species to species (Harvey and Bradbury 1991), but I will use the following generalized example to put the above argument in more concrete terms. Hamilton and Zuk (1982) have argued that some plumage displays may signal the degree to which males vary in the amount of heritable resistance to parasites. Accurate knowledge of this varying level of heritable resistance, which is not directly observable, is of benefit to females since it can be useful information in terms of choosing the best mates. Males benefit from females perceiving them to be of high quality in relation to other males. But the production of ornate plumage reduces other components of the males' fitness. For example, its production incurs significant energetic costs to produce and may increase susceptibility to predation as well (Harvey and Bradbury 1991: 219–220). The argument, then, is that males with lower parasite loads (i.e., higher quality) can "afford" to advertise more than males of lower quality and, hence, that the costly plumage display is an "honest" signal of their underlying heritable resistance to parasites.

What is the "Quality" Being Advertised, and Why?

I have argued above that conspicuous consumption as defined by Veblen may constitute a kind of costly display aimed at reinforcing social status. The immediate problem at hand in developing this idea further is that social status does not itself constitute a "quality" possessed by an individual independently of its being perceived by others—one can't have (or not have) social status by oneself! Hence, for the purposes of the following argument, I propose that "social status" is a relatively vague and general term which relates to a definable quality possessed by an individual such that it is perceived and acted upon by others in a social group in ways that affect the fitness of all the involved parties. This quality can be more specifically termed *social power*, defined as the capacity to extract and defend access to fitness-affecting benefits in the form of goods and services produced by members of a social group individually or through various forms of collaboration.

Social power in turn derives from two general classes of strategic capacities that determine how fitness-affecting benefits produced within the group will be allocated differentially among individuals: the capacity to impose costs and the capacity to dispense benefits (Noë et al. 1991). The first capacity is what is referred to in the behavioral ecological literature as "resource holding power" or RHP. Advertising RHP can potentially resolve a contest before it escalates to a costlier conflict level, but it must be reliable in order to work at all in this manner. For example, someone trying to overthrow or usurp another would benefit from knowing exactly what their "target's" power to resist such an attempt would be. The target would likewise benefit from knowing how big of a threat the potential usurper actually is. But both target and usurper would benefit from the other perceiving his or her power to resist or usurp to be as high as possible. There is, then, a conflict in interest over the honesty of their respective displays of power to impose costs that can be resolved through the handicap principle.

The second strategic capacity associated with social power is the capacity to dispense benefits to potential allies, clients, mates, and other conspecifics who may be in a position to assist an individual in achieving his or her fitness-related goals. For example, if an individual is considering forming an alliance or coalition with another, he or she would benefit from knowing something about the potential ally's ability to hold up their end of the deal, and vice versa. Of course, the power to dispense benefits goes beyond simple alliance formation and other cooperative endeavors. It can quickly develop into the power to acquire and defend benefits at the expense of others when an individual can gain control or possession of a key resource or service that others in the group need. In order to get access to this resource or service, the non-possessors must then render services or pay some kind of rent to the controller. If the commodity is sufficiently limited or rare in relation to the number of individuals that need it, then a "market effect" will emerge, such that the suppliers can act as monopolists and set up a process of "bidding" over access to it. Under the conditions of such an auctioning game, monopolists dispense their services to the highest bidders, often leaving the bidders worse off than when they started. Hence, monopolists are technically "dispensing a benefit," but at a price.

In ongoing social interaction, these two classes of capacities exist as *potentials* which are not often directly observable, except in very small groups where individuals may be able to keep track of each other's power on the basis of past interactions. As group size increases and as the range of variation in social status increases, power needs to be advertised in order to be effective. Furthermore, the amount of power held by individuals or families may increase or decrease as their fortunes rise and fall. For example, in agricultural societies, fortunes may rise and fall each year—hence, periodically scheduled festivals in which individuals / families / coalitions are called upon to display their current "condition" are quite common. The scheduling of such festivals is often itself strategic. For example, in temperate climates, winter solstice festivals—such as the Kwakiutl Winter Dance (Suttles 1991) or, in European societies, Christmas—represent, among other things, a context in which stored foods and other resources are expended at the onset of a long, unproductive winter and spring. Similarly, at mid- and late summer festivals, such as the eastern North American green corn ceremony or western European harvest festivals, freshly harvested crops are expended in feasting rather than stored for the upcoming winter and spring. In both cases, these ceremonies seem to signal various families' or kin groups' capacity to afford to waste stores of food that might otherwise carry them through future hard times.

To visualize more clearly how critical the signaling of social power might be, imagine a society characterized by differences in power on the level of, say, Iron Age Greece or medieval Europe. Yet in this society, everyone wears exactly the same clothes, lives in exactly the same houses, speaks in the same manner, and so forth. How would people know how to interact with each other? At most, the situation would be chaotic; at the very least, people would constantly be making costly mistakes regarding competitive and cooperative interactions—costlier than the signals which function to avoid them.

The question now arises as to the testability of the handicap model. I suggest the following criteria must be met if a costly display is functioning as a handicap aimed at insuring honesty in advertising: (1) individuals will choose with whom to interact and how, when, or how much to interact with them on the basis of varying levels of extravagance of displays; (2) the process of choosing will itself be costly, or making an error in the choice thereof will be costly; (3) costly displays must constitute a fitness-related handicap to the displayer; (4) costly displays must reveal in some way an underlying, unobservable character in the sender that is of interest to the receivers; and (5) the revelation of this underlying character will cause receivers to behave in a way that increases a component of the fitness of the sender other than the one that is being handicapped.

Not all forms of signaling will conform to these criteria. As Maynard Smith (1991, 1994) has shown, costly signaling as a reliability indicator operates specifically when there are conflicts of interest between senders and receivers. If senders and receivers have overlapping interests, signaling should cost no more than what is required to get the message across. Hence, not all signaling is "costly" in the specific Zahavian sense. Many, if not most, forms of the more familiar kind of advertising present in market economies with mass communication probably do not conform entirely to the Zahavian model. Here, some or all of the costs of advertising may take the form of an investment in reaching as many receivers as possible, thereby capturing a larger market share, not necessarily in convincing buyers of underlying quality. Signaling in other areas of human activity may take this form as well.

This last point is related more generally to another critical underlying assumption of the handicap principle, which is that signalers cannot change their level of q—the underlying quality. Taken at face value, this assumption may seem to be very restrictive. Certainly, people gain or lose power during their lifetimes, and increasing or defending power will require pursuing strategies that necessitate the expenditure of time and energy, or taking risks, in and of themselves. I take the main import of this assumption to be that senders cannot change their quality by changing the level or intensity of the advertisement. This follows directly from the logic of the handicap principle: If you don't already possess the underlying quality, you cannot produce an honest signal of it. Furthermore, if it were the case that increasing the cost of the signal increased the level of the underlying quality, the display would no longer be operating as a handicap; it would be like any other fitness-enhancing activity in which time or energy is expended in return for some net fitness-related benefit.

This is a subtle but critical point. It is relatively common in the anthro-pological literature, for example, to view such displays as the Kwakiutl potlatch as a form of competition in which potlatch-givers fight for higher position by giving away property—that individuals gain social status by potlatching (Drucker and Heizer 1967: 134, 146; Suttles 1960). Drucker and Heizer argue that it is in fact the other way around: "The Southern Kwakiutl and their neighbors liked to say, in effect, 'So-and-so was a great chief because he gave many potlatches.' Close analysis reveals that the reverse was actually true: So-and-so gave many potlatches because he was a great ('highly ranked') chief" (1967: 134). They further argue that a potlatch was in fact given to confirm the possession of certain hereditary rights, privileges, and possessions, and "had as its purpose the identification of an individual as a member of a social unit and the defining of his social position within that unit" (Drucker and Heizer 1967: 8, citing Barnett 1938). The Kwakiutl had among them a total of 795 separate "titles" or crests which were tied to specific use rights to fishing, hunting, and gathering locales and other privileges (Suttles 1991). Drucker and Heizer (1967) emphasize that it was not always clear who had possession of what at any given time, and that potlatches constituted a way in which possessors of such privileges could announce what was theirs and validate their claims to it. Hence, there is good reason to believe that potlatches actually did function as costly signals in the Zahavian sense.

On the other hand, it seems quite possible that a behavior that starts out as a costly signal in the Zahavian sense may change through time to become a strategy that increases social status in and of itself. For example, Thorstein Veblen famously argued that higher education was prestigious *precisely because* it was "of no use" (Fershtman et al. 1996: 109), and indeed, universities in the nineteenth century were largely attended by individuals who had independent sources of wealth, such that entering university was a signal that one was not required to enter a trade. Currently, however, number of years of education has become one of the best predictors of lifetime income (Tucker n.d.), such that increasing one's level of education can in fact change one's social status. A similar line of argument might be used to explain the traditional prestige relationship between "pure" and applied research.

What is the Utility of Status?

The key idea in strategic handicap theory is that a signal is effective and reliable because it lowers one component of a signaler's fitness while raising another through the production of the display. I have argued above that social power is an underlying, usually unobservable quality that must be signaled or advertised in order to be effective. The central question now becomes, what is the ultimate utility or benefit to an individual of reinforcing higher social status in relation to others, such that expenditures in the form of costly signaling, which in some cases may be quite large, are justified in some cost-benefit context? The question can be put in more concrete terms as follows. Most current evolutionary anthropological theory (e.g., Betzig 1986; Vining 1986) attempts to couch the fitness benefits of social status in terms of momentary or ongoing increased access to wealth, wherein wealth is by various means converted into larger numbers of direct offspring. Hence, the number of surviving offspring is the measure of fitness, and access to resources is argued to be the principal determinant of this measure. Under this view, socioeconomic status is typically defined in terms of annual income or the amount of wealth that an individual has accumulated at any given time. Hence, an individual with higher socioeconomic status has more wealth than one of lower status and, as a result, will accrue higher fitness. Implicit in this view is the idea that social status is relative and positional because fitness is relative, and, at least in the context of past selective environments, one can never have too much wealth.

The problem with this formulation with respect to understanding con-spicuous consumption as form of costly signaling is this: If the evolutionary point of high socioeconomic status is ongoing, increased access to resources, which are then converted directly into lifetime reproductive success in the form of increased numbers of surviving offspring, then what is the point of wasting time and energy or taking unnecessary risks to signal one's level of wealth to others? Why not just convert the resources expended in signaling directly into offspring?

There are at least two distinct avenues towards resolving this apparent paradox. First, it may be that conspicuous consumption simply does not constitute a costly signal in the Zahavian sense. For example, it may be that increasing the level of conspicuous displays somehow increases access to resources in some manner analogous to mass advertising, as discussed above. Whether or not a display constitutes a strategic handicap could be tested using the criteria developed in the previous section. The second possibility, and one I will pursue in more detail below, is that the primary fitness benefit associated with high social status is not ongoing access to wealth per se, but some other, more long-term component of fitness. If this were the case, then conspicuous consumption could be understood as a costly signal that lowers one component of fitness associated with momentary or ongoing access to wealth, while raising another component associated with long-term fitness.

In the discussion below, I will argue that status reinforcement behavior has evolved as a flexible, state-dependent behavioral strategy that increases probability of survival through relatively infrequent, but recurrent, demographic bottlenecks by determining individual or familial *priority of access* to resources accumulated, produced, or defended by the social group during infrequent but serious shortages, and that what is usually passed off as an occasional perquisite of social status may in fact be its evolutionary raison d'être. Perhaps the simplest way to conceptualize what I am suggesting that status "does" is to consider the following example. Imagine a corporation that is engaged in the manufacture of some product. The corporation is made up of a hierarchically organized group of individuals who not only have different functional roles, but markedly different powers to decide how production is undertaken and how proceeds from sale of the product will be allocated among the workers; such powers would include the power to hire and fire employees below them. Imagine now that the corporation experiences an economic downturn such that it can no longer afford to pay everyone. Who gets laid off first?

In this kind of scenario, social status is relative and positional precisely because it ranks individuals in order of priority of access to what is left in the face of adversity. That social ranking functions in this manner has been fairly well documented for the Tikopia (Firth 1936), the traditional Hopi (Levy 1992), and the Tuareg (Baier and Lovejoy 1977; for a full discussion of these cases see Boone and Kessler 1999).

Since the survivors of a population crash form the population base for the next period of growth, it stands to reason that natural selection might favor individuals or lineages with some heritable capacity to survive (or to have offspring that will survive) infrequent crashes at a higher probability than others in a population. Imagine a population consisting of two variants that employ distinct reproductive strategies. One variant, which we will call the K-strategists, employs a strategy that increases the probability of survival through a bottleneck but which, in order to implement, requires producing fewer offspring than the second variant, which we call r-strategists, who have a higher intrinsic rate of increase than the K-strategists during periods of growth, but lower probability of survival through infrequent crashes. As I show below, the long-term success of one strategy relative to the other will depend on both the severity and the frequency of population crashes, as well as the intrinsic rate of growth of the K-strategists relative to that of the r-strategists.

Below, I present a formal model that evaluates the conditions under which a costly strategy might yield higher long-term rates of increase in the face of random population crashes even if it results in reduced fertility (see Boone and Kessler 1999 for a fuller exposition). The model employs three key variables: r (intrinsic rate of growth during periods between crashes), S (severity of the crash, measured by proportion of the population surviving the crash), and p (probability of a crash occurring during time t).

To evaluate and compare the long-term success of these two strategies under varying environmental conditions, the model employs a variation on the well-known formula for geometric population growth ($N_t = N_0 * e^{rt}$), which includes a term that accounts for the effects of recurrent crashes:

Expected $N_t = N_0 * e^{t[r - p(1-S)]}$

where N_0 and N_t are the starting and ending population sizes; e is Euler's constant, or the base of natural logarithms; t is time in years; r is the intrinsic rate of growth *between* crashes; p is the probability that a crash will occur in any particular year; and S is the proportion of the population surviving the crash. Thus the formula gives the expected population size after time t under conditions in which the population experiences periods of growth punctuated by randomly occurring crashes. By giving the K- and r-strategists different values for r and S, and assuming that both populations grow over the same period and experience crashes simultaneously, we can compare their relative long-term success under varying conditions of crash frequency, crash survivorship, and fertility expressed as r—the strategy that gives the highest overall rate of growth would be favored by selection

Figure 13.2 shows a set of simulated growth trajectories under conditions of p and S that are comparable to the frequency and severity of population crashes resulting from famines in nineteenth-century India and Ireland, where famines occurring at intervals of between 25 and 50 years caused population reductions of between 20% and 25% (Crawford 1989; McAlpin 1983). In the simulation shown here, crashes occur randomly at a mean rate of one every fifty years ($p = 1/50 = .02$). Survival rates are 78% for the r-strategists and 99% for the K-strategists (recall that these figures apply only to excess mortality resulting from the conditions that caused the crash, such as starvation). The rate of intrinsic growth during normal periods is .02 (or 2% per year) for the r-strategists; r for the K-strategists is only 80% as high, at .016 (or 1.6%). Even so, after 200 years, the K-strategists have the highest overall growth rate.

Thus, the above example shows that under fairly plausible conditions, selection could favor a strategy that involves the diversion of resources from the production of offspring if it sufficiently increased survivorship through recurrent bottlenecks. Note that this model essentially reflects what might occur in a costly signaling context: time and energy expenditure in signaling compromises one component of fitness (numbers of surviving offspring) while raising another (probability of survival through bottlenecks). This delayed-benefit aspect of status reinforcement is of particular significance to the issue of the evolution of "altruistic" displays.

Figure 13.2

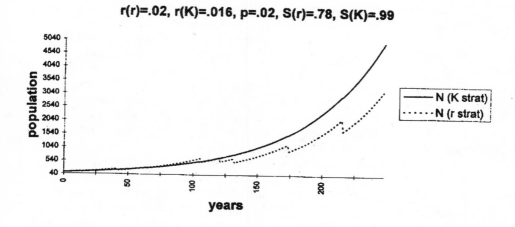

Results of a simulation showing that "K-strategists" (as defined in the text) have a higher overall rate of increase than "r-strategists" even when the costs of higher survivorship result in lower fertility.

Bread or Circuses? Magnanimity as Status Reinforcement

An additional aspect of strategic handicap theory of central interest in this article is the idea that costly signaling can take the form of "altruistic" behavior, such as allofeeding or food sharing and other forms of aid-giving behavior (see in particular Carlisle and Zahavi 1986 and Zahavi 1990 on allofeeding as status reinforcement in Arabian babblers). In this sense, expenditure of time or energy in altruistic behavior signals the sender's ability to bear the short-term costs of cooperation or sharing, wherein some long-term increase in another component of the "altruist's" fitness compensates for this temporary loss in fitness. In other words, "altruism" viewed as costly signaling is essentially a form of delayed benefit altruism, albeit one that does not necessarily depend on reciprocation by the recipient of the altruistic act. In this sense, costly sig-naling in the form of altruism, or what I will call *magnanimity*, corresponds most closely to what is often termed by-product or no-cost mutualism (Con-nor 1986; Dugatkin 1997b: 31-33). It is "no-cost" because the altruist receives a benefit whether the act is reciprocated or not. The altruist may receive an additional short-term benefit in the form of some return payment by a subordinate, who might "invest" in the mutualism by offering loyalty and support, but the long-term benefit of the altruistic act is not itself contingent on return payment by the recipient.

The idea that costly display can take the form of altruism can be expanded to shed light on Mancur Olson's (1971) concept of the "privileged group." A privileged group is one in which one individual in a group has ultimately more to gain from the proceeds of a cooperative enterprise than others in the group; hence, he or she is willing to pay the short-term costs of enforcing cooperation, detecting and punishing cheaters, or any other kind of strategy necessary to maintain cooperation in the group. Boyd and Richerson (1990) have shown that punishment (or enforcement) enables the evolution of cooperation (or any other kind of behavior) in large groups. The problem that then arises is: Who will pay the costs of enforcement? I suggest that if the enforcers gain some long-term benefit in the form of social status in return for the short-term costs they pay in enforcement, the problem is solved. Since significant status differences between individuals and families can exist even in the most egalitarian societies, it may be that all human social groups are in some sense "privileged." Ironically, then, social status reinforcement behavior may turn out to be a critical and essential ingredient of the social glue that results in human "ultrasociality," and a great deal of human helping and sharing behavior may be explained without any special appeal to group selection processes.

Taken at its simplest level, costly signaling theory requires only that the signal constitute a handicap to the sender in order to guarantee its honesty. Is it possible that the form of a costly signal is in some way determined by the underlying quality being advertised? This issue is currently being vigorously debated (Guilford and Dawkins 1995; Johnstone 1997; Zahavi 1993). Zahavi (1977) argues that the handicap must have some functional connection to the underlying quality being advertised in order to be honest. For example, advertisement of nutritional and energy reserves (as in the plumage displays of male birds) must in turn be energetically costly, whereas the capacity for a prey to escape its predators should be advertised by increased risk of capture (as in antelope stotting [Grafen 1990a]; see also Johnstone 1997). However, Johnstone argues against a correspondence, at least a very close one, between handicap and quality: the simple fact that "a signal should entail a certain kind of cost does not impose much of a restriction on its design" (1997: 173). The second objection is that costly displays may be "conventional" in the sense that the "appropriate receiver response (imposing whatever kinds of costs are necessary for the maintenance of honesty) can be elicited by a display of any kind." For example, the social power of an individual may be signaled in any number of ways that display the capacity to "waste" time or energy or take unnecessary risks.

Still, the question of whether a status reinforcement display takes the form of an "altruistic" display, such as a feast or act of philanthropy, or whether it takes the form of a purely self-aggrandizing display that might be determined by specific kinds of political, economic, or ecological conditions seems worth asking. Why are humans such inveterate signalers? Under what evolutionary ecological circumstances might costly signaling take the form of magnanimity?

Zahavi is not very clear about the general conditions under which costly signaling of status would take the form of altruism, but I would suggest the following. First, there should be relatively intense feeding competition within the group; if there weren't, the sharing of food would not constitute a handicap. Second, the primary benefit to grouping should derive from some factor other than cooperation in foraging or food production, such as cooperative territory defense; if the primary benefit were cooperation in food acquisition, there would not be enough conflict of interest between group members to justify costly signaling. These conditions describe fairly accurately Zahavi's Arabian babbler case, and I would suggest that they conform to the majority of ethnographically known human middle-range societies (i.e., tribal and chiefdom societies) in which production and consumption occurs principally at the house-hold level, but levels of intergroup warfare are high. A third necessary condition is that there must be some long-term or delayed benefit to some component of the altruist's fitness that compensates for the short-term loss that is sustained in transferring resources or other benefits to a con-specific. Above, I have argued that such a benefit may take the form of increased priority of access to resources during infrequent shortfalls or other calamities.

Zahavi (1990; see also Carlisle and Zahavi 1986) suggests two reasons why status competition takes the form of competitive altruism rather than open aggression against rivals in social groups in which collaborative activity is critical to individual maintenance and reproduction. First, aggression towards group members is more costly than competitive helping behavior because "any display of aggression which is not successful is witnessed by all group members, and any injury or weakness may be exploited by rivals that are constantly present in the group, waiting for an opportunity to change their rank" (Zahavi 1990:127). Second, individuals investing in the welfare of the group by undertaking costly helping behavior are more likely to attract other individuals as collaborators. In other words, in cooperative groups, dominants may need collaborators to become and remain dominant; by advertising their willingness to be magnanimous, individuals can attract supporters, as well as mates, and demonstrate their capacity to rally support in defense against outsiders.

However, magnanimous displays, in contrast to non-altruistic costly displays, have some distinctive, emergent, group-level characteristics that complicate the picture. First, magnanimous displays differ from other kinds of costly signals in that the display often actually takes the nascent form of an economic transaction; once such a display is iterated a number of times, it may become a regularly expected exchange of goods and services. It is worth at least speculating whether exchange itself might not have evolved from some earlier form of magnanimous display in hominid social groups. For example, note that tipping, which can be seen as a form of magnanimous display directed perhaps as much toward an individual's dining companions as toward the waiter, has evolved in many contexts into a *de rigueur* form of payment. This could be tested by determining whether individuals routinely tip more in the presence of companions, or certain kinds of companions, than alone.

Second, a display in the form of a transfer of goods and services has the effect of decreasing the fitness differential (at least in the short term) between the sender and the receiver by *raising* the receiver's fitness relative to that of the sender. In contrast, in other forms of conspicuous consumption, as for example when resources are channeled into elaborate clothing or housing, the signal may actually have the ultimate effect of *depressing* the fitness of subordinates by taking out of circulation resources that could have been used for their somatic maintenance or reproduction (i.e., the "Let them eat cake" effect).

Of course, in the context of cooperative territory defense, raising the fitness of subordinates may not be an altogether bad thing, at least during other than famine conditions. Consider, for example, a scenario in which a region has filled with autonomous social groups, and competition over territory in the form of intergroup warfare and raiding has begun to intensify. Victorious groups typically have larger group sizes than the losers. At the same time, status competition within each group is increasing as well. At this point it becomes clear that additional socially subordinated group members may become critical contributors to the dominants' long-term survival, and vice versa, and that negotiations over rights to access to resources held in "common," in both the short and long term, would begin to take on the complexity and subtlety observed in real human social groups. One might foresee a two-way market opening up among patrons and clients (Boone 1992b) wherein patrons compete for clients who demand the least (or who are best able to take care of themselves, although of course at a lower long-term level of food security) and clients scramble for patrons who will offer the best terms and/or who have the most power to protect them from the depredations of competing groups. Since how much each would have to offer in the long term may not be directly or immediately observable, it is in this context that we would expect to see honest signaling of these underlying capacities develop. For example, patrons may need to display their capacity to dispense benefits in the form of protection, their ability to form and maintain alliances with other powerful patrons, as well as RHP in the face of would-be outside aggressors.

Now imagine that, in some of these groups, elites signal their power by piling up their year's agricultural surplus in the plaza and burning it up in front of their subordinates. In other groups, elites engage in status displays by staging elaborate feasts and handing out gifts to their subjects. After several generations of intense warfare, which type of display behavior is likely to survive in the population? One might expect that the "feasters" would be much more successful at attracting supporters than the "burners."

If different kinds of costly displays have varying long-term fitness effects on the population as a whole, it seems possible that different kinds of social, political, and economic conditions might select for specific kinds of costly displays by dominants. One might draw the rather obvious conclusion that feasting displays are more effective for maintaining within-group competition, and burning displays are more effective for displaying competitive ability to other groups. For example, Neiman (1997), in what is probably the first application of costly signaling theory in anthropology, has argued that Classic Maya pyramids were built to signal the competitive strength of individual polities engaged intense warfare with their neighbors. This would be an example of a "burning" type of display, since the resources marshaled to build the pyramid then become unavailable for other purposes.

The question then becomes whether *(a)* individual signalers can actually track or predict the long-term group-level effects of a particular form of display (which in turn will affect their own fitness one way or another) and, hence, act strategically, or *(b)* individuals cannot effectively track long-term fitness effects, and selection acts at the group level on more or less random choices regarding the form of displays made by dominants. In some contexts, it must be quite apparent to dominants which form of display would attract the most loyal supporters. In other contexts it may be that lower-status individuals actually vicariously enjoy observing the conspicuous consumption of the elite even when they do not seem to get any direct benefit from it—possibly because the capacity of the elite to display demonstrates competitive strength to outsiders. In any case, this issue does not seem to be altogether clear-cut, and further comparative study would be informative. Furthermore, many forms of conspicuous consumption have much more subtle, indirect, and delayed effects on the economy and the resulting viability of a social group. For example, it seems clear that a very significant proportion of the economy of western society derives from the manufacture and marketing of luxury goods and services, which in turn provides employment for millions of people. Yet surely it is not the intention of those who purchase and consume these goods and services to engage in this behavior for the good of society!

La Dolce Vita: Benefit or Cost?

The question of how far individuals are able to track the long-term, emergent group effects (or indeed, the shorter-term effects on their own fitness) of a particular display behavior is complicated by the peculiar reward structure that seems to underlie conspicuous consumption and other costly status reinforcement displays. To put it simply, when we acquire and consume luxury goods, what may feel to us like benefits should, in the evolutionary perspective developed here, be counted as costs! To illustrate, consider the following simple question: why do people like to eat food? The proximal answer is: because they like the taste, and it gives them a feeling of satisfaction to have eaten. The ultimate answer is, of course, that selection has favored the propensity to feel the pleasurable sensations associated with eating and having eaten because regular food consumption enhances survival and reproduction. It is now possible to restate one of the central questions raised in this article in an analogous way: Why do people want and need status, and why are they willing to pay such high costs to get and maintain it? I have argued above that the primary evolutionary raison d'être of social status is in its effect on long-term lineage survival through infrequent, unpredictable demographic bottlenecks. Yet because such bottlenecks are unpredictable, and because of the impracticality of implementing priority-setting strategies only when the dire need has already arisen (cf. von Schantz's 1984 argument regarding flexible vs. obstinate territory defense strategies), status needs to be continuously reinforced. However, unlike the example given above, where the need for nutrients is more or less continuous, it is quite possible for an individual to live an entire lifetime without ever experiencing the selective conditions that gave rise to the costly signaling behaviors he or she enacts every day. During this time, such an individual will be faced with shorter-term needs relating to other components of fitness (nutrient acquisition, avoidance of environmental hazards and pathogens, parental investment, mating effort, etc.) that require the expenditure of time and energy as well. How might the human psyche have evolved in order to allocate time and energy expenditures effectively between short- and long-term components of fitness? One possibility is that short-term emotional and physiological rewards and punishments (for example, pride and shame) have become attached to forms of energy expenditures that have long-term fitness effects, such that under conditions of increasing status competition, humans will actually prefer to invest time and energy in behaviors that will have (or, at least in their selective history, have had) positive effects on long-term fitness at the expense of behaviors that would appear to have greater short-term, but ultimately ephemeral, effects on the same.

Aside from the general theoretical interest of such a scenario, the position set out above has some more immediate and practical implications for the way we analyze the relationship between socioeconomic status and fitness. Socioeconomic status (SES) is currently almost always defined in terms of total annual income or some other measure of total accumulated wealth. This definition implicitly assumes that accumulated wealth is channeled into lifetime reproductive success in the form of numbers of surviving offspring (holding back some relatively constant amount that goes into growth, development, and somatic maintenance), and it further seems to imply that wealth and social status are completely isomorphic. What I have tried to show in this article is that social status reinforcement is an ongoing, state-dependent strategy that entails significant costs as well as benefits, in both the short- and long-term. It seems at least possible that a group of individuals or families that have identical annual incomes or accumulated wealth might expend widely variable proportions of income on *(a)* conspicuous consumption and other status reinforcement displays, *(b)* the production and rearing of offspring (Kaplan 1996), and *(c)* conservation of resources that can be passed on to offspring in the form of bequests (Rogers 1991). How, why, and under what socioeconomic conditions wealth is variably allocated between these and other components of long- and short-term fitness would seem to be an important avenue for future research.

I thank Paul Andrews and an anonymous reviewer for extensive comments on the manuscript. Hillard Kaplan, Charles Keckler, Karen Kessler, Jane Lancaster, Lance Lundquist, and others provided useful comments and suggestions in the development of the ideas presented in this article. Any errors of fact or interpretation are my own.

14

Evolutionary Economics
of Human Reproduction

Alan R. Rogers

The "Leslie matrix" of demography is extended to deal with categories of
wealth, rather than age, and is used to build an evolutionary model of the
effect of heritable wealth on reproductive decisions. Optimal reproductive
strategies are assumed to be those that maximize the long-term rate of growth
in the numbers of one's descendents. In poor environments, the optimal strategy is to maximize the wealth inherited by each offspring, which requires
limiting their numbers. In rich environments, on the other hand, it pays to
maximize the number of offspring. Strong positive correlations between
wealth and the number of offspring are predicted only in rich environments. Therefore, evidence that the rich reproduce more slowly than the
poor is not inconsistent with the hypothesis that reproductive strategies have
been shaped by evolution.

Introduction

There is no agreement yet about the kind of theory that is appropriate for
understanding human reproductive behavior. In evolutionary theory, reproduction is viewed as part of a "strategy," designed by natural selection, for getting
one's genes into the next generation. The problem with this view, according
to Vining (1986) is that the rich have fewer offspring than the poor, whereas
evolutionary theory would seem to predict the opposite. Perhaps selection
has not yet had time to produce adaptations to the modern environment, and
perhaps, therefore, evolutionary theory is not relevant to the behavior of modern
humans.

Evolutionists have not received this argument meekly (see the commentary
published with Vining's paper). They have pointed out how evolutionary

Reprinted from *Ethology and Sociobiology* (now *Evolution and Human Behavior*) V11: Rogers, A.
"Evolutionary Economics of Human Reproduction," pp. 479-495 © 1990 Elsevier Inc., with permission from Elsevier.

models can be useful even if our adaptation is out of date (Turke 1989), and they have pointed to the substantial evidence for a *positive* relationship between wealth and reproduction (Simon 1977; Mealey 1985; Irons 1979; Daly and Wilson 1983; Turke and Betzig 1985; Essock-Vitale 1984; Borgerhoff Mulder 1989). Yet none have objected to the claim that natural selection must tend to make the rich reproduce faster than the poor. As we shall see, however, this assumption is questionable in populations where wealth, as well as genes, can be inherited.

There is a tradeoff between transmitting genes to the next generation, and transmitting wealth. Wealthy parents must decide between producing many relatively poor offspring, or a few relatively rich ones. Several authors (Kaplan and Hill 1986; Turke 1989) have observed that if selection can ever favor the latter choice, then data such as Vining's may make evolutionary sense. But can it? To find out, we must incorporate inheritance of wealth into an appropriate evolutionary model. Relevant models can be found in the literature of evolutionary biology and economics.

Although evolutionary biologists have not written extensively about human wealth, they have considered a closely related question: How much resource should a parent devote to increasing the number of its offspring, and how much to improving their "quality" as measured by health, size, or nutrition? This question has been approached by assuming that selection will favor the strategy that maximizes either the number of grandchildren, or the number of offspring that survive to reproductive maturity (Smith and Fretwell 1974; McGinley and Charnov 1988). This seems sensible for animals and plants, because the bequests that such parents leave are perishable, and cannot be passed on to later generations. For example, providing a baby bird with extra food might have a large effect on the baby's chance of surviving the first winter, but have little effect on its ability to provision its own offspring in later years. Thus, giving extra food to a baby bird is more likely to affect the number of grandchildren than their quality, and a theory that ignores the quality of grandchildren is appropriate. In many human societies, on the other hand, the situation is different. By leaving large bequests to offspring, a human parent may substantially increase the expected wealth of grandchildren, and the optimal strategy will still depend on the tradeoff between quantity and quality. Clearly, we need a method that can account for the effects of bequests on distant generations.

Economists have developed such methods, though they are not concerned with natural selection. Becker and Barro (1988), for example, assume that individuals will strive to maximize a *dynastic utility function,* which is a weighted sum of the *utility,* or satisfaction, arising from their own consumption and that of all their descendants. The weights are chosen so that one's own consumption matters more than that of one's children, which in turn matters more than the consumption of grandchildren, and so on. Unfortunately, little is known about utility, so the analysis must proceed from assumptions that are more or less arbitrary.

The model developed here combines elements of both approaches. It accounts for effects of reproductive decisions on distant generations, but sidesteps our ignorance about utility by asking which reproductive decisions are favored in the long run by natural selection.

Long-Term Effects of Reproductive Decisions

The model developed here is the simplest one that I have been able to find that incorporates a) wealth differences among individuals, b) an effect of wealth on reproductive success, c) inheritance of wealth, and d) inheritance of reproductive strategies. Because it is simple, it will be relatively easy to understand. It will not, however, describe any real population accurately. Nonetheless, it will provide a convenient "laboratory" within which we can begin to study how inheritance of wealth affects the process of natural selection. Work on more realistic models is in progress, and a two-sex version of the model discussed here is described in the companion paper (Harpending and Rogers 1990).

I assume that wealth occurs in discrete units, that generations do not overlap, and that there is but one sex. Genetic inheritance is assumed to be clonal, so that each individual is genetically identical to her single parent. Reproductive strategies are inherited genetically, so that one's descendants will all share the same reproductive strategy. In the absence of density-dependent population regulation, selection will favor the clone that can sustain the highest rate of growth.

Each individual will divide her wealth into two components, a *fertility allocation,* which is spent on activities that increase the number of her off-spring, and a *bequest,* which is inherited by the offspring that she produces. The evolutionary problem that she faces is to make this division in a fashion that will maximize the representation of her genes in future generations. A "reproductive strategy" is a set of rules determining the fertility allocation as a function of total wealth. The optimal reproductive strategy is the one that maximizes the long-term rate of growth in the numbers of one's descendants.

The growth trajectory of a single clone can be described using a formalism closely related to the "Leslie Matrix" of demography. If the individuals in a single clone can be subdivided into a series of categories based on, for example, wealth or age, we can define a matrix **G** whose ijth entry is the expected number of individuals in category i produced in one unit of time by each individual in category j. For example, if

$$
\begin{array}{c}
\begin{array}{ccc} 0 & 1 & 2 \end{array} \\[4pt]
\begin{array}{c} 0 \\ G=1 \\ 2 \end{array}
\left(\begin{array}{ccc}
0 & 3 & 0 \\
0 & 2 & 3 \\
0 & 1 & 2
\end{array}\right)
\end{array}
\qquad (1)
$$

then individuals in category 1 each give rise to 3 individuals in category 0, 2 in category 1, and so forth. If the categories in question are age categories, and the unit of time much less than a generation, then **G** is the familiar "Leslie matrix" of demography. We can apply this same formalism to the problem at hand by defining **G** in terms of wealth categories rather than age categories, and using the generation as a unit of time. The ijth entry of **G** then becomes the expected number of offspring of wealth i born to each parent of wealth j.

Unlike the Leslie matrix, many of whose elements are zero by definition, any or all of the elements of **G** may be positive. Nonetheless, several prop-erties of Leslie matrices carry over without change to the matrix **G**, provided that its leading eigenvalue, λ, is unique. These properties are listed below and are derived in the appendix.

1)   *The long–term growth rate of the clone is equal to λ, the dominant eigenvalue of **G** (if one exists);*
2)   *The distribution of wealth converges eventually to a "stable wealth distribution" that is proportional to u, the leading column eigenvector **G**;*
3)   *The leading row eigenvector, w^T, contains the* relative long-term fitnesses *of the wealth categories.* (Here and elsewhere, the superscript "T" is used to denote the matrix transpose.) This statement means that the ratio of the expected numbers of descendants, after τ generations, of indi-viduals in wealth classes i and j approaches w_i/w_j as τ approaches infinity. Long-term fitnesses are generally called "reproductive values" in demography. This term is avoided here because it is conventionally defined as a function of age rather than wealth.

Table 14.1
Results for the Example Matrix (1)

Wealth category	Darwinian fitness	Long-term	
		Fitness	Frequency
		w	u
0	0	0.00	0.34
1	6	0.79	0.42
2	5	1.37	0.24

The long-term growth rate, $\lambda = 3.732$, is the dominant eigenvalue of G, relative long-term fitness is w^T (the corresponding row eigenvector), and the long-term frequency distribution is u (the corresponding column eigenvector). Note that wealth Class 2 has the highest long-term fitness even though Class 1 has the highest Darwinian fitness. The long-term fitnesses are normalized so that the average long-term fitness of the fertile wealth Classes (1 and 2) is 1.0. Thus, individuals in wealth Class 1 can expect to leave 0.79 times as many descendants as the average fertile individual.

For the matrix in the example above, $\lambda = 3.732$, and the leading column and row eigenvectors are given in Table 14.1. They show that the long-term fitness of an individual with wealth 2 is nearly twice that of an individual with wealth 1, even though her Darwinian fitness is lower.

Our interest is in using these results to search among reproductive strategies for the one whose long-term rate of growth (λ) is greatest. As an aside, it seems worth mentioning that these results could be used in other ways as well. If **G** were estimated from data, we could calculate the long-term growth rate, wealth distribution, and fitnesses: they are given by λ, u, and w, respectively. In addition, we could use long-term fitness as a "currency" in optimization models. By comparing u with the observed distribution of wealth, we could test the hypothesis that the population is at equilibrium, and consists of individuals with a single reproductive strategy.

Optimal Allocation of Wealth

How many offspring should one produce, if the goal is to maximize the representation of one's genes in future generations? Presumably, fertility is costly, and increasing fertility allocations yield increasing fertility according to some law of diminishing returns. An individual must decide how much wealth to allocate to fertility, and how much to leave as a bequest. The set of rules that govern such decisions may be described by a *strategy vector,* whose ith entry is the fertility allocation of parents with wealth i. Our problem is to find the strategy vector that maximizes the rate at which one's descendants will increase. The results of the preceding section reduce this problem to two steps: 1) find a way to determine **G** from a strategy vector and 2) search among all strategy vectors for the one that maximizes λ.

Determining G from a Strategy Vector

It seems plausible to assume that no offspring can be produced unless some wealth is allocated to fertility. The threshold $s,$ below which no offspring are produced, is referred to as the "starvation threshold." Above this threshold, increasing fertility allocations yield increasing fertility according to some law of diminishing returns. I assume that the number of offspring, $m_x,$ born to parents with fertility allocations x $(\geq s)$ is the nearest integer to $m_+ (1 - e^{-(x-s)})$ where m_+ is the maximum number of offspring. The shape of m_x when $m_+ = 3$, and $s = 4$ is shown in the upper panel of Figure 14.1. To improve realism, it might be preferable to define m_x as the expected number of offspring, and allow it to take noninteger values. However, this makes the computer program too slow to be useful.

I assume that bequests are divided as evenly as possible among offspring. For example, when five units of wealth are divided among four offspring,

Figure 14.1

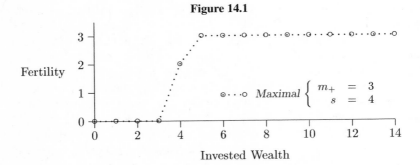

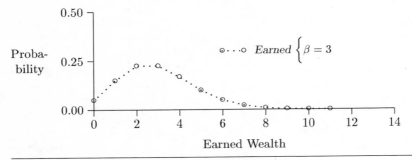

Maximal fertility, m_x, and the probability distribution, p_x, of earned wealth.

three of them will inherit one unit and the other one will inherit two units. It will be interesting, in future work, to investigate alternative assumptions such as that of primogeniture.

To ease the computational burden, I also assume that there are only K possible wealth categories, corresponding to individuals with wealth 0, 1, . . ., $K - 1$, respectively. A model allowing infinite wealth was also developed, but required too much computer time to explore numerically.

In addition to inheriting wealth, individuals also "earn" some on their own. Initial earnings are independent Poisson random variables, truncated so that earnings cannot exceed $K - 1$ units of wealth. Formally, this distribution is $p_x = C\beta^x e^{-\beta}/x!$, where β, is (roughly) the mean of the distribution, and C is a constant chosen so that $\sum_{x=0}^{K-1} p_x = 1$. The bottom panel of Figure 14.1 illustrates the form of this distribution when $\beta = 3$. After each individual has received both earned and inherited wealth, her total wealth is reduced if necessary to ensure that total wealth does not exceed $K - 1$.

The Search Problem

Given a way of mapping reproductive strategies into **G**, we still need to search among all feasible strategies for the one that maximizes λ. Unfortunately, with K wealth categories, there are $(K - 1)!$ feasible strategies, and each eigenvalue calculation takes on the order of K^3 floating point operations. Thus, computer time required by a exhaustive search of the strategy space increases as $(K - 1)! \times K^3$, which is feasible only when K is small. The results in Figures 14.2 through 14.4 were obtained by a local optimization algorithm that begins with an initial strategy and tries all possible "one-step perturbations" (those obtained by adding 1 or −1 to one element of the strategy vector). Each time a strategy vector is found that increases λ, that vector is accepted and the algorithm begins again. The algorithm stops when it finds a strategy vector that cannot be improved by any one-step perturbation. To increase the likelihood that the local optima thus obtained are also global optima, the algorithm was started from 100 random initial strategies for each set of parameters. This generates numerous local optima, and the one with the greatest value of λ is chosen. In practice, the local optimum with the highest value is found far more often than any other. This suggests that high peaks in the surface are also broad, so that our chances of finding the global optimum in this way are quite good. Exhaustive searches have been done with K as large as 15, and have always led to the same optimum as the local optimization algorithm.

Figure 14.2

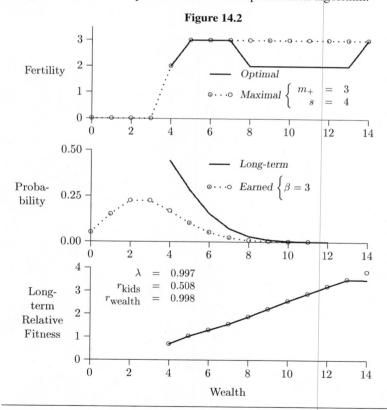

Optimal reproduction in an environment of intermediate quality.

Results

Figure 14.2 presents results for the assumptions illustrated in Figure 14.1. The curves describing maximal fertility (m_x) and the distribution (p_x) of earned wealth are indicated by circles connected by dots, and are identical to those in Figure 14.1. The solid lines pertain to the optimal reproductive strategy. They begin at the starvation threshold, s, because wealth classes below s are sterile and have been neglected for computational convenience. The solid line in the upper panel shows how optimal fertility varies with wealth. If a clone of descendants follows this fertility schedule, it will converge towards the frequency distribution of wealth shown in the middle panel of Figure 14.2. The solid line in the bottom panel of this figure gives the long-term fitness of individuals in each wealth class. These long-term fitnesses are normalized so that the average individual (excluding sterile wealth categories) has a value of 1. Thus, Figure 14.2 tells us that individuals with 14 units of wealth will eventually have more than three times as many descendants as the average fertile individual.

Figure 14.3

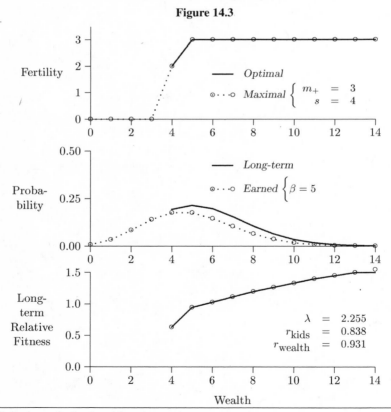

Optimal reproduction in a rich environment.

The most striking features of Figure 14.4 are the nonmonotonic relationship between wealth and optimal fertility, and the linearity of long-term fitness. Fertility increases with wealth only until wealth exceeds five units, and beyond that it wobbles up and down. Consequently, the correlation (r_{kids}) between wealth and optimal fertility is only 0.508. On the other hand, long-term fitness is nearly a perfectly linear function of wealth over the entire range, as is reflected in the correlation, $r_{wealth} = 0.999$. Thus, wealth is an excellent proxy for long-term fitness.

The linearity of long-term fitness breaks down a little in the final wealth class. It was not clear, at first, whether this was an artifact of the restriction that individual wealth cannot exceed $K - 1 = 14$ units. To evaluate this possibility, a run was made using $K = 40$ wealth classes. In models of this size, limitations on the accuracy with which eigenvalues can be calculated make it impossible to find a unique optimal strategy. There are many strategies with indistinguishable eigenvalues. To cope with this problem, optima were calculated in two different ways: once by resolving all ties in favor of the strategy specifying more investment (to give an upper bound on optimal investment) and

Figure 14.4

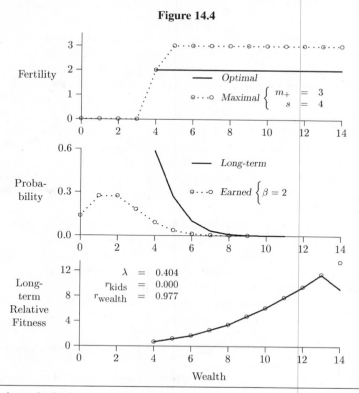

Optimal reproduction in a poor environment.

once resolving ties in favor of the lesser investment (to give a lower bound on optimal investment). The upper and lower bounds coincide until wealth exceeds 15, as do the long-term fitnesses that they imply. These long-term fitnesses are plotted as open circles in the lower panel of Figure 14.2, and show that the deviation from linearity at the right edge of that figure is indeed an artifact of the fixed upper limit on wealth. Relaxing the upper bound on wealth makes long-term fitness almost perfectly linear.

Figure 14.3 shows the effect of enriching the environment by increasing mean earned wealth (β) from 3 to 5. The optimal reproductive strategy is now to have as many offspring as possible. Nonetheless, wealth is still the better proxy for long-term fitness, as is shown by the correlations, $r_{\text{wealth}} = 0.931$ and $r_{\text{kids}} = 0.838$. Some curvature is now apparent in long-term fitness, but not much.

Figure 14.4 shows the effect of a deterioration of the environment: mean earned wealth (β) has dropped to 2. Now the optimal strategy is to allocate as little to fertility as possible, maximizing the wealth of offspring but minimizing their number. Consequently, optimal fertility is the same for all wealth categories above the starvation threshold, and the correlation (r_{kids}) between optimal fertility and wealth is zero! Once again, wealth is a far better proxy for long-term fitness than is fertility.

The open circles in the bottom panels of Figures 14.3 and 14.4 are analogous to those in Figure 14.2, and refer to models with 40 wealth classes. The congruence between the results for $K = 40$ and $K = 15$ suggests that the results are, with one exception, insensitive to assumptions about maximal individual wealth. The exception is the comparatively low long-term fitness of the highest wealth category, which is clearly an artifact of the artificial upper limit on individual wealth.

A broader view of the model's behavior is provided by the perspective drawings in Figure 14.5, which cover a wide range of values of β and s. In each drawing, the quality of the environment improves as one moves from the left corner to the right corner. At the left corner, the starvation threshold (s) is high, and expected earned wealth (β) is low, so offspring are unlikely to reproduce unless they inherit wealth. At the right corner, the opposite conditions prevail, so inherited wealth does not limit reproduction. Consider first the upper drawing, which illustrates the behavior of r_{wealth}. At the left (in poor environments), $r_{\text{wealth}} = 1$, which means that wealth is an excellent predictor of long-term fitness. As one moves to the right (into more favorable environments) r_{wealth} gradually declines. Still, it appears that wealth is a fair proxy for long-term fitness in all but the richest environments. Neither of the two surfaces continues all the way to the corner on the right. This is because, in these richest of environments, there is no variation in either long-term fitness or immediate reproductive success, and r_{wealth} and r_{kids} are therefore undefined. The surface for r_{kids} shows a very different pattern. r_{kids} is zero in poor environments, and suddenly jumps to values near unity when the environmental quality exceeds a threshold marked by the "cliff" in the middle drawing in Figure 14.5.

Figure 14.5

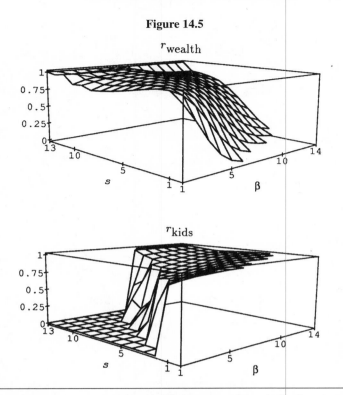

$r$wealth

$r$kids

Sensitivity of results to variation in s, and β. The two surfaces depict the values of r_{wealth}, and r_{kids} over a wide range of values of s and β. In all three drawings, $m_+ = 3$.

The variation in these surfaces is mainly confined to a single dimension running from left to right, which is measured by $\beta - s$, and can be interpreted as "environmental quality." Figures 14.6–14.8 plot r_{kids} and r_{wealth} against the measure of environmental quality. Note how little detail is lost by collapsing the surfaces in Figure 14.5 onto the two-dimensional scatter plot in Figure 14.7. The "cliff" in the r_{kids} surface is still apparent, and similar cliffs can be seen in Figures 14.6 and 14.8, which refer to different values of the parameter m_+. These cliffs reflect a sudden shift in reproductive strategies that occurs roughly where $\beta = s$. When $\beta < s$, the optimal reproductive strategy maximizes the wealth inherited by each offspring, which makes $r_{kids} = 0$ as in Figure 14.4. On the other hand, when $\beta > s$, the optimal strategy maximizes the number of offspring, generating higher values of r_{kids}, as in Figure 14.3. Intermediate strategies, such as that in Figure 14.2, occur only along the face of the cliff in Figure 14.5.

Figure 14.6

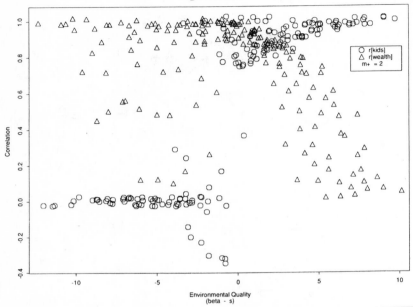

Plot of correlations against environmental quality, assuming $m_+ = 2$.

In populations where $r_{wealth} > r_{kids}$, increasing one's wealth is more important than increasing the number of one's offspring. This suggests that selection may favor economic motivations over motivations relating to mating and reproduction. In such a population, an economic theory of wealth maximization would predict behavior better than the fitness maximizing models that behavioral ecologists have preferred. On the other hand, where $r_{wealth} \ll r_{kids}$, a model of fitness maximization should work best. Figures 14.6–14.8 show that $r_{wealth} \geq r_{kids}$ over a wide range of parameter values.

These figures also show that the results are affected in only minor ways by variation in m_+. The most important effect is on strategies, such as that shown in Figure 14.2, that are intermediate between maximizing the wealth of offspring, and maximizing their number. Such strategies occur in a small region along the face of the cliff when m_+ equals 2 or 3, but are absent entirely when $m_+ = 6$. Because these intermediate strategies occur only in a restricted range of parameter values, it is unlikely that they are important in nature.

In summary, the assumptions of the model studied here imply that the optimal strategy is to maximize the wealth of offspring when offspring can expect to earn less than they need, but to maximize the number of offspring when expected earnings are larger. In the former case, wealth is a better proxy for long-term fitness than is immediate reproductive success. In the latter case, immediate reproductive success is the better proxy.

Figure 14.7

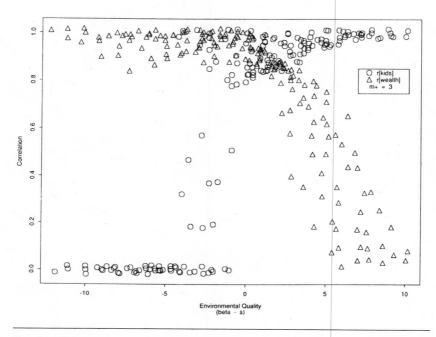

Plot of correlations against environmental quality assuming $m_+ = 3$. Environmental quality is measured by $\beta - s$, the expected excess of individual earnings above the starvation threshold. The position of each point in Figures 14.6–14.8 have been perturbed by a small random amount to reduce the number of points that lie atop one another.

Discussion

The contrast between Figures 14.3 and 14.4 exposes a tradeoff that is analogous to that embodied in the theory of r- and K-selection (MacArthur and Wilson 1967). In an environment with scarce resources, selection favors parents who maximize the quality of their offspring. In a richer environment, it is better to maximize the quantity of offspring instead. But in the standard theory, quality is favored only to the extent that it increases the number of offspring that survive to maturity. The model in Figure 14.4 favors quality even at the expense of the number that survive.

The novel features uncovered here are caused by the simultaneous action of two modes of inheritance: genetic inheritance of reproductive strategies, and cultural inheritance of wealth. Other studies (Boyd and Richerson 1985; Rogers 1988) have shown that nongenetic transmission can drastically alter the equilibria favored by natural selection, and the present findings provide another example of this principle.

Figure 14.8

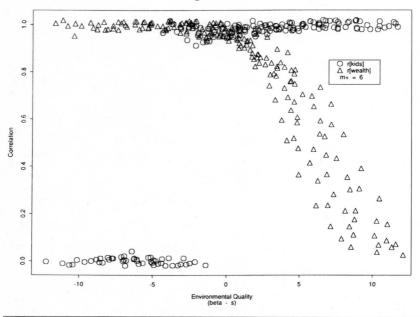

Plot of correlations against environmental quality, assuming $m_+ = 6$.

In the simplest evolutionary models, selection favors those who produce the largest number of surviving offspring. If, in addition, one assumes that the ability to rear offspring is constrained by wealth, then it follows that the rich should reproduce faster than the poor. However, this simple argument does not hold when wealth is heritable. Those who produce many offspring may lose out, in the long run, to those whose offspring are fewer, but wealthier. Selection favors small broods of wealthy offspring in poor environments, but large broods of poor offspring in rich ones.

This weakens Vining's (1986) claim that the "fundamental premise of sociobiology" is refuted by the modern data on wealth and reproduction. Figure 14.2 shows that the relationship between wealth and optimal production of offspring can be negative over at least a part of the range of wealth categories. It is only in rich environments that a strong positive relationship is predicted between wealth and offspring production. Therefore, there is no necessary contradiction between the assumption that reproductive strategies make evolutionary sense in modern environments, and the evidence that fertility does not increase wealth.

The most striking feature of the results presented above is the strong relationship between wealth and long-term fitness. Over a wide range of parameter values, wealth is at least as good a predictor of long-term fitness as is immediate reproductive success. Thus a desire to maximize wealth might serve the interests of long-term fitness at least as well as a desire for sex, or for offspring. This suggests that selection may often favor economic motivations over reproductive ones. Furthermore, the linearity of the graph of long-term fitness against wealth seen in Figures 14.2–14.4 suggests that the marginal value of wealth may be nearly as great for the wealthy as for the poor: the 14th dollar increases fitness nearly as much as the 4th. Consequently, selection may favor economic motivations that are equally strong for individuals of all wealth classes.

These conclusions, of course, refer to the hypothetical population described by the model, and should be related to the human condition only with caution. There is no reason to believe either that patterns of inheritance of wealth have been stable long enough for the model described here to have reached equilibrium, or that the model describes recent human evolution even approximately. One of the least satisfying assumptions is that of clonal inheritance, which implies that even distant descendants are genetically identical. In a diploid, sexual population, distant descendants will share few genes with any given ancestor. Consequently, increasing the wealth of descendants should help less in a diploid model than in the clonal model developed here. Therefore, we may expect long-term fitness to be more closely related to immediate reproductive success than the present model suggests.

On the other hand, there is another unrealistic feature with an opposing effect. I have assumed earned wealth to be independent of inherited wealth. In reality, the heir of a large estate is likely to earn more than a pauper. Now if, by leaving my child a dollar, I can improve her chances of earning another dollar, then the inheritance is worth more than a dollar. A wealthy parent faces a tradeoff between transmitting genes to the next generation, and transmitting dollars. Making an inherited dollar more valuable will bias this tradeoff in favor of dollars. Thus, a positive correlation between earned and inherited wealth should favor reproductive restraint among the wealthy. If this effect should happen to cancel that of sexual reproduction, the predictions of the present model may turn out to be realistic after all.

Summary

A model of clonal inheritance and density-independent population growth is used to study the effect of heritable wealth on reproductive strategies. The optimal strategy in poor environment is to maximize the wealth inherited by each offspring, which requires minimizing their numbers. In such environments, the correlation between wealth and long-term fitness is high, while that between immediate reproductive success and long-term fitness is near zero. The optimal strategy changes when the expected earnings of an offspring are large enough for it to survive and reproduce. It is then best to maximize the number of offspring, thereby minimizing the wealth of each. Long-term fitness becomes highly correlated with immediate reproductive success, but weakly correlated with wealth.

This paper has benefitted from the comments of Nick Blurton Jones, Monique Borgerhoff Mulder, Robert Boyd, Elizabeth Cashdan, Eric Charnov, Colin Clark, Malcolm Dow, Henry Harpending, Kristen Hawkes. B. Miličić, Simon Tavaré, Paul Turke, and David Wilkie. I thank James Crow for pointing out the similarity between my **G** matrix and the Leslie matrix of demography, George C. Williams for pointing out the probable effect of diploid inheritance, and Kim Hill for pointing out the probable effect of a correlation between earned and inherited wealth. This work was supported in part by National Institutes of Health grant No. MGN I R29 GM39593—01.

Appendix

This appendix derives the properties noted above for the matrix **G**, all of which are shared by the Leslie matrix of demography. However, published derivations of these properties generally refer to the entries of the Leslie matrix that are zero by definition (see, for example, p. 48 of Pollard 1973). Because no such restrictions apply to **G**, it is necessary to derive these properties in the present more general context.

The results below all depend on the assumption that **G** has a unique dominant eigenvalue. The conditions under which this is so are given by the theorem of Perron and Frobenius (see any text on positive matrices) and are satisfied provided only that, except for "sterile" wealth classes (those producing no offspring), each wealth class produces some offspring who remain in the same wealth class as their parents, and descendants of each wealth class may eventually reach any other wealth class, though not necessarily in a single generation.

To justify claim (2), let

$N(t)$ = the total clone size in generation t;

$n(t)$ = a vector whose ith entry is $n_i(t)$, the number of individuals in wealth class j in generation t.

The expected number of offspring in each wealth category is the sum of the expected contributions from each parental wealth category: $n_i(t) = \sum_j g_{ij} n_j(t-1)$, or

$$n(t) = Gn(t-1) = G^t n(0). \qquad (2)$$

As $t \to \infty$, $G^t \to \lambda^t u w^T$, provided that λ is unique. Thus, $n(t)$ will eventually become proportional to u. This justifies claim (2).

Once the stable wealth distribution has been reached, $n(t+1) = \lambda n(t)$, and therefore $N(t+1) = \lambda N(t)$. This justifies claim (1).

The third of the properties of **G** can be verified by considering what I shall call "τ-generation fitnesses." The τ-generation fitness of an individual in wealth class i is denoted by is $w_i(\tau)$ and refers to the expected number of her descendants after τ generations. For example, $w_i(2)$ is the expected number of grandchildren, it $w_i(1)$ the expected number of children, and $w_i(0) = 1$, since we can count an individual as her "own descendant" zero generations hence. An individual's τ-generation fitness is the sum of the $(\tau - 1)$-generation fitnesses of her offspring, i.e., $w_j(\tau) = \sum_i w_i(\tau - 1)g_{ij}$, or

$$w^T(\tau) = w^T (\tau = 1)\mathbf{G} = w^T(0)\mathbf{G}^T, \qquad (3)$$

where $w(\tau)$ is a vector of τ-generation fitnesses, and $w(0)$ is a vector with each element equal to unity. This equation is analogous to (2), and implies that $w(\tau)$ will converge to w, the dominant row eigenvector of $\mathbf{G}$, as $\tau \to \infty$. This shows that w can be interpreted as a vector of "relative long-term fitnesses." The ratio $w_i(\tau)/w_j(\tau)$ converges to $w_i(\tau)/w_j$ as τ increases. This justifies (3), the third and final claim about the properties of $\mathbf{G}$.

15

More Status or More Children? Social Status, Fertility Reduction, and Long-Term Fitness

James L. Boone and Karen L. Kessler

Over the past two decades, there been several general kinds of approaches to explaining reduced fertility in modern industrialized populations from an evolutionary perspective and, more specifically, why socioeconomic status currently seems to be negatively correlated with lifetime reproductive success in modern Western contexts (Borgerhoff Mulder 1998). One is that reduced fertility is simply a maladaptive strategy that plays out in a novel environment characterized by high levels of production, which allows for unprecedented accumulations of wealth, such that wealth is no longer consistently channeled into the production of correspondingly large numbers of offspring (see, for example, Hill 1984; Vining 1986).

A variant of the novel environment hypothesis, and one that isolates specific proximal mechanisms for reduced fertility in a more rigorous manner, has been presented by Kaplan (1996), who sees contemporary fertility reduction associated with the demographic transition as a response to competitive wage labor markets that characterize modern industrial and postindustrial societies. In Kaplan's view, as competition for high-paying jobs increases, parents are motivated to invest more and more in education and job training, particularly when increased levels of training result in higher incomes for offspring. Under conditions of continuing intergenerational reinvestment of dividends from investment in offsprings' earning power—what Kaplan refers to as embodied capital—parents are predicted to reduce numbers of offspring to a lower limit of two. But because offsprings' earning power is never actually converted into higher lifetime reproductive success, it appears that this strategy of trading off offspring quality (measured in terms of earning power) against quantity does not necessarily lead to higher fitness in the long term. Rather, fertility reduction

Reprinted from *Evolution and Human Behavior* 20. James L. Boone and Karen L. Kessler, pp. 257-277, Copyright © 1999, with permission from Elsevier.

is seen as a response to a novel environment—the competitive wage economy of industrialized society—in which the costs of producing high-quality offspring are atypically high.

Rogers (1990b, 1991, 1995) and Harpending and Rogers (1990, 1991) have developed a quantity versus quality approach that focuses on the idea that, in human social systems, wealth is often heritable. Rogers (1991) argues that "by leaving large bequests to offspring, a human parent may substantially increase the expected wealth of grandchildren" (p. 481), which, in turn, might lead to higher levels of reproductive success in succeeding generations. Hence, in stratified societies, parents may be faced with a tradeoff between quantity—the total number of offspring—and quality, defined as the amount of wealth that can be bequested to each offspring. Rogers has developed an elegant approach to analyze the intergenerational effects of trading off quantity against quality. This model has produced mixed results. In the one-sex (clonal) model assuming density-independent population growth, wealth was a better predictor of second- and third-generation reproductive success than was overall fertility (Rogers 1991). In a later two-sex model, Rogers (1995) was unable to find conditions under which reduced fertility would result in greater long-term fitness. Harpending and Rogers (1990, 1991) considered the possibility that under conditions of density-dependent population regulation, such as might exist in a stratified society in which the lowest social stratum constitutes a demographic sink, a strategy that favored offspring quality over quantity might result in higher long-term fitness. We will return to this last issue later in this article.

One of the problems with understanding the relationship between social status and fertility has always been that, in virtually all previous formulations, socioeconomic status is analyzed not in terms of an ongoing strategy, but rather as an *outcome* of a strategy. Typically, it is defined in terms of a variable such as annual income or total accumulated wealth. In optimal foraging theory terms, it is as if annual income or accumulated wealth constituted a *net return rate*—which is then applied in its entirety to the production and rearing of offspring. Hence, these approaches attempt to couch the fitness benefits of social status in terms of increased access to resources, wherein resources are converted into higher lifetime reproductive success. Even approaches that have attempted to look at long-term fitness, such as that of Rogers (1995) (discussed earlier), have not really departed from this basic position. For example, the logic behind Rogers's bequesting model is that parents might restrict the number of offspring in one generation in order to conserve resources that can be passed down to offspring, such that their descendants might convert such resources into more offspring in a future generation. Such analyses have provided valuable insights into the cost-of-children end of the tradeoff. In the end, however, it is clear that parents do not invest everything they have into the rearing of offspring—expenditures associated with status reinforcement need to be accounted for as well (Boone 1998). To do this, we need to analyze social status striving as an ongoing strategy with both costs and benefits, not just as the outcome of a strategy.

In this article, we will explore the possibility that lower than expected fertility of humans can be explained as part of an evolved strategy to maximize long-term fitness in the face of relatively infrequent, but severe, calamities that result in significant demographic crashes. Three conditions must be met for this model to be plausible: (1) human population history has been characterized by local periods of growth punctuated by recurrent crashes caused by calamities such as climatically induced resource shortfalls; (2) a strategy is available to individuals that increases the probability of survival through a crash, but that, to implement, requires diverting resources away from producing more offspring; and (3) long-term fitness benefits to increased survivorship through a crisis must outweigh or equal the fitness benefits that would accrue to putting the same resources into higher fertility during periods of growth. We present a model that shows that increases in survivorship can outweigh the benefits of higher fertility even if crises are neither very frequent nor particularly severe.

The idea that a lower mean number of offspring can constitute a tradeoff against high variance in a fluctuating environment is not a new one. Field studies of birds have shown that the fitness benefits of reduced reproduction may only be apparent during infrequent bad years (Stearns 1992:155). For example, the study by Tutor (1962) (Daly and Wilson 1983:27) of boat-tailed grackles in Texas showed that, during a good breeding season, the average number of fledglings per nest increased monotonically with clutch size, peaking with the maximum clutch size of five. Yet the modal clutch size observed in all years was only three. In a subsequent bad year characterized by severe flooding, larger clutch sizes actually produced fewer fledglings, and the average number of fledglings per nest peaked at a lower clutch size of three. Hence, what appeared to be a less than optimal clutch size in good years was, in fact, the most successful clutch size when unpredictable climatic factors affecting rearing of hatchlings was taken into account. For example, in the grackle example just cited, reducing clutch size was sufficient to increase survivorship of fledglings through a bad year. Larger clutches show greater variance in fitness because they suffer lower reproductive success (i.e., numbers of fledglings) during bad years than do individuals laying smaller clutches. In this case, stabilizing selection appears to produce a modal, optimal clutch size, in which higher clutch size is weeded out during bad years.

This is the logic behind what are called *bet-hedging strategies,* wherein higher clutch size is traded off against lower variance in the number of successful fledglings (Boyce and Perrins 1987; Gillespie 1977; Winterhalder and Leslie unpublished). In our version of the bet-hedging model, the number of offspring produced is not itself the key variable affecting survival through a crash, but rather the effect of a continuous, on-going tradeoff between the energetic costs of a higher survivorship strategy—social status maintenance—and the number of offspring produced, as well as with other components of fitness.

Catastrophic Population Dynamics

Classic models of logistic population growth assume that animal populations grow until they reach a stable state at or near K. To the extent that competition is a force shaping the evolutionary process, Darwin's initial formulation of the theory of evolution by natural selection assumed that populations are usually stable and at or near K. Under such conditions, individuals with the highest number of surviving offspring (i.e., the highest lifetime reproductive success) are favored by selection. Recent empirical work in population ecology, however, has shown that many animal species, including large mammals, are characterized by periods of growth interrupted by relatively infrequent crashes brought about by starvation, disease, and other environmental factors that may be, to varying extents, density independent (Dunbar 1987:77; Mangel and Tier 1993; Young 1993). For example, Young (1993) surveyed a series of 80 documented natural die-offs in large mammals, including herbivores, carnivores, and primates, which were characterized by peak-to-trough reductions in population of at least 25%. Within the sample, Young found that the modal degree of population reduction during crashes fell in the range from 70% to 90%, although he suggested that this may be partly due to underreporting of die-offs of lesser severity. There was also a pronounced paucity of die-offs involving reductions of greater than 90%, which Young suggested may be due to the fact that natural selection has eliminated populations susceptible to total extinction (p. 414). Interestingly, all three classes of mammals (herbivores, carnivores, and primates) were equally represented in the 70% to 90% range, although carnivore die-offs were more likely to be caused by disease, whereas herbivore and primate die-offs were more likely to be caused by habitat decline and starvation.

Our interest in this phenomenon focuses on the issue of whether it is possible that natural selection can favor state-dependent strategies (McNamara and Houston 1992) that increase an individual's or lineage's probability of survival through a crash relative to others in a population. Selective pressures for such strategies may be particularly strong in species such as hominids, which are characterized by long lifespans (which increases the probability of actually experiencing a disaster during the lifetime), and extended juvenile dependency periods with high levels of parental investment (which would make reproductive failure during a crisis relatively costly).

The question then arises as to what extent human populations have been subject to periodic die-offs in their evolutionary history. Several lines of indirect evidence lead to the conclusion that infrequent, but severe, population crashes may have been a chronic condition of human population history. First of all, it is difficult to reconcile the high rates of growth (and the potentially high reproductive capacity) observed in modern human populations, including modern hunter-gatherers, with the extremely low overall rates of growth estimated throughout most of human history unless one argues that periods of relatively high growth were counterbalanced by periodic population crashes. Hassan (1982:254), using the geometric growth equation in connection with plausible estimates of population sizes at different points in the European and Middle Eastern prehistoric sequence, has estimated that, almost, overall growth rates up until the Neolithic transition ranged between 0.00007% and 0.011% per year. Yet even the modern !Kung, who live in a relatively marginal foraging environment, have a calculated growth rate of 0.7% ($r = .007$), whereas a tropical foraging group, the Ache (Paraguay), sustained a precontact growth rate of 2.5% ($r = .025$) (Keckler, 1997). The problem of the lack of fit between estimated growth rates in prehistory and observed growth rates in modern hunter-gatherers can be more clearly visualized by taking the !Kung growth figure of $r = .007$ and applying it to a starting population of 10,000; such a population would grow to the present population of the Earth in 1,849 years.

A second line of evidence for population bottlenecks in human population history comes from mortality profiles of prehistoric skeletal populations (Keckler 1997). Such populations often exhibit a high proportion of prime adult deaths, resulting in a flattened mortality profile, compared to the usual U- or J-shaped profile (with higher mortality among the young and old) that typically characterizes mammal populations. These flattened morality profiles do not easily fit any known model life table. Using a series of simulations using life table characteristics of two hunter-gatherer populations, the !Kung and the Ache, as well as the Coale and Demeny West Model 1 life-tables, Keckler (1997) showed that by adding random population crashes involving reductions ranging from 10% to 54% at a mean rate of one every 30 years, he could generally produce both the near-zero overall population growth rates estimated by Hassan as well as the flattened mortality profiles observed in prehistoric skeletal populations. Keckler concluded that human population growth in the past is likely to have been characterized by what he refers to as a biphasic pattern: relatively long periods of growth counterbalanced by short intervals of catastrophic decline.

Finally, Belovsky (1988:343) and Winterhalder et al. (1988:309), have independently presented optimal foraging theory based models of hunter-gatherer population growth that predict stable limit cycles of growth and crash with a periodicity of about 50 years under varying conditions. These are predator-prey density-dependent models and therefore differ from models that assume crashes occur as the result of density-independent factors.

Social Status and Differential Survival

The second element in our model posits that a strategy is available to individuals that increases the probability of survival through a crash relative to others in a population, but which is energetically costly to implement. Here we argue that in ranked and stratified human societies, social status is an important factor determining the probability of lineage survival through crashes. Such might be the case where a population is differentiated in terms of priority of access to resources or other benefits produced within the group that affect survival, maintenance, and reproduction. In times of shortage, lower-ranking families will be the first to be adversely affected; higher-ranking families would be affected last. High social status is clearly costly to acquire and maintain: the lavish sumptuary displays associated with status competition in ranked and stratified middle-range societies are well documented in the ethnographic literature. Much less common are documented cases of how social status plays out in terms of survival during calamities such as famines.

A particularly stark example of the effect of status on survival is provided by Sacks (1996:36-37), who describes the effects of a typhoon that swept the Micronesian atoll of Pingelap in the Caroline Islands in 1775. Of a total population of nearly 1,000, 90% were killed outright in the storm, and most of the survivors starved to death within a few weeks. In the end, only 20 or so survivors were left, including the hereditary chief and members of his household. Here we present further ethnographic and historic examples that show that social status is an important factor in determining who survives through catastrophic food shortages and other kinds of disasters.

Tikopia

Tikopia is a small, isolated Polynesian outlier consisting of approximately 4.6 km² of land and 2 km² of reef located in the Santa Cruz Islands group, about 12.5 degrees south of the equator. Its nearest neighbor is the even smaller island of Anuta, 112 km across the open ocean. In 1929, Raymond Firth recorded a population of 1,281 people, or approximately 400 persons per square mile. The population of Tikopia was divided into four patrilineal clans, each with its own clan chief, or *ariki*. The population was further divided into two general classes: chiefly families and commoner families (Firth 1965:187-236). Although theoretically the clan chiefs had ultimate authority and ownership of clan lands, all families had relatively permanent use rights to individual plots of land consisting of orchards and other prepared fields within the clan lands. Fishing beds around the shores of the island were not individually owned, but rights to fish them were closely managed by the clan chiefs.

An extensive ethnographic study of Tikopia was carried out in 1928-1929 by Raymond Firth (1957, 1965). In 1952, Firth and an assistant, James Spillius, revisited the island shortly after a hurricane had devastated the island's crops.

Coconut and breadfruit trees were flattened, and taro and other root crops were ruined by inundation by salt water and salt spray. A second hurricane the following year brought further devastation, and the result was a severe famine that lasted more than a year. Firth (1959) and Spillius (1957) provide about the most extensive documentation of the sociopolitical and demographic effects of a natural disaster on a small-scale society that is available in the ethnographic literature.

One of the most arresting aspects of the social dynamics that ensued during the crisis was the privileged status of the chiefs and their families in terms of their "right to survive" through the disaster. Spillius writes:

> What was communicated explicitly here was that the chiefly families would be the last to die. The implication was that special groups, namely commoners, would have to be sacrificed to guarantee the survival of the chiefs and the *maru* [the chiefs' overseer-enforcers, usually their brothers or close relatives]. In fact, everyone, commoners included, felt that it would be unthinkable for the chiefs to die. Theoretically, the chiefly families own the land and there is a mystical connection between the well-being of the chiefs and the well-being of the land. But in spite of consensus on the necessity of preserving the chiefs, the commoners had no intention of sacrificing themselves. (Spillius, 1957:13)

Be that as it may, Spillius points out that some *maru* "suggested that those who had no food should not steal but should sit in their houses and die, and others suggested that if people had no food, then there might be a 'land for them beyond the sea,' and they should take to their canoes" (Spillius 1957:13). In fact, as the famine grew worse and discord on the island increased, caused largely by widespread theft of food from gardens, five canoes full of people attempted to leave the island on *forau,* or suicide voyages, although they were eventually persuaded to return. During the 3-month long peak famine period, the death rate rose to about three persons per week due to excess deaths from starvation, or four times the normal (Firth 1959:62), even though nearly 20 tons of rice and other relief foods were brought in from outside.

Unfortunately, we are not given a breakdown of the survivorship by social status, which in Tikopian society was determined largely by genealogical position within a descent group. But there is ample evidence that chiefly households had a great deal more social and political power than the commoners with respect to accessing critical resources, and good reason to believe that in times of famine, chiefly households would have priority of access to whatever food resources were left. Indeed, the clan chiefs had the power to banish individuals who were reduced to stealing food from chiefly gardens. Chiefly houses were generally distinguished by having a great deal more authority in everyday life. Chiefs also could intercede in feuds or disputes over land. Clan chiefs and ritual elders held titular ownership of the few freshwater springs that run on the island, thus controlling both drinking water and water for irrigation during droughts (Kirch and Yen 1982:18). The chiefs could impose *tapus* on the planting or harvesting of various crops and in this manner could control which plots could be planted or left to fallow. In this sense, the clan chiefs could control how much land was under production at any given time and prevent overexploitation of productive lands. Tapus also could be imposed against fishing along designated portions of the reef for indeterminate periods of time, probably in response to overfishing.

Members of chiefly houses could give orders, but did not take them from commoners. A chiefly member might strike a commoner with impunity, but if a commoner struck a chiefly house member, he might be forced to undertake a *forau,* or suicide voyage, possibly along with the rest of his household, unless a chief interceded with an invitation to remain. Banishment through forced voyage also could be a punishment meted out by a clan chief for crimes such as persistent theft, insulting the chief or one of his close relatives, engaging in an intrigue with the chief s daughter, or disregarding the chief's or his family's welfare in a crisis (Kirch and Yen 1982:135). In the 1952-1953 famine, theft from gardens became rampant and created a great deal of discord on the island. Anxiety over impending banishments led one individual to complain:

> I who sit here, look on my children, who are starving. I get my canoe ready to go. If a chief comes to block me I say "Shall I stay here to steal from your orchards? Or are you prepared to feed me?" Then he is silent and lets me go." (Firth 1959:66).

Hopi

Another example of how lineage status can affect survivorship through famine is found among the 19th century Hopi, a population of maize cultivators living in northeastern Arizona. Although the Hopi often are considered to be a good example of an "egalitarian" society, Hopi matrilineal clans were, in fact, ranked in terms of the size and quality of cultivable fields located on alluvial fans watered by Spring runoff (Levy 1992). High-ranking clans held the most extensive and well-watered fields, whereas low-ranking clans often had no land at all and were forced to farm common fields controlled by the village chief, who was typically the brother of the matriarch of the top-ranking clan. Furthermore, within each clan there was a distinction between a "blue-blood" primary line and secondary matrilineages, which also affected access to farming lands and grain stores.

High-ranking clans were responsible for underwriting elaborate katsina dances and other ceremonials that were energetically costly and time consuming to prepare. The capacity to underwrite these ceremonials appears to have been an important element in a clan's justification for greater access to cultivable lands (Levy 1992:25, 46). Droughts were a significant cause of shortages of stored maize; sharing of stored maize during shortages occurred almost exclusively within clans. In a recent reassessment of Hopi social structure, Levy (1992:56) argues that "a restricted and tenuous resource base required that Hopi society structured itself on an inequitable distribution of land" such that methods were "devised to 'preserve the core' of the land controlling descent groups by sloughing off the excess population in an orderly manner during times of scarcity." Similarly, Eggan (1966:125) remarks that "in case of drought, all resources are concentrated for the preservation of the central clan core, and other clansmen may be forced to migrate or starve." For example, in the mid-18th century, 40 Hopi families were forced to leave their village during an extreme drought and attempted to take refuge among the Navajo in the Canyon de Chelly area. In this case, the Hopi men were killed and the women and children taken in (Levy 1992:108).

Tuareg

Yet another example of the relationship between social status and survival through famine is found among the Tuareg, a stratified pastoralist population in Saharan West Africa. Baier and Lovejoy (1977) have argued that the intricately graded system of status differentiation among the Tuareg and their dependents, which had previously been argued to have little political significance, actually operates to determine priority of access to remaining resources during periods of extreme famine caused by droughts: "[social status distinctions]—which had little importance in prosperous times—came into play in periods of scarcity, when they provided a pattern, for sloughing off excess population. The social system offered a clearly delineated blueprint of the order of precedence, from nobles down to *iklan* [slaves]" (Baier and Lovejoy 1977:404).

These examples are of relatively small-scale ranked and stratified societies in which priority of access to resources is justified largely by rules of historical precedence and genealogical ties to putative founders, or "first-comers," and in some cases descent from mythological or supernatural beings. For example, Eggan (1966:124-125) argued that the position of Hopi clans within villages was established, at least theoretically, by the order of arrival of the group into the area and maintained by the possession and use in periodic ceremonies of ceremonial objects and clan fetishes:

> 26 at the head of the prestige hierarchy is the Bear Clan, the members of which arrived first in the Hopi region and made a compact with Masu'u, the god of life and death, in which he gave the Hopi lands and crops in exchange for carrying out the proper rituals. Late-comers received portions of this estate in exchange for the performance of ceremonies for rain for the crops, but the "last" arrivals often possessed no rituals and offered their services as guards. Their position was marginal, and usually they were not assigned clan lands.

Eggan's discussion emphasizes the fact that the ability to engage in energetically costly ceremonies and rituals—what we discuss in more detail later as "social advertising"—was a key factor in justifying and maintaining a clan's position in the group. Although the Hopi origin myth defines "late-comers" in terms of an in-migration process, the marginalization of secondary lineages might just as easily result from intrinsic growth through the operation of some kind of rule of primogeniture, as is the case in Polynesian social structure (Boone 1992b:334).

Differential Survival in Complex Societies

Genealogical ties and rules of historical precedence continue to be important factors determining priority of access to resources in complex societies, but in stratified societies characterized by wage labor and a developed market system, market forces during periods of scarcity typically also will determine who will survive a famine. When a shortage presents itself, scarcity drives food prices up past what the poorest can afford, and starvation ensues. This appears to be one of the most important factors affecting differential mortality in famines in social systems operating on a cash market economy (Sen 1981). The effects of differential access to resources on survival is exemplified in the mortality data presented in Table 15.1, collected during a famine that occurred in Bengal in 1943. Although Table 15.1 does not tell us much about the actual social status of the occupational groups in this sample, it does show that there

Table 15.1
Destititution and Mortality Rates by Occupation During the
1943 Bengal Famine, Ordered by Percent Mortality
(Based on Five Surveyed Villages in Faridpur, Bengal, India)

Occupation	Percentage destituted	Precentage "wiped off" (died)
Landlord	0.0	0.0
Crop-sharing landlord	6.3	0.0
Office employee	10.0	0.0
Peasant sharecropper	18.4	6.4
Artisan	35.0	10.0
Petty trader	31.8	14.0
"Unproductive"	44.4	16.7
Priest and petty employee	27.3	27.3
Agricultural labor	52.4	40.3
Total	28.5	15.2

Source: Sen 1981:74.

was significant differential survival among the groups, which seems to have been determined by either access to basic food resources or more-or-less drought-resistant sources of revenue. That is, individuals with direct access to lands (landowners and share-croppers) and local officials (office employees) had the lowest mortality rates.

Although famines are a very common cause of local crashes in which mortality rates are differentiated by social status, other kinds of natural disasters may have similar effects. The Titanic disaster is a somewhat idiosyncratic case, but is graphically illustrative of the effects of social status on survivorship in a disaster. In the case of political disasters and warfare, wealth and social prestige often allow high-status individuals to buy protection or to escape an area of conflict altogether.

[This is a shortened version of the original and here some text was omitted.]

Status Competition and Fertility Reduction

This model (omitted; please refer to original) shows that, under plausible conditions, a strategy that increases the probability of survival through infrequent crashes could be selected for even if it entailed reduction of fertility in relation to others in the group with a lower probability of survivorship. Put another way, this model captures the specific environmental conditions under which reduced fertility could be adaptive. These conditions include the existence of demographic bottlenecks of sufficient frequency and severity through which there is variation in the population with regard to probability of survival through the crash. Here we discuss in more detail the nature of the tradeoff between social status maintenance and fertility.

Because an individual or family's energy budget is finite, it may be that under conditions of high levels of status competition, the costs of status maintenance are high enough to compromise the ability to produce more offspring i.e., there is a tradeoff between status maintenance and numbers of offspring. One of the implications of this viewpoint on the relationship between social status and fertility is that high status families may have *more* or *fewer* offspring than lower status families, depending on the costs and benefits of status reinforcement relative to the costs and benefits of having more offspring in a given ecological context.

Although social status is often conceptualized as a characteristic or quality that a particular individual possesses, it is perhaps better understood as a quality of an individual that resides in the perceptions of others in a social group and their resultant behavior toward that individual. What a high-status individual actually possesses is what might be called *social power,* which we define here as the ongoing capacity to extract or defend a disproportion of the fitness affecting benefits in the form of goods and services produced collectively by members of a social group. Boone (1998) argued elsewhere that social power relates to two classes of capacities: (1) the capacity to impose costs on potential usurpers, and (2) the capacity to dispense benefits to potential allies, supporters, clients, etc. When others in a social group perceive the level of these capacities, they will behave accordingly with regard to the way negotiations over the way resources and other benefits acquired, produced, or defended within the group are allocated.

This way of thinking about social status allows us to see that there are really two running accounts in the energy budget of social status reinforcement behavior. First, individuals may invest time and resources into increasing social power—i.e., the capacity to impose costs and dispense benefits, as well as in specific skills, education, training—qualities that Kaplan (1996) has referred to as embodied capital. But possession of these qualities is often useless in and of itself unless others in the social group perceive how much power the individual in question actually has and, as a result, act accordingly in ways that benefit that individual. Hence, a second running account in the status-striving energy budget consists of a social advertising budget. In highly competitive contexts, social advertising can become particularly costly when the advertising takes the form of a strategic handicap. That is, because advertisers will always benefit from "consumers" perceiving their power to be greater than it actually is, while consumers will benefit from having as accurate an estimate as possible of the advertiser's power, a conflict of interest over the reliability of the advertisement invariably arises. This conflict can be resolved by the evolution of costly signals—advertisements so costly that they can't be faked without a net loss to the advertiser's fitness (Boone 1998; Frank 1988:96113; Grafen 1990a; Zahavi and Zahavi 1997).

To summarize, the costs incurred in status striving fall into two categories: (1) time and energy invested in increasing social power, and (2) time and energy invested in advertising social power. We turn now to the issue of the economic benefits of social status striving. There is good reason to believe that acquiring status can lead to higher rates of gain in ongoing economic transactions that occur within a social group. For example, Ball and Eckel (1996, 1998) have shown that acquiring status puts individuals in a stronger negotiating position in simple bargaining games. Specifically, in ultimatum games where some players were artificially furnished with higher status, Ball and Eckel showed that players were more likely to accept an uneven split from individuals who had been given the higher status distinction. This is an interesting result because it seems to capture the essence of the value inherent in gaining higher social status, and it suggests that high-status individuals may enjoy an ongoing economic benefit in terms of their ability to extract a disproportion of the resources and other benefits produced collectively within a social group. However, we must now ask: what happens when individuals have to pay the costs of acquiring, maintaining, and advertising social status themselves? How much should they be willing to pay for it?

Status reinforcement appears to be a state-dependent strategy (McNamara and Houston 1992, 1996), which means that how much an individual or family can invest in status versus other components of fitness such as reproduction will depend on their current budget and the benefits of investing in status relative to other activities. Status is also a positional good (Frank 1985), which means that the value of status is always relative to others in the social group. If status has some value and is in limited supply, it stands to reason that the absolute costs of status acquisition and maintenance should increase with social status itself—that is, higher status individual pay more for their positions than lower status individuals. Whether the amount invested in status, *proportional* to what is invested in other components of fitness such as reproduction, increases, decreases, or remains constant as social status increases is an interesting question for future empirical research. There appear to be at least three main possibilities, which we outline here.

If, as suggested earlier, the value of social status relates solely to increasing rates of gain that are a function of relative position within the social group, then the *net* economic benefit to status (i.e., the total rate of gain obtained as a function of status minus the costs of acquiring and maintaining it) should always increase with status. Under these conditions, the amount paid for status *proportional* to other expenditures such as somatic maintenance and reproduction (in this case, including the amount invested in passing status onto offspring) might well be the same for all individuals in the population. In other words, investment in status might act a lot like energetic investment in somatic maintenance, and the increasing absolute costs of status reinforcement would simply be a scalar phenomenon. Under this scenario, because the net economic returns to acquiring and maintaining status could then be invested in activities such as reproduction, there would be no reason to expect that higher-ranking families would ever have to sacrifice fertility for status. The complicating factor is, of course, that if status is worth having as a parent, it is worth passing on to children as well, which itself will be costly. Under this view then, the problem of fertility reduction in relation to social status might simply revert to a cost-of-children issue.

The second possibility, which we suggested in the long-term differential survival hypothesis, is that status has, or had in past selective environments, some significant value to fitness above and beyond the continuous net economic benefits it brings to ongoing economic transactions in a social group. Under this view, status reinforcement behavior would constitute a distinct component of fitness and would not simply act as an engine to produce more or better children through greater access to resources. If this were true, then it would be worth paying more for status than one might expect if its primary function were tied to increased short-term net rates of gain. We might expect to see humans behave as if they desired wealth because it helps them buy more status, rather than desiring status because it generates more wealth, and they should behave as if they were willing to sacrifice at least some lifetime reproductive success to acquire or maintain more status (see discussion of "status anxiety" later). Under the long-term survival scenario, the costs of status *proportional* to an individual's or family's total budget should vary across a given status distribution in a social group. For example, the amount paid for status proportional to the total budget might first increase and then perhaps decrease again as one moves up the social hierarchy. If this were so, then we might expect fertility to first decrease with status, and then increase again at the very top of the social hierarchy, at least in situations where the costs of maintaining and advertising status have become very great.

The third possibility is that the long-term survival benefit associated with social status is a predictable outcome of the short-term benefit effect. That is, social power that leads to higher ongoing rates of gain in the short run translate into higher priority of access during periods of adversity. Note that in both scenarios, the value of status lies in relative position within the group. The possibility of such a dual role for social status is intuitively appealing because it suggests that status could produce higher reproductive success during periods of growth (i.e., when resources are plentiful) —including enhanced capacity to disperse and to fill empty niches left by crashes—*and* higher probability of survival through demographic bottlenecks, when resources are in short supply. Hence, status may have a dual function that corresponds to the biphasic growth and crash population dynamic discussed at the beginning of this article.

Under the dual function scenario, humans might still appear to be willing to pay more for status than it appears to be worth in terms of ongoing net economic returns for the following reason. As stated earlier, the absolute costs of status increase as status increases. Because status is positional, each time an individual of lower status is added to a group the costs of status reinforcement increase for each individual above them—that is, additions below create a domino effect caused when each higher status individual defends his position against the individual just below him. When the total range of social statuses begins to approach that of traditional and modern complex societies, the costs of maintaining social status, including the costs incurred in social advertising, can become very high indeed.

Now, as the costs of maintaining high status increase as the result of the addition of lower-status individuals to group, it may develop that higher-status individuals along part or all of the ranked distribution will begin to pay such high costs for status maintenance that they actually get less net benefit (i.e., benefits that could be invested in other components of fitness such as reproduction) than individuals below them in the ranked distribution. But because status is a positional and a nondivisible good (it has to be furnished in its entirety at any given position to have any utility at all), it may not be possible to pay a little less to adjust the net benefit back into the higher benefit region below without losing one's position altogether. If it is true that relative position also affects probability of survival through a crash, humans might behave as if the losses incurred in paying less for status and losing one's position altogether are greater than the losses sustained in maintaining status at an ongoing net economic loss relative to individuals lower in the ranked distribution. Under these conditions, humans might appear to behave as if they were committing the Concorde Fallacy (Dawkins 1976:162; i.e., the fallacy that the amount already invested in a project should determine whether to continue with it, even if prospects for future gains look dim), at least with respect to status maintenance behavior.

Selection would then favor individuals who "hang tough" and accept decreases in net economic returns to status maintenance, which may, after all, turn out to be temporary. For example, it may be that the human psyche is designed to respond to gross rates of economic gain, and to ignore or at least valuate at a different rate reductions to net return that result from paying more for social advertising. Boone (1998) has argued elsewhere that some of the costs of social advertising associated with conspicuous consumption—i.e., social advertising in the form of acquisition and display of luxury goods and special privileges—seem to "feel" like benefits rather than costs.

In any case, at the population level, the total amount of resources put into social status reinforcement relative to other activities (i.e., reproduction) should increase as the total range of social status within the social group increases. Hence, a useful way of looking at fertility reduction episodes like the demographic transition might be to consider them as a special case of the process captured in the logistic equation: as populations grow up against some finite limit in resource availability, the effects of status competition produce a dampening effect on reproductive rates at the top of the social scale, and higher mortality towards the bottom. Although our model has emphasized the effects of a situation in which there are infrequent catastrophic down-turns in the availability of resources, similar effects of social ranking on survivorship can occur in richer environments where populations are expanding faster than the productive potential of the environment. Under these circumstances, the lower social stratum of a stratified social system may act as a demographic sink. Harpending and Rogers (1990, 1991) have argued that such may have been the case in early modern Europe. It is notable that perhaps the best known case of fertility reduction among upper classes—that of the demographic transition—occurs in precisely this geographical and temporal context. There is good evidence that this pattern of fertility reduction had its origins among the landholding nobility in the 16th and 17th centuries, and that it spread "top-down" through the lower classes later on (Livi-Bacci 1979).

The mystery surrounding the demographic transition then becomes: why do individuals with more resources reduce their fertility first? We seem to be faced with at least the possibility that, in highly competitive contexts, when individuals are faced with a tradeoff between investing resources into more status (and the material trappings it entails) versus more offspring, they will prefer more status. Yet the only way in which this preference could evolve is if status positively affects some critical component of fitness *other than* number of surviving offspring or even grandoffspring.

Johansson (1987:463) has argued that fertility reduction in the early stages of the demographic transition was the outcome of high levels of "status anxiety" among elites. She defines status anxiety as "a strong commitment among married couples to the preservation of the material basis of their own high social status and to the transmission of that status to their children." Now, the idea that high-status families might sacrifice numbers of offspring to preserve or increase social status would seem antithetical to the position that the primary fitness benefit of high social status is increased lifetime reproductive success. If, however, the advantage of high rank is more closely tied to a separate component of fitness, that of long-term lineage survival through calamities, as we argued earlier, the effect of "status anxiety" is much easier to understand. The term "status anxiety" is actually quite apt and prompts a very interesting question from the evolutionary psychological point of view. If the adaptive logic behind status striving is to acquire wealth and produce as many offspring as possible from it, how then could a psychological mechanism like status anxiety become fixed, such that it actually seems to trigger reduced fertility? If, on the other hand, status anxiety is tied more closely to a fear of falling into a less secure social milieu, then it becomes a logical candidate for a psychological mechanism that negotiates the tradeoff between investing more resources in social status versus investing in more offspring.

Our argument that fertility reduction may be a generalized response to increasing status competition in complex social systems would be strengthened if it could be shown that secular declines in fertility in high status social contexts are not limited to the case of the demographic transition in Western Europe. As Davis and Daly (1997:431) have pointed out, modern Western society may not be unique with respect to secular trends in fertility reduction—Imperial Roman populations experienced what looks very much like a demographic transition as well. We expand on their point to emphasize that, as in the case of early modern Europe, fertility reduction occurred first and foremost among the wealthy classes.

There is good evidence that the Roman upper classes were limiting the number of their children through abortion, infanticide, abandonment, and contraception beginning just before the time of Christ and continuing for several centuries (Brunt 1971:140-154; Dixon 1988:71-97; Dixon 1992:119-123; Hopkins 1965; Parkin 1992:115-123). The Emperor Augustus instituted laws in 18 BC and AD 9, which were designed to reward couples who had children and to penalize those who did not. The *Ius liberorum* (the right of children), a law that, among other things granted access to political office at an earlier age to males with children and disallowed childless and celibate women and men from receiving portions of their inheritances, was offered to citizens of the city of Rome who had at least three children, to parents of four children in the rest of Italy, and parents of five children in the provinces (Brunt 1971:563-565). This suggests that Roman upper class families were in the practice of having less than three children (Brunt 1971:154). The laws were clearly aimed at the senatorial and equestrian classes, but were extended to freedmen, the descendants of whom were not reproducing fast enough to replace themselves by the time of the early Empire (Duff 1958:191). Even in the contemporary sources, the limitation of births by upper classes is seen as essentially deriving from the same motivations as those cited by Johansson (1987) in late medieval and early modern Europe, that is, the desire to maintain a high standard of living and to avoid the diminution of the family estate. Children were in any case seen as financially crippling and a burden (Dixon 1988:96); classical authors typically viewed family limitation as a function of choice and cited greed, vanity, and desire for material wealth as the cause (Dixon 1988:102). The classical historian Polybius, for example, wrote somewhat hyperbolically that the result of this pattern of family limitation was "universal childlessness and a dearth of men, the desolation of cities and failure in production, though we have not been in the grip of continuous wars or pestilences" (Polybius xxxvi; cited in Brunt 1971:141). As to the proximate cause of this fertility decline, Polybius suggested that "ostentation, the love of money, and the habits of indolence have made men unwilling to marry, or if they do, not to raise the children born, except for one or two at most out of a larger number, whom they desire to leave rich and bring up in self-indulgence."

Conclusion

We have argued that what is often passed off as an occasional perquisite of high social status—higher probability of survival through famines and other disasters that create demographic bottlenecks—may in fact be the evolutionary *raison d'être* of status striving. Under these circumstances, humans might be willing to pay more for status than it appears to be worth in terms of net economic return in the short run. Some psychological aspects of behavior associated with social striving appear to support this hypothesis, but it is difficult at this time to point to any solid empirical evidence, because aside from the work by Frank (1988) on the positional nature of status and the simple experiments described on the effects of status on the outcome of ultimatum games (Ball and Eckel 1998), virtually nothing is known about the energetics and behavioral ecology of status reinforcement behavior. In the future, empirical studies in the tradition of optimal foraging studies, which keep track of actual expenditures and economic returns to social status acquisition and maintenance behavior, should shed more light on this intriguing issue.

We thank Martin Daly, Jennifer Nerissa Davis, Henry Harpending, Monique Borgerhoff Mulder, Hillard Kaplan, Jane Lancaster, Fraser Neiman, Eric Smith, Troy Tucker, Eckart Voland, Margo Wilson, and Carla Wofsey for comments on earlier versions of this paper.

16

The Demographic Transition: Are We Any Closer to an Evolutionary Explanation?

Monique Borgerhoff Mulder

Demographic transitions normally entail two features. First, there is a radical decline in the number of offspring that parents produce, despite overall increases in the availability of resources[1]. This is particularly true of the 19th century European transition, where marital fertility halved in less than 30 years in some countries (Figure 16.1), but also characterizes some 20th century transitions in the developing world, such as in Thailand.[2] Second, rich (and often aristocratic) families reduce their fertility earlier, and often more markedly, than the rest of the population,[3] such that the positive correlations commonly found between wealth and fertility in predemographic populations disappear: censuses from England and Wales in 1911 (Ref. 4) demonstrate this point (Table 16.1).

The fact that people in an increasing number of societies worldwide voluntarily reproduce at lower levels than would apparently maximize their lifetime reproduction poses a major challenge to evolutionary anthropologists. Indeed, the sociologist Daniel Vining identified this puzzling trend as the ultimate challenge to evolutionary approaches to human behaviour.[4] In some senses, this view is naive: many social scientists fail to appreciate that natural selection favours different optimal allocations of effort to reproduction in different ecological and social environments and hence that lower-than-maximum levels of fertility might be optimal. Nevertheless, behavioural ecologists are puzzled by the dramatic nature of the decline. They are also deeply puzzled by the emerging negative correlations between wealth and reproduction, when evidence that the wealthy outreproduce the poor is so prevalent in predemographic transition populations.[5] Consequently, a range of hypotheses are now being explored to explain why parents with access to plentiful resources choose low fertility rates (Box 1).

Reprinted with permission from *Trends in Ecology and Evolution* vol. 13, no. 7 July 1998.

Figure 16.1

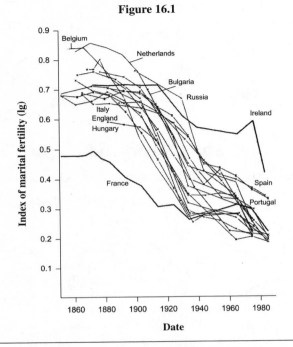

Index of Marital Fertility (I_g) in Selected National Populations of Europe. *From Ref. 1, with permission.*

Three Hypotheses for Demographic Transition

In the first hypothesis, evolutionary anthropologists suggest that, in modern societies, lowered fertility rates are optimal (with respect to fitness) because of the competitive environment in which offspring are raised—a world in which high levels of parental investment are critical to a child's success[9] and costly to the parent.[10] This hypothesis closely reflects historical demographers' recognition that the 19th century fertility decline coincided with two processes: the devaluation of child labour in cottage industry and its banning in factories, and the growing benefits associated with the education of children for employment in an increasingly competitive open-market economy.[11] It is also linked to the observation that parents in land-limited agrarian communities selectively abandoned later-born sons to reduce the number of potential inheritance claims on the estate,[12] thereby avoiding impoverishment and loss of family rank. This hypothesis poses the quantity and/or quality trade-off (first pioneered for clutch size variation by Lack[6]) that has structured much subsequent thinking in life history analysis, including work on humans.[13, 14]

Table 16.1
Surviving children per married couple (where wife's age exceeds 45 years)
classified by social status, 1911, England and Wales[a]

Social class	Surviving children per married couple
Professional	2.94
Lower white collar	3.38
Skilled manual	3.82
Semiskilled manual	3.79
Unskilled	3.88
Textiles	3.31
Coal mining	4.45
Agricultural labourers	4.57

[a] *From Ref. 4, with permission.*

Attractive as this hypothesis is, it has no direct support. Kaplan *et al.* tested its central prediction that the number of grandchildren would peak at an intermediate level of fertility.[15] Using data on contemporary men in the state of New Mexico (USA), they did not find a curvilinear relationship (Figure 16.2). Indeed, although it is well known that higher parental fertility in modern societies is linked to lower investment in individual children, the low income of children from large families does not decrease their fertility. In short, although birth rates can respond almost instantaneously to changes in the costs of children (such as was seen in the precipitous decrease in fertility in East Germany once state benefits contingent on children were withdrawn at reunification[16]), the overall fitness benefits of such adjustments remain elusive.

Until recently, attempts to support this hypothesis analytically have also failed. Investigators modelling the competitive environments in which high levels of parental investment are critical to offspring success have been unable to simulate the classic features of demographic transitions.[17] Although they can show that low fertility is favoured when the costs of raising children are high[18] and when the effects of parental investment on offspring quality diminish slowly,[19] they cannot simulate situations in which high-income groups maximize their fitness by producing fewer offspring than low-income groups.[17,20] The failure of both empiricists and modellers to support Lack's hypothesis is disappointing, as it had seemed such a good candidate.

Figure 16.2

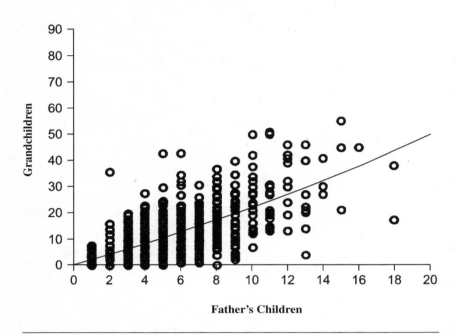

The relationship between the fertility of Albuquerque men (number of offspring produced) and their production of grandchildren. If the number of grandchildren peaks at an intermediate level of fertility (as in a tradeoff model), a second order polynomial regression should yield a positive linear term and a negative squared term. No such effect is found. The model (reduced to include only significant effects) shows that the number of children produced is by far the strongest predictor of variation in the numbers of grandchildren produced. In other words, men with the most children have the highest number of third generation descendants, contrasting sharply with the observed modal fertility of two children in this population. *From Ref. 15, with permission.*

The second hypothesis (Box 1) is that lowered rates of fertility are a consequence of darwinian, but nongenetic, mechanisms of inheritance. This view stems from the recognition that, as a result of cultural evolutionary processes, not all human traits and institutions need to be seen as serving individual reproductive interests[8]. The argument is that traits that do not necessarily enhance genetic fitness can spread through a population as a result of imitation, especially if these traits are expressed by people who are otherwise very successful. Because, in a competitive market economy, childless individuals (or people with fewer than the average number of children) are often highly successful in careers as teachers, communicators or politicians, these people serve as models for others in the population. In the model, if a suite of traits are imitated indiscriminately, low fertility might spread. Boyd and Richerson[8] have used mathematical models to show how this process (which they call indirect bias) might arise through ordinary evolutionary processes if the costs associated with reaching adaptive solutions are high.

This idea is intriguing. It may well account for the rapid spread of fertility-limiting behaviour through populations (as, for example, in Sweden[21]) and is intricately linked to the notion, now popular among demographers and social scientists, that it is changes in ideas (rather than changes in the economy) that cause fertility transitions[22]. However, the cultural evolutionary hypothesis poses problems. First, there is no clear reason why influential trendsetting individuals choose lower fertility in the first place. Granted, there will be trade-offs between seeking socioeconomic status and reproducing early and often. But why reproduction is sacrificed to such extremes still needs to be explained or at least raises questions about how such status-seeking becomes stable in a population where there may be countervailing selection for high fertility. Second, theorists working on cultural inheritance build their models with very different assumptions concerning the mechanisms of evolutionary processes than do behavioural ecologists. Such abandoning of the basic organic evolutionary model may still be premature, although the potential importance of such mechanisms is pointed to in newer work, outlined here.

The third hypothesis (Box 1) is that low fertility is maladaptive—a by-product of changes in our environment that serves no adaptive function. The commonest cause cited here is the availability of cheap and efficient birth control methods. Thus, Perusse[23] has shown that wealthier men in his Canadian sample achieve higher copulation rates than their poorer counterparts: without the availability of contraceptives, the wealthy would outreproduce their less wealthy competitors. This is an interesting finding but not a sound explanation for the demographic transition. The European transition started long before effective birth control technology was available. Furthermore, in many contemporary African nations the transition fails to occur, despite the availability of contraceptives. Finally, this perspective does not explain why some sectors of the population adopt contraceptive use before others. Most importantly, the hypothesis that low fertility is a maladaptive by-product of rapid social change is not an explanatory theory at all because it fails to specify precisely what has changed in the environment, why these changes lead to lowered fertility and even what kinds of evolved mechanisms might underlie this response.[15] The overall justification of an evolutionary approach is given in Box 2.

Box 1
Evolutionary Hypotheses for Demographic Transitions

Lowered rates are:

(1) **Optimal because of the competitive environment in which offspring are raised.** In conditions where high levels of parental investment are critical to offspring and costly to parents, parents optimize fitness by producing a few children with high levels of investment rather than many with less investment *per capita*. This hypothesis implies a trade-off between offspring quantity and quality.[6,7]

(2) **A consequence of darwinian but nongenetic mechanisms of inheritance.** Boyd and Richerson[8] suggest that traits (such as low fertility) associated with successful individuals are preferentially imitated by others in the population. Indiscriminate imitation of the traits of successful models might evolve through ordinary organic adaptive processes if the costs to an individual reaching the local optimal value for traits through experimentation and individual learning are high.

(3) **A maladaptive by-product of rapid environmental change that has no adaptive value.** Owing to the radical changes in the social, economic, political and ecological conditions that modern humans now experience, evolved mechanisms (either psychological or physiological) no longer generate appropriate responses to external conditions. As a result, maladaptive levels of fertility are observed.

Box 2
Why an Evolutionary Theory of Fertility
Transitions is Required?

A theory of fertility should specify what causes both the overall reduction in fertility and the observed relationships between independent variables (such as wealth) and fertility. It must also account for why these changes produce a demographic transition on the basis of a broader model of fertility determinants. Demography, as a discipline, has little theory of its own; accordingly, Keyfitz[24] observes that demography "has withdrawn from its borders and left a no-man's land which other disciplines have infiltrated". Evolutionary theory is one such intruder[25]. Another is economics, but recent overviews suggest that economic theories of fertility suffer from several conceptual weaknesses. Most notably, demand-orientated models ("how many children do people want?") are poorly linked to broader explanations for why humans behave as they do[26].

So what can evolutionary theory offer? Most fundamentally, it provides a justification for why organisms, including humans, are designed by natural selection to maximize fitness—economics offers no such causal closure. As regards fertility, models can be formulated of how fertility is likely to be affected by extrinsic features of the environment, in conjunction with evolved physiological and psychological mechanisms[25,27]; this work is often informed by work on other species[28]. Although it is increasingly recognized that not all aspects of human behaviour can be characterized as adaptations, variability in fertility is one trait for which simple adaptive models are likely to be valuable, at least as a starting point [Box 1 (1)], because fertility differentials are so exposed to selection. Where such models fail, evolutionary, but less explicitly adaptationist, perspectives are being investigated [Box 1 (2), (3)].

Enter Evolutionary Psychology

Perhaps because of difficulties with each of these hypotheses, attention has turned recently to the psychological mechanisms underlying reproductive behaviour. Can mechanisms be identified (and we are considering here primarily psychological mechanisms) that would have been favoured in pretransition human populations and can also account for the fertility patterns seen in modern societies?

Rogers[20] became interested in identifying the conditions in which material motivations might be selected over reproductive motivations, an important issue considering that consumption of material goods seems to compete with childbearing. His first model showed that, in environments where wealth is heritable, an individual's wealth is a better predictor of the number of second and third generation descendants produced than is fertility.[17] Under such conditions, material motivations could be selected over pure reproductive motivations through conventional darwinian processes. Unfortunately, a more realistic diploid version of the model[20] could simulate no such equilibrium.

Kaplan also focuses on psychological mechanisms.[25] He retains the assumption that, as in all organisms, human fertility is the outcome of an allocation decision between the number and the quality of offspring and, therefore, a function of the effects of resources on reproductive success. However, because optimizing fertility relative to the impact of parental investment on the reproductive outcomes of the next generation is such a difficult problem to solve, simple decision making rules ("rules of thumb") could have evolved.[15] Specifically, psychological mechanisms might have been designed by natural selection to maximize the sum of the energetic resources stored by individuals and their descendants. Kaplan and Lancaster are starting to test elements of this model using empirical data from their sample of contemporary New Mexico men, with considerable success.[29]

But what about pretransition societies? Social anthropologists and historical demographers commonly argue that parents exhibit fertility schedules that serve to protect the integrity of a family's material property. For example, primogeniture and unigeniture (whereby the complete inheritance is given to the first or last born son, respectively) have long been interpreted as strategies for preventing impoverishment and loss of family rank, and infanticide has been interpreted as a means of shaping appropriate sets of heirs.[30, 31] However, in highly pronatal communities, such as many parts of contemporary Africa (and indeed most pre-transition populations), it is generally assumed that people have as many children as they can (or at least as many as women can subject to the trade-off between offspring survival and interbirth interval[13]). A recent collaborative project uses empirical data to explore whether reproductive or material motivations better predict the marital and reproductive de-cisions of men in a highly pronatal Kenyan Kipsigis community.[32] The results (Box 3) show that material motivation is pre-eminent, even though no obvious heir-shaping strategies are evident in this society. In summary, there are now intriguing arguments and scattered seeds of evidence that a highly materialistic human psyche might be evolutionarily stable and deeply rooted in our past, and not just an aberrant outcome of modern society.

Box 3
Dynamic State Models: Do Kipsigis Men
Marry for Money or Love?

Dynamic state models[33] are increasingly used in studies of human fertility[18,34] to identify how environmental factors affect reproductive decisions. In a study of a Kenyan Kipsigis community[32], a dynamic model is used to determine the precise nature of the terminal payoff (wealth or children) individual men appear to be maximizing, as judged by their life history decisions.

A model was built of men's marital careers, with the aim of predicting when they should take subsequent wives. In the model, men are characterized by different states—their capital (livestock and land), the number of their wives and the number of their children. Their economic productivity depends on their capital assets—the labour of their wife (or wives) and various stochastic factors. Net food surpluses, after family consumption, are invested in livestock. Children are born and entail costs (illnesses, education) that are met by livestock sales. Marrying a wife is costly (the bride-wealth payment) and each year every man gets the option to choose another wife (there is no divorce in the population).

The analysis was in three steps. First, an array of terminal functions (or maximands) were constructed, based on differently weighted combinations of offspring and wealth-per-offspring (capturing a continuum of reproductive to material motivation). Second, using backward runs of the model,[33] the optimal marital career was determined for each weighting of the terminal function. This showed that, given the same initial starting conditions, men concerned with maximizing wealth per child married fewer wives than men concerned with maximizing the numbers of their children. The former also had a higher wealth threshold at which they took additional wives. The third and key step entailed comparing each set of optimal strategies (predicated on different terminal functions) with what Kipsigis men actually do. The figure shows that the best fit between the model and the observed behaviour occurs where the terminal function is heavily weighted towards accumulating a lot of wealth for each child. In other words, a Kipsigis man's behaviour seems optimally designed to accumulate material resources over the lifespan, resources that are divided among his sons at death.

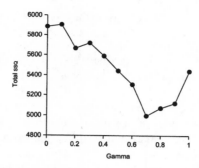

The total sum of squares (total SSQ) measures the error between the optimal strategy predicated on a particular terminal function and the observed behaviour of real men. The terminal function (gamma) represents the conflicting "motivation" to maximize children (gamma=0) or wealth per child (gamma = 1). The values of gamma are shown by filled circles and the line is interpolated for ease of viewing. Results proved robust to sensitivity analyses. *From Ref. 32, with permission.*

Closing in on an Explanation

So why, within societies, do richer people have fewer children? Several studies are closing in on an explanation for this second puzzle posed by the demographic transition.

Assuming there is a general human psychology designed to maximize the sum of the incomes of all descendants produced, Kaplan *et al.* argue that rich, highly skilled parents have fewer children than poorer, less skilled parents be-cause the time and resources the former put into their off-spring are intrinsically more valuable (i.e. produce more skills) than the time and resources invested by the latter.[15, 25] It is widely recognized that the rate at which a child learns depends on the knowledge and skills that it already possesses. If such a nonlinear effect is real, then the opportunity costs of producing an additional child are greater among the rich than they are among the poor, driving the negative or curvilinear relationships between parental wealth and fertility noted by Vining. Mace[35] models a similar process in which different strata in society optimize fitness (with different fertility levels) as a result of following different decision rules. Therefore, although wealth and fertility may be positively correlated within strata, the relationship can break down between strata. Finally, Rogers, who was until recently unable to simulate an environment in which optimal fertility decreases with wealth,[20] now reports that in environments in which inheritance greatly boosts an individual's ability to earn an income (each dollar inherited generates on average two dollars of earned income) wealthy parents can attain higher long-term fitness at equilibrium than poorer parents by producing fewer children (A.R. Rogers, unpublished).

In summary, the rich and skilled may produce fewer children than the poor and unskilled because different groups within a society vary in the learning trajectories of their off-spring, in the effects of inherited wealth on earning power, and in wealth. This outcome arises purely from mechanisms shaped by ordinary adaptive processes. Furthermore, once such fertility reductions appear among the rich, they can potentially spread to other social classes, even if the appropriate conditions do not exist in these groups, by the process of indirect bias posited by cultural evolutionists [Box 1(2)].

Conclusion

The demographic transition that has, or is, affecting so many parts of the world is a highly complex phenomenon with no single form and no common set of causes[36]. At least two phenomena need to be explained: the reduced levels of fertility despite relative material plenty; and the erosion of the widespread correlations between resources and fertility characterizing predemographic transition populations. The contribution of behavioural ecology to this perplexing issue consists of disentangling the notion of "the transition" and tackling its various components with appropriate theory. Although rapid declines in fertility may ultimately prove to be a maladaptive by-product of massive changes in the market place (because no data currently attest to long-term fitness payoffs), our explanations increasingly specify precisely what has changed in the environment and what kinds of evolved mechanisms might underlie this phenomenon. More broadly, these developments point to how evolutionary psychology, behavioural ecology and cultural inheritance models—three approaches once characterized as mutually incompatible—each stand poised to contribute to an evolutionary explanation for the complexities of human behaviour.

Acknowledgements

Thanks to Tim Caro, Sarah Hrdy and Eric Smith for comments and discussion, and to Alan Rogers for permission to cite an unpublished result.

Notes

1.   Coale, A.J. and Treadway, R. (1986).
2.   Knodel, J. et al. (1982).
3.   Livi-Bacci, M. (1986).
4.   Vining, D.R. (1986).
5.   Cronk, L. (1991b).
6.   Lack, D. (1947).
7.   Smith, C.C. and Fretwell, S.D. (1974).
8.   Boyd, R. and Richerson, P.J. (1985).
9.   Irons, W.G. (1983).
10.   Turke, P. (1989).
11.   Lesthaeghe, R. and Wilson, C. (1986).
12.   Voland, E. and Dunbar, R. (1995).
13.   Blurton Jones, N. (1987).
14.   Hill, K. and Hurtado, A.M. (1996).
15.   Kaplan, H.S. et al. (1995).
16.   Conrad, C., Lechner, M. and Werner, W. (1996).
17.   Rogers, A.R. (1990b).
18.   Beauchamp, G. (1994).
19.   Pennington, R. and Harpending, H. (1988).
20.   Rogers, A.R. (1995).

21.   Carlsson, G. (1966).
22.   Watkins, S.C. (1990).
23.   Perusse, D. (1993).
24.   Keyfitz, N. (1984).
25.   Kaplan, H. S. (1996).
26.   Robinson, W.C. (1997).
27.   Ellison, P.T. *et al.* (1993).
28.   Wasser, S.K. (1994).
29.   Kaplan, H.S. and Lancaster, J.B. (2000b).
30.   Wrigley, C. (1978).
31.   Hrdy, S.B. and Judge, D.S. (1993).
32.   Luttbeg, B., Borgerhoff Mulder, M., and Mangel, M. (2000).
33.   Mangel, M. and Clark, C.W. (1988).
34.   Mace, R. (1996).
35.   Mace, R. (2000b).
36.   Low, B.S. (1994).

17

Biophilia and the Conservation Ethic

Edward O. Wilson

Biophilia if it exists, and I believe it exists, is the innately emotional affiliation of human beings to other living organisms. Innate means hereditary and hence part of ultimate human nature. Biophilia, like other patterns of complex behavior, is likely to be mediated by rules of prepared and counter prepared learning—the tendency to learn or to resist learning certain responses as opposed to others. From the scant evidence concerning its nature, biophilia is not a single instinct but a complex of learning rules that can be teased apart and analyzed individually. The feelings molded by the learning rules fall along several emotional spectra: from attraction to aversion, from awe to indifference, from peacefulness to fear-driven anxiety.

The biophilia hypothesis goes on to hold that the multiple strands of emotional response are woven into symbols composing a large part of culture. It suggests that when human beings remove themselves from the natural environment, the biophilic learning rules are not replaced by modern versions equally well adapted to artifacts. Instead, they persist from generation to generation, atrophied and fitfully manifested in the artificial new environments into which technology has catapulted humanity. For the indefinite future more children and adults will continue, as they do now, to visit zoos than attend all major professional sports combined (at least this is so in the United States and Canada), the wealthy will continue to seek dwellings on prominences above water amidst parkland, and urban dwellers will go on dreaming of snakes for reasons they cannot explain.

Were there no evidence of biophilia at all, the hypothesis of its existence would still be compelled by pure evolutionary logic. The reason is that human history did not begin eight or ten thousand years ago with the invention of agriculture and villages. It began hundreds of thousands or millions of years

---

ago with the origin of the genus *Homo*. For more than 99 percent of human history people have lived in hunter-gatherer bands totally and intimately involved with other organisms. During this period of deep history, and still farther back, into paleohominid times, they depended on an exact learned knowledge of crucial aspects of natural history. That much is true even of chimpanzees today, who use primitive tools and have a practical knowledge of plants and animals. As language and culture expanded, humans also used living organisms of diverse kinds as a principal source of metaphor and myth. In short, the brain evolved in a biocentric world, not a machine-regulated world. It would be therefore quite extraordinary to find that all learning rules related to that world have been erased in a few thousand years, even in the tiny minority of peoples who have existed for more than one or two generations in wholly urban environments.

The significance of biophilia in human biology is potentially profound, even if it exists solely as weak learning rules. It is relevant to our thinking about nature, about the landscape, the arts, and mythopoeia, and it invites us to take a new look at environmental ethics.

How could biophilia have evolved? The likely answer is biocultural evolution, during which culture was elaborated under the influence of hereditary learning propensities while the genes prescribing the propensities were spread by natural selection in a cultural context. The learning rules can be inaugurated and fine-tuned variously by an adjustment of sensory thresholds, by a quickening or blockage of learning, and by modification of emotional responses. Charles Lumsden and I (1981, 1983, 1985) have envisioned biocultural evolution to be of a particular kind, gene-culture coevolution, which traces a spiral trajectory through time: a certain genotype makes a behavioral response more likely, the response enhances survival and reproductive fitness, the genotype consequently spreads through the population, and the behavioral response grows more frequent. Add to this to the strong general tendency of human beings to translate emotional feelings into myriad dreams and narratives, and the necessary conditions are in place to cut the historical channels of art and religious belief.

Gene-culture coevolution is a plausible explanation for the origin of biophilia. The hypothesis can be made explicit by the human relation to snakes. The sequence I envision, drawn principally from elements established by the art historian and biologist Balaji Mundkur, is this:

1.   Poisonous snakes cause sickness and death in primates and other mammals throughout the world.
2.   Old World monkeys and apes generally combine a strong natural fear of snakes with fascination for these animals and the use of vocal communication, the latter including specialized sounds in a few species, all drawing attention of the group to the presence of snakes in the near vicinity. Thus alerted, the group follows the intruders until they leave.

3.   Human beings are genetically averse to snakes. They are quick to develop fear and even full-blown phobias with very little negative reinforcement. (Other phobic elements in the natural environment include dogs, spiders, closed spaces, running water, and heights. Few modern artifacts are as effective—even those most dangerous, such as guns, knives, automobiles, and electric wires.)

4.   In a manner true to their status as Old World primates, human beings too are fascinated by snakes. They pay admission to see captive specimens in zoos. They employ snakes profusely as metaphors and weave them into stories, myth, and religious symbolism. The serpent gods of cultures they have conceived all around the world are furthermore typically ambivalent. Often semihuman in form, they are poised to inflict vengeful death but also to bestow knowledge and power.

5.   People in diverse cultures dream more about serpents than any other kind of animal, conjuring as they do so a rich medley of dread and magical power. When shamans and religious prophets report such images, they invest them with mystery and symbolic authority. In what seems to be a logical consequence, serpents are also prominent agents in mythology and religion in a majority of cultures.

Here then is the ophidian version of the biophilia hypothesis expressed in briefest form: constant exposure through evolutionary time to the malign influence of snakes, the repeated experience encoded by natural selection as a hereditary aversion and fascination, which in turn is manifested in the dreams and stories of evolving cultures. I would expect that other biophilic responses have originated more or less independently by the same means but under different selection pressures and with the involvement of different gene ensembles and brain circuitry.

This formulation is fair enough as a working hypothesis, of course, but we must also ask how such elements can be distinguished and how the general biophilia hypothesis might be tested. One mode of analysis, reported by Jared Diamond [New Guineans and Their Natural World, Chapter 8 of *The Biophilia Hypothesis*], is the correlative analysis of knowledge and attitude of peoples in diverse cultures, a research strategy designed to search for common denominators in the total human pattern of response. Another, advanced by Roger Ulrich and other psychologists, is also reported [in *The Biophilia Hypothesis*]: the precisely replicated measurement of human subjects to both attractive and aversive natural phenomena. This direct psychological approach can be made increasingly persuasive, whether for or against a biological bias, when two elements are added. The first is the measurement of heritability in the intensity of the responses to the psychological tests used. The second element is the tracing of cognitive development in children to identify key stimuli that evoke the responses, along with the ages of maximum sensitivity and learning propensity. The slithering motion of an elongate form appears to be the key stimulus producing snake aversion, for example, and preadolescence may be the most sensitive period for acquiring the aversion.

Given that humanity's relation to the natural environment is as much a part of deep history as social behavior itself, cognitive psychologists have been strangely slow to address its mental consequences. Our ignorance could be regarded as just one more blank space on the map of academic science awaiting genius and initiative, except for one important circumstance: the natural environment is disappearing. Psychologists and other scholars are obligated to consider biophilia in more urgent terms. What, they should ask, will happen to the human psyche when such a defining part of the human evolutionary experience is diminished or erased?

There is no question in my mind that the most harmful part of ongoing environmental despoliation is the loss of biodiversity. The reason is that the variety of organisms, from alleles (differing gene forms) to species, once lost, cannot be regained. If diversity is sustained in wild ecosystems, the biosphere can be recovered and used by future generations to any degree desired and with benefits literally beyond measure. To the extent it is diminished, humanity will be poorer for all generations to come. How much poorer? The following estimates give a rough idea:

- Consider first the question of the *amount* of biodiversity. The number of species of organisms on earth is unknown to the nearest order of magnitude. About 1.4 million species have been given names to date, but the actual number is likely to lie somewhere between 10 and 100 million. Among the least-known groups are the fungi, with 69,000 known species but 1.6 million thought to exist. Also poorly explored are at least 8 million end possibly tens of millions of species of arthropods in the tropical rain forests, as well as millions of invertebrate species on the vast floor of the deep sea. The true black hole of systematics, however, may be bacteria. Although roughly 4,000 species have been formally recognized, recent studies in Norway indicate the presence of 4,000 to 5,000 species among the 10 billion individual organisms found on average in each gram of forest soil, almost all new to science, and another 4,000 to 5,000 species, different from the first set and also mostly new, in an average gram of nearby marine sediments. Fossil records of marine invertebrates, African ungulates, and flowering plants indicate that on average each clade—a species and its descendants—lasts half a million to 10 million years under natural conditions. The longevity is measured from the time the ancestral form splits off from its sister species to the time of the extinction of the last descendant. It varies according to the group of organisms. Mammals for example, are shorter-lived than invertebrates.
- Bacteria contain on the order of a million nucleotide pairs in their genetic code, and more complex (eukaryotic) organisms from algae to flowering plants and mammals contain 1 to 10 billion nucleotide pairs. None has yet been completely decoded.

- Because of their great age and genetic complexity, species are exquisitely adapted to the ecosystems in which they live.
- The number of species on earth is being reduced by a rate 1,000 to 10,000 times higher than existed in prehuman times. The current removal rate of tropical rain forest, about 1.8 percent of cover each year, translates to approximately 0.5 percent of the species extirpated immediately or at least doomed to much earlier extinction than would otherwise have been the case. Most systematists with global experience believe that more than half the species of organisms on earth live in the tropical rain forests. If there are 10 million species in these habitats, a conservative estimate, the rate of loss may exceed 50,000 a year, 137 a day, 6 an hour. This rate, while horrendous, is actually the minimal estimate, based on the species/area relation alone. It does not take into account extinction due to pollution, disturbance short of clear-cutting, and the introduction of exotic species.

Other species-rich habitats, including coral reefs, river systems, lakes, and Mediterranean-type heathland, are under similar assault. When the final remnants of such habitats are destroyed in a region—the last of the ridges on a mountainside cleared, for example, or the last riffles flooded by a downstream dam—species are wiped out en masse. The first 90 percent reduction in area of a habitat lowers the species number by one-half. The final 10 percent eliminates the second half.

It is a guess, subjective but very defensible, that if the current rate of habitat alteration continues unchecked, 20 percent or more of the earth's species will disappear or be consigned to early extinction during the next thirty years. From prehistory to the present time humanity has probably already eliminated 10 or even 20 percent of the species. The number of bird species, for example, is down by an estimated 25 percent, from 12,000 to 9,000, with a disproportionate share of the losses occurring on islands. Most of the megafaunas—the largest mammals and birds—appear to have been destroyed in more remote parts of the world by the first wave of hunter-gatherers and agriculturists centuries ago. The diminution of plants and invertebrates is likely to have been much less, but studies of archaeological and other subfossil deposits are too few to make even a crude estimate. The human impact, from prehistory to the present time and projected into the next several decades, threatens to be the greatest extinction spasm since the end of the Mesozoic era 65 million years ago.

Assume, for the sake of argument, that 10 percent of the world's species that existed just before the advent of humanity are already gone and that another 20 percent are destined to vanish quickly unless drastic action is taken. The fraction lost—and it will be a great deal no matter what action is taken—cannot be replaced by evolution in any period that has meaning for the human mind. The five previous major spasms of the past 550 million years, including the end-Mesozoic, each required about 10 million years of natural evolution to restore. What humanity is doing now in a single lifetime will impoverish our descendants for all time to come. Yet critics often respond, "So what? If only half the species survive, that is still a lot of biodiversity—is it not?"

The answer most frequently urged right now by conservationists, I among them, is that the vast material wealth offered by biodiversity is at risk. Wild species are an untapped source of new pharmaceuticals, crops, fibers, pulp, petroleum substitutes, and agents for the restoration of soil and water. This argument is demonstrably true—and it certainly tends to stop anticonservation libertarians in their tracks—but it contains a dangerous practical flaw when relied upon exclusively. If species are to be judged by their potential material value, they can be priced, traded off against other sources of wealth, and—when the price is right—discarded. Yet who can judge the *ultimate* value of any particular species to humanity? Whether the species offers immediate advantage or not, no means exist to measure what benefits it will offer during future centuries of study, what scientific knowledge, or what service to the human spirit.

At last I have come to the word so hard to express: spirit. With reference to the spirit we arrive at the connection between biophilia and the environmental ethic. The great philosophical divide in moral reasoning about the remainder of life is whether or not other species have an innate right to exist. That decision rests in turn on the most fundamental question of all: whether moral values exist apart from humanity, in the same manner as mathematical laws, or whether they are idiosyncratic constructs that evolved in the human mind through natural selection. Had a species other than humans attained high intelligence and culture, it would likely have fashioned different moral values. Civilized termites, for example, would support cannibalism of the sick and injured, eschew personal reproduction, and make a sacrament of the exchange and consumption of feces. The termite spirit, in short, would have been immensely different from the human spirit—horrifying to us in fact. The constructs of moral reasoning, in this evolutionary view, are the learning rules, the propensities to acquire or to resist certain emotions and kinds of knowledge. They have evolved genetically because they confer survival and reproduction on human beings.

The first of the two alternative propositions—that species have universal and independent rights regardless of how else human beings feel about the matter—may be true. To the extent the proposition is accepted, it will certainly steel the determination of environmentalists to preserve the remainder of life. But the species-right argument alone, like the materialistic argument alone, is a dangerous play of the cards on which to risk biodiversity. The independent-rights argument, for all its directness and power, remains intuitive, aprioristic, and lacking in objective evidence. Who but humanity, it can be immediately asked, gives such rights? Where is the enabling canon written? And such rights, even if granted, are always subject to rank-ordering and relaxation. A simplistic adjuration for the right of a species to live can be answered by a simplistic call for the right of people to live. If a last section of forest needs to be cut to continue the survival of a local economy, the rights of the myriad species in the forest may be cheerfully recognized but given a lower and fatal priority.

Without attempting to resolve the issue of the innate rights of species, I will argue the necessity of a robust and richly textured anthropocentric ethic apart from the issue of rights—one based on the hereditary needs of our own species. In addition to the well-documented utilitarian potential of wild species, the diversity of life has immense aesthetic and spiritual value. The terms now to be listed will be familiar, yet the evolutionary logic is still relatively new and poorly explored. And therein lies the challenge to scientists and other scholars.

Biodiversity is the Creation. Ten million or more species are still alive, defined totally by some 10^{17} nucleotide pairs and an even more astronomical number of possible genetic recombinants, which creates the field on which evolution continues to play. Despite the fact that living organisms compose a mere ten-billionth part of the mass of earth, biodiversity is the most information-rich part of the known universe. More organization and complexity exist in a handful of soil than on the surfaces of all the other planets combined. If humanity is to have a satisfying creation myth consistent with scientific knowledge—a myth that itself seems to be an essential part of the human spirit—the narrative will draw to its conclusion in the origin of the diversity of life.

Other species are our kin. This perception is literally true in evolutionary time. All higher eukaryotic organisms, from flowering plants to insects and humanity itself, are thought to have descended from a single ancestral population that lived about 1.8 billion years ago. Single-celled eukaryotes and bacteria are linked by still more remote ancestors. All this distant kinship is stamped by a common genetic code and elementary features of cell structure. Humanity did not soft-land into the teeming biosphere like an alien from another planet. We arose from other organisms already here, whose great diversity, conducting experiment upon experiment in the production of new life-forms, eventually hit upon the human species.

The biodiversity of a country is part of its national heritage. Each country in turn possesses its own unique assemblages of plants and animals including, in almost all cases, species and races found nowhere else. These assemblages are the product of the deep history of the national territory, extending back long before the coming of man.

Biodiversity is the frontier of the future. Humanity needs a vision of an expanding and unending future. This spiritual craving cannot be satisfied by the colonization of space. The other planets are inhospitable and immensely expensive to reach. The nearest stars are so far away that voyagers would need thousands of years just to report back. The true frontier for humanity is life on earth—its exploration and the transport of knowledge about it into science, art, and practical affairs. Again, the qualities of life that validate the proposition are: 90 percent or more of the species of plants, animals, and microorganisms lack even so much as a scientific name; each of the species is immensely old by human standards and has been wonderfully molded to its environment; life around us exceeds in complexity and beauty anything else humanity is ever likely to encounter.

The manifold ways by which human beings are tied to the remainder of life are very poorly understood, crying for new scientific inquiry and a boldness of aesthetic interpretation. The portmanteau expressions "biophilia" and "biophilia hypothesis" will serve well if they do no more than call attention to psychological phenomena that rose from deep human history, that stemmed from interaction with the natural environment, and that are now quite likely resident in the genes themselves. The search is rendered more urgent by the rapid disappearance of the living part of that environment, creating a need not only for a better understanding of human nature but for a more powerful and intellectually convincing environmental ethic based upon it.

Notes

I first used the expression "biophilia" in 1984 in a book entitled by the name (*Biophilia,* Harvard University Press). In that extended essay I attempted to apply ideas of sociobiology to the environmental ethic.

The mechanism of gene-culture coevolution was proposed by Charles J. Lumsden and myself in *Genes, Mind, and Culture* (Harvard University Press, 1981), *Promethean Fire* (Harvard University Press, 1983), and "The Relation Between Biological and Cultural Evolution," *Journal of Social and* Biological structure 8(4) (October 1985): 343-359. It represents an extension of theoretical population genetics in an effort to include the principles of cognition and social psychology.

Balaji Mundkur traced the role of snakes and mythic serpents in *The Cult of the Serpent: An Interdisciplinary Survey of Its Manifestations and Origins* (State University of New York Press, 1983).

Jared Diamond's study of Melanesian attitudes toward other forms of life and Roger S. Ulrich's review of psychological research on biophilia are presented elsewhere in this volume (i.e., Kellert & Wilson 1993).

I have reviewed the measures of global biodiversity and extinction rates in greater detail in *The Diversity of Life* (Harvard University Press, 1992).

In evaluating the environmental ethic I have been aided greatly by the writings of several philosophers, including most notably Bryan Norton (*Why Preserve Natural Diversity?,* Princeton University Press, 1987), Max Oelschlaeger (*The Idea of Wilderness: From Prehistory to the Age of Ecology,* Yale University Press 1991), Holmes Rolston III (*Environmental Ethics: Duties to and Values in the Natural World,* Temple University Press, 1988), and

Peter Singer *(The Expanding Circle: Ethics and Sociobiology,* Farrar, Straus & Giroux, 1981).

18

Human Behavioral Ecology: 140 Years without Darwin is Too Long

Gordon H. Orians

Introduction

The science of ecology developed in the North Temperate Zone, a region highly modified by human activities. Nevertheless, ecologists typically chose to study what appeared to be more "natural" environments, because they believed that the study of those environments would better reveal the complex dynamics of ecological systems. In addition, many basic ecologists looked down on their colleagues who attempted to solve environmental problems, a rather surprising state of affairs given that theoretical ecology has been powerfully influenced and stimulated by applied concerns throughout its history (Kingsland, 1985).

Fortunately, emphasis on the over-inflated differences between basic and applied ecology has greatly diminished. Today, basic and applied ecologists, to the extent that this distinction still has meaning, work comfortably together and regularly exchange roles. Associated with this change has been an increased willingness of ecologists to study human-modified environments and to consider humans as components of functioning ecosystems.

This is, of course, a highly desirable development, but the study of human ecology and of humans as components of ecosystems concentrates on the effects that humans have *on environments* rather than the effects of *environments on us*. The profound influences of humans on biogeochemical cycles, species extinctions, habitat fragmentation, landscape modification, and disease ecology, to name a few of the most striking examples, make such an emphasis natural and important, indeed almost inevitable.

Nevertheless, the relative neglect of the strong influences that environments have on people should be of central concern to ecologists. Humans have strong emotional responses to living organisms and natural environments. Involvement with nature is strongly motivating to people in many ways. These powerful motivations influence how we respond to nature, how we attempt to manipulate nature, and why we care about it. Thus, a better understanding of environmental influences on human behavior, in addition to its intrinsic interest, should provide insights into why we manipulate nature as we do.

The central goal of behavioral ecology is to explain the ways in which individual organisms make decisions about habitat, shelter, food, and mates. Humans make choices among these features of the environment, but human perceptions of and responses to environments are imbedded in a complex nexus of symbolisms and cultural memories (Schama, 1995) that have no counterpart in the responses of other organisms. These complexities make it difficult to untangle the respective influences of our cultural and evolved responses to environments. Nevertheless, techniques exist that are yielding considerable progress.

Here I examine some important components of human behavioral ecology from an evolutionary perspective. I concentrate on habitat selection because a rich body of research exists on this topic and because it has been the focus of most of my own research efforts. Time does not permit me to delve into such fascinating and equally important topics as selection of food and mates. I also explore why we have been so reluctant to extend evolutionary thinking to human behavioral ecology, and I discuss some broader implications of studies of evolved human responses to nature.

Evolution of Human-Environment Interactions

The belief that contact with nature is good or beneficial to people is an ancient one. The gardens of ancient Egypt, the walled gardens of Mesopotamia, and the gardens of merchants in medieval Chinese cities indicate that for centuries people have gone to considerable lengths to maintain and enhance their contacts with nature. More recently, the belief that exposure to nature fosters psychological and physical health has formed part of the justification for providing parks and other natural settings in cities and for preserving wilderness (Schama, 1995).

These ideas were extended by E. O Wilson in his book *Biophilia* (1984), in which, in addition to supporting the view that contact with nature is valuable, he asserted that there is a partial genetic basis for the positive responses of humans to nature. Wilson did not claim that there is a hereditary program hard-wired into human brains. Rather, he suggested that our responses and learned reactions to nature are biased in certain directions by our evolutionary history.

Why human responses to nature might have affected fitness is obvious. Our ancestors lived in environments devoid of modern comforts and conveniences.

Their survival, health, and reproductive success depended on their ability to seek and use environmental information wisely. They had to know how to interpret signals from the animate and inanimate environment and how to adjust their behavioral responses to them. They needed to understand relationships between habitats and resources and how to evaluate habitats. Such responses could, of course, develop ontogenetically via learning, but efficiencies can be achieved if an organism selectively retains certain information while ignoring or paying less attention to other types of environmental information.

Because emotional responses drive decisions, the study of emotions is a central focus of human behavioral ecology (Pugh 1977, Pulliam and Dunford 1980). Evolutionary biologists expect emotional responses to evolve in response to conditions that influence fitness. Put in the simplest terms, those of our ancestors who did not enjoy food and sex, for example, were more poorly represented genetically in future generations than those who did enjoy and, hence, seek out, food and sexual partners. Similarly, individuals who selected inferior environments in which to live should have been less well represented genetically in future generations than individuals who made better habitat choices.

Because humans, like all other organisms, use environmental information primarily to make decisions, a problem-solving perspective is a useful way of viewing our interactions with environments. This perspective suggests that, for biological as well as cultural reasons, our interactions with nature are likely to be quite complex. Fear and loathing as well as pleasure and joy characterize our experiences with the natural world because that world is a source of both resources and danger. Which resources are needed from environments change over time, and so do the abilities of environments to provide them.

The genetic bases for responses to organisms and landscapes should express themselves in what may be termed *biologically prepared learning*. As initially proposed by Seligman (1970, 1971), biologically prepared learning theory holds that evolution has predisposed humans, as well as other animal species, to learn easily and quickly, and to retain those associations or responses that foster survival when certain objects or situations are encountered (Pulliam and Dunford 1980). Moreover, even if modern societies have, as a result of large-scale transformations of environments, largely eliminated the real danger posed by the objects of fears and phobias, fear and avoidance responses may nonetheless persist because selection against those responses may be very slow (Wilson 1984).

Biologically prepared learning should be evident only for stimuli that have had significant influence on survival and reproductive success during our evolutionary history. Prepared learning theory does not postulate that adaptive responses to natural stimuli must appear spontaneously or in the absence of learning, although such spontaneity is not ruled out. Rather, it suggests that responses should appear quickly when the right condition-ing stimuli appear, and that these responses should be unusually resistant to extinction or forgetting.

Emotional responses to environments are major components of the field of aesthetics. The study of aesthetics dates back to the work of Edmund Burke (1757), but attention has been directed almost exclusively to human artistic creations rather than to nature (Hepburn 1976, Rose 1976). The neglect of nature is a by-product of the view that has dominated Western thinking for centuries—namely that cultural symbols and art forms create the aesthetic experience. Therefore, the study of aesthetics was viewed as the domain of artists and philosophers. Any attempt to explore the biophysical bases of aesthetic responses to the environment was regarded as futile and ideologi-cally dangerous (Cosgrove 1984). Although no-one could seriously doubt the strong influence of culture on the way humans see the environment and the symbolisms we attach to natural objects (Appleton 1990, Schama 1995), it would be remarkable if evolutionary biology had nothing to contribute to the subject.

A central contribution of evolutionary biology to the field of aesthetics is the recognition that *beauty* is not an intrinsic property of the objects that we call beautiful, but is rather the product of interactions between traits of objects and the human nervous system. Investigators who believe that beauty is an intrinsic property of objects seek correlations between the characteristics of objects and aesthetic responses. Such analyses are useful but an evolutionary perspective suggests that features of environments and objects in them should be viewed in functional rather than morphological terms because humans evaluate environ-ments, not necessarily consciously, in terms of the opportunities they provide for pursuing activities that contribute positively to survival and reproductive success (Appleton 1975).

Appleton identified three functional concepts—prospect, refuge, and haz-ard—upon which he built his theory of environmental aesthetics. Appleton's framework, which focuses on the initial evaluation and exploration of unfamiliar environments, has stimulated much activity among students of landscape and architecture (Jakle 1987, Hildebrand 1991). However, it does not include consideration of the types of information humans use to assess the habitability of environments or the time period over which environmental information is relevant (Orians and Heerwagen 1992, Heerwagen and Orians 1993).

The Types of Environmental Information

The environment bombards animals with a vast array of information. Some of this information is highly relevant to survival and reproductive success, but much of it is not. Moreover, which information is relevant changes over time and with the needs of the organism. To guide our thinking about responses to such complex inputs, classifications of environmental information are useful.

One classification divides information into categories based on the type of objects being identified. Some of these objects are resources, such as food, water, and protection from the physical environment. Information about these resources should motivate liking and stimulate approach. Other objects, such as potentially dangerous animals, competitors, human enemies, and hazardous physical objects (cliffs), should motivate fear and avoidance.

Another classification of environmental information is based on the time frame over which the information is relevant. Time is a continuous variable, but it is heuristically useful to divide time into four rough categories that correspond to the time frames of major types of decisions that animals make (Table 18.1). Some environmental information signals current events whose significance will soon pass. Responses to such information must be rapid and decisive (Plate 1).

Other information signals seasonal environmental changes that are associated with variations in the kinds of resources that are or will be available and where they will be found (Plate 2). This information is relevant to different kinds of decisions, and such decisions can be, and usually are, made deliberately after considerable reflection, evaluation, and discussion.

A third type of information comes from features of the environment, such as vegetation succession and changes in the courses of rivers over periods of years to decades. Finally, information arrives from relatively permanent objects in the environment. Mountains do erode and lakes fill in, but these features change very slowly with respect to human lifetimes and the durations affected by human decisions. These structures may be regarded as fixed features of the environment from the perspective of the lives of the people.

Table 18.1
Time Frames of Environmental Information

Temporal category	Examples affected	Decisions affected
Short-Term (minutes to hours)	Weather changes (thunder, clouds, wind)	Seeking shelter, initiating outdoor activities
	Appearance of dangerous animals, enemies	Immediate defensive actions
	Appearance of prey	Immediate hunting activities
	Illumination changes	Moves to appropriate locations for spending the night
Seasonal	Day-length changes, vegetation growth, flowering, precipitation changes	Shifts of hunting sites, planting and harvesting of crops
Multi-year changes (decadal)	Vegetation succession, erosional changes (river meanders, lake sedimentation)	Shifts of hunting sites, movement of villages
Long-Term (decades to centuries)	Topography	

Biophobia

I begin my overview of human emotional responses to nature with negative responses–biophobia–because its experimental underpinnings are stronger than those for positive responses. A major reason for the difference is that positive conditioning studies are usually more difficult to perform than aversive conditioning experiments. Producing an immediate, strongly positive unconditioned stimulus in the laboratory is more difficult than providing an immediate negative unconditioned stimulus.

The most powerful evidence for a genetic basis for phobias comes from research on human twins, which has yielded convincing evidence that genetic factors play major roles in a wide range of human traits, such as obesity, personality, and physiological reactivity. Recent twin studies have yielded data consistent with biologically prepared learning theory because "individual-specific" aversive or traumatic events are highly specific for generating animal phobias (Kendler et al. 1992) and fear of open spaces (Moran and Andrews 1985).

Environmental psychologists, particularly in Sweden and Norway, have carried out a series of imaginative experiments that provide interesting insights into how exposure to stimuli influence both the acquisition and retention of fears. Most of these experiments have used variations on an approach pioneered by Swedish psychologist Arne Öhman. The experiments typically involve comparisons between defensive or aversive responses that are conditioned (learned through repeated exposure or experience) to slides of fear-relevant and fear-neutral stimuli. Responses are typically assessed by recording autonomic nervous system indicators such as skin conductance and heart rate.

Initially defensive responses are conditioned by showing either fear-relevant stimuli (snakes or spiders) or neutral stimuli (geometric figures) and pairing each slide presentation with an aversive stimulus (the "unconditioned" stimulus), usually an electric shock that mimics a bite. This phase of the experiment makes it possible to compare fear-relevant and neutral stimuli with respect to the speed and magnitude of acquisition of defensive/aversive responses.

Following the initial acquisition phase, the same stimuli are presented 10-40 additional times but without reinforcement of an electric shock. This "extinction" phase allows comparison of the fear-relevant and neutral stimuli in terms of resistance to forgetting the defensive/aversive response acquired earlier. The general result of these studies is that conditioned responses are usually, but not always, acquired more quickly, but responses to snakes and spiders are always more resistant to extinction than responses to neutral stimuli (McNally 1987).

These results could, of course, be due to prior cultural reinforcement. To test this possibility, investigators have used similar conditioning experiments that exposed individuals to snakes and spiders and to far more dangerous, strongly culturally conditioned, modern stimuli such as handguns and frayed electrical wires. Conditioned aversive responses to these modern dangerous stimuli extinguished more quickly than responses to snakes and spiders (Cook, Hodes, and Lang 1986; Hugdahl and Karker 1981), a result incompatible with a social reinforcement explanation. In addition, aversive responses to fear-relevant natural stimuli can be acquired merely by telling a person that shock will be administered. Aversive responses to fear-irrelevant natural stimuli cannot be elicited in this manner (Hugdahl 1978).

In still another study, subjects were exposed to an allegedly phobic experimenter/actor who reacted fearfully to slides or fear-relevant natural stimuli (snakes, spiders, rats) or fear-irrelevant natural stimuli (berries). People acquired much more persistent defensive reactions when watching the experimenter's reactions to fear-relevant than to fear-irrelevant stimuli (Hygge and Öhman 1978). Similar results have been obtained with rhesus monkeys exposed to fear-relevant stimuli (toy snakes and crocodiles) and to fear-irrelevant stimuli (toy rabbits) (Mineka et al. 1984, Cook and Mineka 1989, 1990).

Even more striking, from an evolutionary point of view, are the results of "backmasking" experiments in which slides are displayed subliminally (15-30 milliseconds) before being "masked" by a slide of another stimulus or setting. Even though the subjects are not consciously aware of having seen the stimulus slide, presentations of natural settings that contain snakes or spiders elicit strong aversive/defensive reactions in nonphobic persons (Öhman 1986, Öhman and Soares 1993). For people already with biophobias, a masked subliminal presentation is sufficient to elicit defensive responses to the feared stimulus (Öhman and Soares 1994).

Thus, experimental results show that aversive responses can occur without recognition or awareness of specific natural threat stimuli, but no such responses exist to neutral or fear-irrelevant stimuli. These results are inconsistent with a purely cultural/learned interpretation of phobias. They strongly indicate a genetic basis for the responses. Moreover, the nature of the responses makes adaptive sense.

Habitat Selection

A crucial step in the lives of most organisms, including humans, is selection of a habitat. Habitat selection depends on the recognition of objects, sounds, and odors to which an organism responds *as if* it understood their significance for its future survival and reproductive success. Responses are initially emotional feelings that lead to rejection, exploration, or a certain use of the environment. If the strength of these responses is a key to immediate decisions about where to settle, as empirical data suggest, then the ability of a habitat to evoke these emotional states should be positively correlated with the expected survival and reproductive success of an organism in it. In other words, good habitats should evoke strong positive responses; poor habitats should evoke weak or even negative responses (Orians and Heerwagen 1992). Similarly, specific physical features of an environment or particular kinds of organisms should evoke positive or negative responses depending on the way individuals of the species interact with those features or organisms.

Habitat selection has served as a perspective for a number of studies on human responses to landscape features (Orians 1980, Heerwagen and Orians 1993). Humans confront problems of habitat assessment that are especially difficult, because human offspring are dependent on their parents for long time spans. Rarely did habitats occupied by humans during most of our evolutionary history provide reliable resources long enough to enable permanent occupation of sites. Frequent moves through the landscape were the rule, even though traditional sites might be revisited on an annual basis (Lovejoy 1981, Campbell 1985, Blumenshine 1987). From an evolutionary perspective, few generations have passed since we started to live in mechanized and urban environments. Thus, evolutionarily based response patterns of humans to landscapes are unlikely to have been substantially modified by exposure to the new environments in which we live.

Human responses to environmental cues are highly variable because the needs of people change with age and the immediate situation. A family planning a picnic may view an approaching storm very differently than a farmer whose crops are wilting because of lack of rain, even though both prefer to be in a dry place when the rain begins. A person's physiological state influences the perceived importance of water, food, or shelter. Nevertheless, positive responses to indicators of the presence of food, water, shelter, and protection from predation should be general, as should negative responses to potential hazards, such as inclement weather, fire, dangerous predators, and barriers to movement. Although no direct evidence yet exists for genetic influences on these responses, a number of studies have used selection-thinking to generate predictions of human responses to environmental features. Some of these predictions have been tested experimentally.

The Savannah Hypothesis. A variety of evidence strongly suggested that *Homo sapiens* evolved in African savannahs and only recently has invaded other continents. Because a relatively small number of generations has transpired since humans have occupied temperate habitats, it is reasonable to postulate that landscape features and tree shapes characteristic of high-quality African savannahs (Plate 3) should be especially attractive to humans today (Orians 1980). This hypothesis has been tested by determining responses of people to tree shapes and by examining the features of "aesthetic environments," that is those environments, such as parks and gardens, that are designed to be attractive places in which to spend time (Orians 1980, 1986).

Habitat selection theory suggests that tree shapes characteristic of environments that provided the highest quality resources for evolving humans should be more pleasing than shapes of trees that dominate poor quality habitats. Trees that grow in the highest quality African savannahs have canopies that

Plate 1

Environmental information that has short-term relevence: Dangerous mammals require immediate responses.

Plate 2

Environmental information that has short-term relevence: Flowers signal seasonal changes in resource availability.

Plate 3

Modern African savannah with characteristic Acacia tortilis, surrounded by shrubs on a matrix of herbaceous steppe plants.

Plate 4

Formal European garden: Villa Gamberaia, Settignano, Italy.

Plate 5

Formal European garden: Vaux-le-Vicomte, France.

Plate 6

Advertisers use landscapes to help sell their products.

Figure 18.1

One of the pages of photos of *Acacia tortilis* used by Heerwagen and Orians to test features of trees regarded as attractive. Trees that bifurcate close to the ground were preferred over those trunks split higher up.

are broader than they are tall, trunks that bifurcate close to the ground, and layered canopies (Figure 18.1). People on three continents preferred Kenyan *Acacia tortilis* trees that had highly or moderately layered canopies, lower

Table 18.2
Changes Made by Humphrey Repton in His Designed Landscapes ($N = 18$)

Vegetation Changes	No. landscapes changed
Added scattered trees and copses	9
Removed trees to open distant view	8
Added copses at water's edge	5
Broke up straight pasture/wood edge	5
Opened woods to allow penetration and visual access	4
Changed tree shapes	6
Bifurcated trunks close to ground	(4)
Added canopy layering	(2)
Added flowers and shrubs	2
Added woods	2
Removed shrubs	2

Animals	No. added or removed
Added	
Cows and sheep	89
Deer	86
Removed	
Horses	3
Geese	one flock
Ducks	one flock

trunks, and higher canopy width/tree height ratios than trees with narrow canopies and trunks that bifurcated higher above the ground (Heerwagen and Orians 1993).

The kinds of changes recommended by landscape architects to their prospective customers are another source of data on human responses to landscapes. Humphrey Repton, an 18th-century British landscape architect, presented his clients with "before" and "after" drawings of their estates and bound these drawings and accompanying prose in red covers (Repton 1907). If we assume that Repton wished to encourage rather than discourage potential clients, the changes Repton drew in his "Redbooks" can be used to test predictions from evolutionary hypotheses or human habitat selection. Heerwagen and Orians (1993) predicted that Repton generally would change landscapes by creating more savannah-like scenes, increase visual access and penetrability of closed woods, open up distant views to the horizon, add refuges and cues signaling ease of movement, and add evidence of resource availability, particularly large mammals. All of these predictions were supported by their analysis of 18 before and after drawings (Table 18.2).

Flowers provide environmental signals. A love of flowers is such a pervasive human trait that it has not been regarded as a topic worthy of serious investigation. However, it is not obvious why an omnivorous primate should take flowers to hospitals, bring them to dinner parties and housewarmings, and annually spend billions of dollars on them. An evolutionary perspective suggests that flowers may evoke strong positive feelings because they have long been associated with food resources. Because flowers precede fruits, flowering plants provide excellent cues to timing and locations of future resources. Especially in species-rich environments where it may be difficult to find particular plant species when they are not in a reproductive state, paying attention to flowering plants may enhance resource acquisition abilities in the future. Until the 19th century, honey was the only natural source of sugar, and beekeeping is an ancient human enterprise.

To date no analyses have been made to determine which traits of flowers evoke strong positive feelings. Nonetheless, the obvious changes produced in many species of flowers by artificial selection—such as increased size and duplication of floral parts—appear to result in flowers likely to have been associated with large nectar rewards. Anecdotal evidence suggests that strongly zygomorphic flowers are generally more attractive than symmetrical ones, but whether zygomorphism has been favored during selection under domestication remains to be determined.

Restorative Responses

A functional-evolutionary approach to human emotions has been useful in a number of fields. Researchers in psychiatry (Nesse 1990), medicine (Williams and Nesse 1991), and nutrition (Harris and Ross 1987) have examined modern diseases from the perspective of hunter-gatherer lifestyles. Nesse, for example, argues that negative emotional states such as fear, anxiety, or depression may be adaptive because they focus attention on the situation at hand. If so, medication may, in some circumstances, be truly maladaptive. Lundén and Ulrich (1990) investigated whether exposure to visual stimulation in hospital intensive care units facilitates postoperative recovery of open-heart-surgery patients. At Uppsala University Hospital in Sweden, 166 patients who had undergone open-heart surgery were randomly assigned to a visual stimulation condition consisting of a nature picture (either an open view with water or a moderately enclosed forest scene), an abstract picture dominated by either curvilinear or rectilinear forms, or a control condition consisting of either a white panel or no picture at all. Patients exposed to the open view of water experienced much less postoperative anxiety than the control groups and the groups exposed to other types of pictures. The enclosed forest setting did not reduce anxiety significantly compared to the control conditions. Interestingly, the rectilinear abstract picture was associated with *higher* anxiety that the control conditions. Other indicators, such as blood pressure, frequency of use of pain killers, and length of hospital stay, correlated with these psychological measures.

Findings from more than 100 studies of recreationists in wilderness areas have shown that stress reduction consistently emerges as one of the key benefits (Knopf 1987, Ulrich, Dimberg, and Driver 1991). Most studies on urban parks and other urban natural settings also have found that restoration from stress is a key benefit (Ulrich and Addoms 1981, Kaplan and Talbot 1983, Schroeder 1989). These studies show that certain natural configurations and elements are more effective than others in eliciting restoration. Specifically, such states as "relaxation" and "peacefulness" are evoked by exposure to settings having savanna-like properties or water. University students typically seek natural environments or urban settings dominated by natural elements (wooded urban parks, places offering views of natural landscapes, locations at edges of water) when they feel stressed or depressed (Francis and Cooper-Marcus 1991).

Manipulation of Landscapes Serves Many Purposes

Mimicry of savannah-like features is clearly not the only pattern shown by our manipulations of environments. Indeed, formal European Renaissance gardens differ strikingly from tropical savannahs. These gardens reflect another human tendency, namely to control environments to serve varied ends. Food production is one of these ends, and nobody suggests that a cornfield is designed to be especially attractive. The objective of the manipulation is bushels per acre, not aesthetics.

Environments are also manipulated to exhibit the power of its owner. Formal European gardens are best viewed from the home of the owner, not by walking through them (Plate 4). That they may be a form of territorial dominance display is indicated by interactions between Louis XIV and his superintendent of finances, Nicolas Fouquet.

On 17 August 1661, Louis XIV was hosted by Fouquet at his newly completed elegant chateau Vaux-le-Vicomte (Plate 5). Fouquet arranged for a spectacular display of fireworks, but the more the king saw, the more threatened he felt. Within three weeks, Fouquet was arrested for treason and convicted, and he spent the remainder of his life interned in the terrible Haute-Savoie fortress of Pignerol. Louis XIV then stripped Vaux of its treasures and took its architect (Le Vau), painter (Le Brun), landscape gardener (Le Notre), and the hydraulic engineers to Versailles with a mandate to build an estate superior to what had been constructed at Vaux-le-Vicomte!

Domination of nature and the tendency to assume that natural processes serve human ends takes on other forms, including our tendency to find meaning in stellar constellations and human resemblances in natural formations. We act as though the forces of stellar evolution and erosion operate to replicate our features. If nature doesn't provide enough human-like features to satisfy us, we may carve mountains into human faces, as we did at Mount Rushmore in South Dakota.

Environment and Higher Order Cognitive Functions

For decades it has been known that stress can reduce performance on cognitive tasks (Glass and Singer 1972, Hockey 1983). Most studies have investigated performance declines for "lower order" cognitive tasks such as proofreading that require narrowly focused attention on a restricted set of information and do not require integration of diverse information (Hartig et al. 1991). Exposure to nature can promote recovery from mental fatigue stemming from such work situations that require prolonged, directed, effortful attention (Kaplan and Talbot 1983).

Higher order cognitive functioning, which involves integrating diverse material or associating previously unrelated information or concepts, is required for creative problem-solving. An increased ability for integrating diverse information and for applying previously learned information to new situations must have played a key role in human evolution. Therefore, humans might have a partly genetic predisposition for enhanced *higher order* cognitive functioning when engaged in nonurgent tasks in certain natural settings. Preliminary evidence indicates that natural surroundings foster performance on higher order tasks and that positive emotional states assist recall of information from memory and creative problem-solving (Isen 1990).

The Conceptual Structure of Science

All forms of intellectual activity are manifestations of human behavior, and their patterns reveal the imprint of countless generations of natural selection

(Tiger and Fox 1971). Not surprisingly, science, which is a particular form of human behavior, exhibits many similarities to other forms of human behavior. Some of these similarities are so pervasive and so seemingly self-evident that we seldom stop to think about them or question them.

A pervasive feature of human intellectual activity that has received little attention is the tendency to resolve complex events and inputs into polar categories. This tendency characterizes the conceptual structure of such diverse fields as art, literature, physics, sociology, and biology. Polarity is not simply a feature of "Western" thought, based on Graeco-Roman logic and a language tied to a series of such crucial modifiers as con-dis, post-ante, pro-con. Polar typology is prevalent in many cultures with different languages and different intellectual histories.

The list of polar categories in ecology and evolutionary biology shown in Table 18.3 is not exhaustive, but it illustrates the large number of basic concepts in our discipline that are polar in nature. Ecology's polar concepts are not relics of past thinking. The cutting edge of our science is plowed with polar constructs such as density dependence–density independence, stochastic–deterministic, and source–sink. I vividly recall the hours we spent as

Table 18.3
Polar Constructs in Ecology and Evolutionary Biology

---

attack – escape (ethology)
birth – death
chaos – limit cycles
continental islands – oceanic islands
density dependence – density independence
determinate – indeterminate
discrete population growth – continuous population growth
distribution – abundance
equilibrium – nonequilibrium
exploitation competition – interference competition
host – parasite
immigration – emigration
island – mainland
male – female
origin – evolution
predator – prey
proximate factors – ultimate factors
saturated communities – unsaturated communities
source – sink
stable – unstable
stochastic – deterministic
sympatric – allopatric
vicariance distribution – dispersal distribution

---

graduate students debating density-dependent vs. density-independent population regulation. In retrospect, it seems rather silly.

Thus, humans have a strong tendency to categorize events and concepts in ways that are strongly polar. This applies equally to those situations in which the real world does appear to be polar—life and death, male and female—and to situations where the real world obviously exists as a complex continuum—food, emotions, and scientific phenomena. The pervasiveness of polar constructs, of which I have given only a brief accounting, suggests that a genetically determined property of our nervous system may be involved.

At least two explanations exist for the satisfaction people feel when they have resolved complex situations into simple polar opposites. One is that it is a byproduct of a nervous system that is built upon units that are basically polar in their functioning. The other assumes that polarization of complex situations on average led to better-adapted behavior than more complex interpretations of environmental input. The two explanations are not necessarily mutually exclusive.

Although nervous impulses are an all-or-none phenomenon, there are reasons for not accepting this explanation as the major cause of the prevalence of polar typology in human thought processes. This same nervous system produces such complex human emotions as fear, hatred, love, admiration, pity, and despair. These states tend to be strongly graded and not polar. All of us know by introspection that our feelings at any moment are often a mixture of these intrinsically complex mental states. If our nervous system is capable of producing this range of complex emotions, surely it was not inevitable that our most intense intellectual efforts should produce simple polar resolutions of obviously complex phenomena.

A more likely explanation for our tendency to polarize complex inputs is that this form of intellectual activity was favored by natural selection because, for two reasons, it resulted in adaptive behavior. First, informational input to the brain must be reduced; second, action is fundamentally polar.

For a number of reasons, information falling on our sense organs is inevitably lost during perception (Marler 1961). Some of the loss is a byproduct of amplification, because amplification always involves selection. Some of the loss is the result of the evolution of highly developed sense organs. Natural selection cannot favor the evolution of sense organs that transmit more information than the central nervous system of an animal is capable of receiving and integrating. The human central nervous system is well-designed to handle large amounts of information and, indeed, the hallucinations of humans subjected to conditions of sensory deprivation suggest that the "need" of the brain for information is so great that it "invents" its own information if none is being provided to it.

Nevertheless, limits exist on central nervous processing of information, and not all environmental information can be synthesized and interpreted.

Simplification is possible because not all information is equally important. Failure to detect the presence of a predator or location of food may be fatal, whereas failure to detect a slight change in ambient temperature may be trivial. Therefore, sense organs evolve to transmit information of importance to the organism with greater accuracy and completeness than unimportant information.

No matter how complex the informational input, many decisions animals make are binary. A perceived object is either food to be eaten or it is unpalatable. An object is a suitable mate or it isn't. An object is to be approached or avoided. Not only are the appropriate actions polar, but they must often be made with great rapidity. An organism that ponders the desirability of fleeing from a hungry predator is more likely to be captured than one that flees instantly. A sexually active male that debates excessively the desirability of mating with a receptive female finds another's genes in his place.

Thus, complex animals with elaborate receptors make many important polar decisions while receiving complex inputs that make the resolution of responses difficult. Not surprisingly, the central nervous system appears to function to analyze complex informational inputs in ways that enable rapid polar decisions to be made. Even when we are contemplating issues for which no quick polar responses are required, we nevertheless tend to organize our thoughts in terms of polar constructs. The strongly polar conceptual structure of our science may be a legacy of this product of our evolution.

Why are We So Reluctant to Embrace the Evolutionary Roots of Human Behavioral Ecology?

Rugged mountains, dashing waters, powerful storms with thunder and lightning, the rising and setting of the sun, the waxing and waning of the moon, and seasonal changes in day length have long captured the attention of people and stimulated in them a sense of awe. To account for such natural events, our ancestors invoked gods, goddesses, and spirits. The sun was driven across the sky in a chariot, and serpents consumed the moon. In the West, the development of science gradually replaced mystical and religious explanations for those physical events. Today only natural explanations for physical events are taught in public schools.

The situation is quite different with respect to the marvelous diversity of the living world because of the peculiarities of the phenomenon of adaptation. The oldest, and still most widely-accepted, explanation of adaptation is creation by a knowledgeable designer. It was not until Charles Darwin published *The Origin of Species by Means of Natural Selection* in 1859 that a theory was available to explain adaptation without recourse to a supernatural designer.

Remarkably, 140 years after Charles Darwin articulated one of the most exciting and unifying of all scientific concepts, the fact of evolution is still rejected by a significant fraction of people in the United States. The teaching of evolution continues to be greeted with formidable resistance, the strength of which may be stronger today than at any time during the past several decades.

Why should biological evolution in general and human evolution in particular come up against the last stronghold of belief in design by creation or "intelligent design" as it is usually labeled today? Why is evolution the sticking point when a literal interpretation of Genesis is not only incompatible with biological evidence, but is also incompatible with physics, chemistry, geology, paleontology, anthropology, and astronomy as well? Why are people who accept the findings of science in general unwilling to accept them when they offer explanations of the adaptedness of life?

Many factors probably contribute to the resistance to accepting that natural processes have produced the diversity of life. The remarkable learning abilities of humans and the mystery of how thinking helps us solve problems seem to defy rational explanation. Even more fundamentally, evolution requires us to think in new and different ways about the meaning of human existence and the place of humans in the world. It makes us question the uniqueness of our position and challenges our strongly held belief in free will. In short, it demands much more of us than acceptance of the theory of gravitational attraction.

An insightful perspective on why acceptance of the world into which Darwin led us has been slow was penned by Nobel Laureate Jacques Monod:

> [C]hance alone is at the source of every innovation, of all creation in the biosphere. Pure chance, absolutely free but blind, at the very root of the stupendous edifice of evolution: this central concept of modern biology is no longer one among other possible or even conceivable hypotheses. It is today the sole conceivable hypothesis, the only one that squares with observed and tested fact. . . . . There is no scientific concept, in any of the sciences, more destructive of anthropocentrism that this one, and no other so arouses an instinctive protest from the intensely teleonomic creatures that we are. (Monod 1971:112-113)

A corollary of the difficulty human minds have had in accepting a Darwinian view of the world is the strong Western intellectual belief that there is an unbridgeable gap between humans and animals. This belief is combined with the position that other species were created especially with human needs in mind:

> It was with human needs in mind that the animals had been carefully designed and distributed. Camels, observed a preacher in 1696, had been sensibly allotted to Arabia, where there was no water, and savage beasts "sent to deserts, where they may do less harm." It was a sign of God's providence that fierce animals were less prolific than domestic ones and that they lived in dens by day, usually coming out only at night, when men were in bed. Moreover, whereas other members of wild species all looked alike, cows, horses, and other domestic animals had been conveniently variegated in color and shape, in order "that mankind may the more readily distinguish and claim their respective property." The physician George Cheyne in 1705 explained that the Creator

made the horse's excrement smell sweet, because he knew that men would often be in its vicinity. (Thomas 1983:19)

The fact that human bodily functions are shared with animals has, accordingly, stimulated much concern and led to efforts to find fundamental differences despite the obvious similarity of metabolism in humans and animals. One interesting approach was to suggest that physical modesty about bodily functions was a trait that distinguished humans from beasts. This position is illustrated by a passage in the diary of New England clergyman Cotton Mather for 1700:

> I was once emptying the cistern of nature, and making water at the wall. At the same time there came a dog, who did so too, before me. Thought I; "What mean and vile things are the children of men. . . How much do our natural necessities debase us, and place us . . . on the same level with the very dogs." My thought proceeded. "Yet I will be a more noble creature; and at the very time when my natural necessities debase me into the condition of the beast, my spirit shall (I say at that very time!) rise and soar. . . Accordingly, I resolved that it should be my ordinary practice, whenever I step to answer one or the other necessity of nature to make it an opportunity of shaping in my mind some holy, noble, divine thought."

As science abolished, one after the other, what appeared to be sharp distinctions between humans and other mammals, new ways were found to defend the gap. In the 19th century one of the great arguments against vaccination was that inoculation with fluid from cows would result in the "animalization" of human beings. Why eating other species, a universal practice, did not have the same effect was never made clear. Bestiality became a capital offense in Britain 1534 and, with one brief interval, remained so until 1861. Incest, by contrast, was not a secular crime at all until the 20th century.

These beliefs may seem quaint to us today, but some of today's defenses of the gap between humans and animals will probably seem equally as quaint to our descendants. The fact that we share more than 97 percent of our genes with chimpanzees has done little to change human self-perceptions. The virulence of the attacks on sociobiology, most of which were based on nonscientific arguments, unfortunately including some advanced by evolutionary biologists, shows that defense of the gap is still strong, even among highly educated people. The view that humans are born as blank slates upon which the environment writes the behavioral script is still the dominant belief among professional social scientists (Cziko 1995). The health sciences have been remarkably slow to embrace "Darwinian medicine," an unfortunate state of affairs because its acceptance would result in significant positive changes in medical practices (Profet 1991, 1992, Williams and Nesse 1991, Ewald 1994).

Some Implications

Thus, an increasing variety of evidence shows that nature has a more powerful impact on our emotional and physical health than has been appreciated

(Ulrich et al. 1992, Ulrich 1984). Yet we spend more time and devote more money to determining the habitat needs of animals that we intend to maintain in zoos than we do to our own habitat needs when we design buildings, neighborhoods, and cities. The view that we are born as blank slates and can adapt equally to all types of environments still dominates Western thought and, hence, strongly influences current practices in the design of spaces used by humans. Buildings are still built with many windowless offices, even though research demonstrates that people are psychologically uncomfortable working in them; we attempt to compensate for the lack of windows with wall decorations that mimic views of nature (Heerwagen and Orians 1986). Most architecture students in the United States do not take even one course in psychology!

In marked contrast, the advertising industry has long recognized the importance of human responses to environmental features. A perusal of magazines with striking visual advertisements reveals the extensive use of savannah-like landscapes, savannah tree shapes, and striking geological features to help sell products that bear no relationship to natural environments or sexual attractiveness (Plate 6).

Finally, an understanding of evolved human responses to environments may assist ecologists in our attempts to convince the public of the importance of biodiversity and ecosystems. Typically we base our arguments on the goods and services ecosystems provide. Although there is no question that human societies are fundamentally dependent upon ecosystem goods and services, it is evident that these issues rarely stimulate the level of concern that leads to action. Surveys repeatedly show that Americans do not understand the meaning of the word "biodiversity." Consequently, they attach a low priority to biodiversity preservation. In addition, the loss of a species rarely affects the lives of people other than the few researchers who were attempting to forestall the extinction.

Clearly, additional motivators are needed if people are to develop the sense of commitment to the biological resources of Earth that will be required to generate the sacrifices that must be made if biological impoverization of the planet is to be avoided. There is no magic bullet that can accomplish this task, but we may improve our success if we make more use of the deep psychological and physiological roots of human responses to nature. These roots, which express themselves in powerful emotions, are what change decision-making. When we have walked through many of the doors opened for us by Charles Darwin, we have found rooms rich with rewards. We can no longer afford to resist or ignore the door that opens into the room where we will encounter the concepts and tools to enable us to better understand our own behavioral ecology.

Conclusion: Integrating the Biological and Social Sciences to Address Environmental Problems

Iver Mysterud and Dustin J. Penn

We hope that the papers in this reader have convinced you that evolutionary perspectives on human behavior offer many important insights into our environmental problems—why they arise, and how we might design more effective solutions. In this concluding chapter, we summarize some highlights of the reader, and we discuss some additional points that were not addressed. In particular, we consider how our species' *environmental impact exacerbates our economic, social, and medical problems, in ways not generally appreciated.* Finally, we discuss how evolutionary research on human behavior is helping to synthesize the badly fragmented fields of human knowledge, which is necessary to effectively address our environmental problems.

As we pointed out in the Introduction, the source of our environmental problems is not a scientific mystery. Our species' environmental impact (I) is the product of population size (P), resource consumption per capita or affluence (A), and the impact of technologies used to supply each unit of consumption (T) or $I=P \times A \times T$ (Ehrlich and Holdren, 1971; Holdren and Ehrlich, 1974; Holdren, 1991; Ehrlich and Ehrlich, 1990). Evolutionary perspectives of human behavior offer insights into the first two components in this equation. Part 5 addresses population issues, such as why people have children in general, and in specific situations, and why people are having fewer children in many countries (demographic transitions). Parts 3 and 4 address resource consumption, including conspicuous consumption. In Part 2, we saw that environmental problems are not new to the modern world, contrary to popular belief. Humans have always used resources, and sometimes this has led to extermination of animal species and destruction of habitats. The low environmental impact of traditional and non-Western cultures have provided the inspiration for the ecological noble savage hypothesis. Yet, we have seen that people are not necessarily conservationists (or follow a conservation ethic) just because they have a low environmental impact. Several authors have stressed the need to define conservation independently of societies' impact on their environment:

281

conservation is about passing up or sacrificing short-term benefits of resource consumption. The ecological noble savage is a myth, although as Alvard (1998b, p. 71) states, "The research is moving on from simply debunking myth to sophisticated analyses aimed at understanding the contexts that do and do not favor resource restraint." The challenge is to find out how the humans have been "designed" by natural selection to use resources, and we feel that several of the papers in this reader offer genuinely new and important insights. In Part 3, we saw that the tragedy of the commons provides a general framework for understanding why humans cause many environmental problems. Of course, over-exploitation is not inevitable. In specific contexts, humans have managed common-pool resources in sustainable ways. It is *unmanaged* commons that lead to tragedies. Thus, sustainable management is possible; however, we need a better understanding of the conditions necessary to avoid tragedies of the commons.

The only variable of the I=PxAxT equation that evolutionary research has not illuminated so far is *technology*. It is obvious that the technology in the past was less environmentally destructive than today. It is a novel evolutionary and environmental situation to have such effective and destructive technology (especially when seen together with a large global population and the existence of global markets for trade of resources and products). Having said that, one should not underestimate certain aspects of the technology of these so-called "primitive" peoples. For example, use of fire may be a very efficient way of modifying habitats, used by humans for millennia (Bird, 1995). Still, technology is undergoing an extremely rapid evolutionary process — similar in many ways to genetic evolution (Basalla, 1988) — but our genetic evolution can not possibly keep up with such rapid advancements. Our evolutionary mismatch with technology has major implications for our environmental problems (Ornstein and Ehrlich, 1989), though this perspective is not widely appreciated.

The modern environmental movement was partly based on the realization that technological innovations, such as DDT and other pesticides, provide a double-edged sword: they fix some problems, but they often have undesirable side-effects on humans and the environment. We humans were not designed to perceive or anticipate the harmful effects of our increasingly complex technologies. Many people still under-estimate the harmful side-effects (and over-estimate the potential benefits) of technology, although environmental thinkers often make the reverse mistake. We need to be guarded about new technologies, such as genetically modified organisms, but it is unrealistic to expect people to become luddites. We need to learn more about our species' cognitive limitations for assessing the benefits and risks of technology, and moreover, how the technology we adopt is influenced by social forces (e.g., political and economic interests). For example, we have an increasing number of ecological friendly technologies or "ecotechnologies" whose potential is still unrealized

(Myers, 2000; Hawken et al., 1999). The failure to replace wasteful, polluting technologies due to political and economic struggles is a topic that deserves attention from evolutionary researchers.

We often focus our attention on environmental destruction, and we forget to emphasize the amazing amount of time, energy, and money people invest into surrounding themselves with plants and animals, and preserving natural habitats. Part 6 addresses the idea that humans have an instinctive psychological need for natural environmental stimuli, such as plants and animals (biophilia hypothesis). This includes instinctive biases to learn, and become imprinted on certain environmental stimuli. If this hypothesis holds true—and several lines of evidence support it (see Gardner and Stern, 1996 and Frumkin, 2001 for nice overviews) – it is far from trivial, and definitely not only of academic interest. The absence of natural stimuli in urban environments may be harmful to people's psychological health and well being. Thus, human biophilia gives an additional reason for slowing and stopping human population growth, human consumption, environmental pollution, and the rate of extinction of plant and animal species.

More and more people throughout the world live in drastically modified environments, and as populations grow, more people will live in urban areas. Expanding human settlements continue to consume farmlands, woods, and wilderness areas, and people are increasingly exposed to evolutionary novel environmental conditions like air pollution. If the biophilia hypothesis is correct, a growing number of people will suffer from emotional and other psychological problems. Environmental changes of the modern world are causing numerous other health problems that also need to be considered.

Environmental, Social, and Medical Problems Interact in the Modern World

Our species' environmental impact affects our existing economic, social, and medical problems. For example, environmental changes and degradation exacerbate violence and conflicts over access to scarce resources (Homer-Dixon, 1991, 1999; Gleditsch, 1997). This occurs on every level from local to international. Research on social conflict is increasingly interdisciplinary, and evolutionary researchers are helping to understand violence, and other human subsistence strategies (e.g. Wrangham and Peterson, 1996; Buss and Shackelford, 1997; Low, 2000b). Understanding how humans evolved to use and get access to resources is crucial to better manage increased levels of social problems in coming years. Although evolutionary perspectives on human violence and social conflicts have long been treated with hostility or suspicion by a number of quarters, especially by humanists, Marxists, and social scientists (Montagu, 1968), we expect Darwinism will be at the center of integration processes of natural and social sciences.

In addition, our modern environmental changes are creating new challenges for medicine, especially epidemiology and toxicology. Global warming may lead to changed distribution of infectious diseases and increased incidence of disease and death due to heat (McMichael, 1993). A warmer and more humid climate may lead to growth and increased distribution of microorganisms, vectors and hosts, and a possible outcome is that tropical diseases may increase in incidence and spread north- and southwards. One can for instance imagine that malaria will "move" northwards in Africa and reach northern Africa and southern Europe. Even though a consequence of a reduced ozone layer may be increased production of vitamin D in the skin, several negative consequences may occur: eye problems (e.g. snow blindness, cataracts) and skin problems (cancers). Thus, the link from human resource use and environmental problems to medicine may be direct, and health personnel may in the coming years witness first-hand the effects of human impacts on the environment.

Every field from demography, environmental sciences to toxicology and sciences of human behavior will to an increasing extent have to be integrated to meet the upcoming challenges. Evolutionary approaches will be central in this integration process, and medicine is actually in the process of becoming transformed by evolutionary approaches (Ewald, 1994; Boaz, 2002; Eaton et al., 1988, 2002; Williams and Nesse, 1991; Lappé, 1994; Nesse and Williams, 1994, 1997, 1998; Stearns, 1999; Trevathan et al., 1999). Darwinian medicine is helping to elucidate age old and modern health problems. Evolutionary researchers correctly predicted that pathogens would rapidly evolve resistance to antibiotics if over-prescribed (Ewald, 1994). They also found that fever is a functional defense against pathogens and not a pathological side-effect as had been assumed (Kluger, 1991; Kluger et al., 1996). Similarly, they recognized that sequestering free serum iron in the liver during certain infections is an adoptive defense rather than a pathology (Weinberg, 1984). By removing iron from circulation, the body removes an essential nutrient for bacteria. A failure to understand that a low iron level in certain environmental contexts is a healthy response against infection, led to well-intended but harmful interventions in third world countries (Stuart-Macadam and Kent, 1992). An evolutionary perspective on health is necessary if medicine is to meet the challenges of our modern world.

The environmental sciences and medicine have both resisted applying evolutionary biology to humans. Yet many of our environmental and health problems are due to the fact that most humans live in an evolutionarily novel environment (Penn, 2003). Modern medicine has helped to reduce or eliminate many infectious diseases: however, evolutionary researchers are finding strong evidence that this explains the development of allergies, atopies, and other immunopathologies in developed countries (Whary and Fox, 2004; Wilson and Maizels, 2004). Our sociocultural evolution has been extremely rapid during the past ten thousand years, but our genes have not had time to adapt to many aspects of the modern world. This point has been criticized and improved (Dunbar and Strassmann, 1999), and most agree that there are several novel environmental factors to which we are not adapted. Let us consider two principally different kinds of examples.

The environmental literature often focuses on human food production, and how to meet the demands for an increased human population and that global environmental problems may threaten our food production (e.g., Dyson, 1999; Smil, 2000). It also stresses how eating lower on the food chain would greatly help to reduce our environmental impact, especially in developed countries. Yet, we are only beginning to hear evolutionary perspectives that add insights into the problem of feeding the poor in developing countries, and reducing consumption in rich countries. A problem for rich and poor countries is that we may not be well adapted to eat large amounts of grains and other cereals (Eaton et al., 1997; Cordain, 1999). Consuming large amounts of processed carbohydrates from grains (and later from sugar cane) is relatively novel, first begun by some human groups approximately 10,000 years ago, and by some modern humans only decades ago. This new dietary consumption is linked to various chronic lifestyle diseases, which appear late in life, usually after most people have reproduced. Consequently, selection against such "stone age genes" will be slow (Neel, 1999).

Another problem is that we are exposed to a large number of evolutionary novel chemical pollutants. Synthetic toxins constitute particular problems, not just because some are more dangerous than natural toxins, but because some are chemically different from toxins humans are adapted to handle. For example, we have no enzyme system designed to deal with PCBs or organic mercury complexes. The liver is ready to detoxify many plant toxins, but lack the capacity to handle some novel substances. Humans have no natural inclination to perceive or avoid some toxins. We have been equipped with the ability to smell or taste common natural toxins and the motivation to avoid such smells and tastes (we sense them as repelling). However, we have no such mechanisms to protect us from many artificial toxins that are odorless and tasteless, such as lead, in spite of being bad for us. The same is true of potentially mutagenic or carcinogenic radioisotopes. For example, sugar synthesized from radioactive hydrogen or carbon tastes as sweet as that made of ordinary stable isotopes—we have no way of detecting the difference with our senses (Nesse and Williams, 1994).

The Importance of Biological Mechanisms

Even though it is insightful to view novel environmental changes from an evolutionary perspective, this does not mean that an evolutionary perspective is sufficient. Indeed, it can be argued that an evolutionary perspective is often trivial to the extent that it just states the obvious, i.e., that several features in the environment are new to humans and our evolved senses and defense systems. Proximate approaches that address the underlying mechanistic details are critical. It is equally important to integrate theories, results, principles and ideas from approaches like ecosystem ecology and environmental sciences. Consider the following examples from toxicology.

There is a large and growing number of synthetic chemical pollutants commonly found in our environment and our bodies that may act as or interfere with hormones (endocrine disrupting chemicals or so-called "hormone mimics" or "environmental estrogens") (Colborn et al., 1996). Such substances have become common since the 1940s as the chemical industry began manufacturing them. More than 37 chemicals have been identified as being able to mimic estrogen in the body or to interfere with the various systems that regulate the body's production of estrogen and other hormones (Dibb, 1995). These chemicals include the chlorinated hydrocarbon, DDT, its breakdown product DDE, dioxins, PCBs, several pesticides and fungicides and some chemicals used in detergents and plastics (Dibb, 1995). In addition, scientists are suspecting 3,000 chemicals in our food and in the environment of having endocrine disrupting effects (Dyer, 1995). They cause various problems in animal populations (Dibb, 1995; Gray et al., 1999), and are suspected to play a role in the increasing incidence of reproductive abnormalities in men (Dibb, 1995; Sharpe and Skakkebæk, 1993), quality and number of sperms in semen the last 50 years (Carlsen et al., 1992; Auger et al., 1995), and breast cancer and some infertility problems in women the last 50 years (Davis and Bradlow, 1995). This provides an example of how our bodies lack the ability to deal with environmental changes we are making in the modern world, and how difficult it is to anticipate the harmful effects (For a recent review of endocrine disruptor chemical pollutants on the behavior of humans and other animals see Zala and Penn, 2004).

The problem with eating fish and other species high on the ecological food chain in the modern world is that they often contain dangerously high concentrations of toxic pollutants due to *bioaccumulation* (i.e., the accumulation of ever higher concentrations of substances, like heavy metals and some hormone mimics in the food webs). These substances include synthetic chemicals (DDT and PCBs) released into new environments, or released at concentrations never before encountered in our evolutionary history (heavy metals). Toxic effects from *heavy metals* provide an example of the importance of mechanistic approaches to human behavior and our social and medical problems. Humans have created some novel situations (and controversies) by use of *mercury* in vaccines for children (Downing, 2000; Rimland, 2000), and in filling materials for dental cavities (dental amalgams) (Nylander et al., 1987; Pleva, 1992, 1994). Although the scientific issues here are not settled at the moment, it is a safe bet that controversy will persist until mercury is used more cautiously. Another heavy metal that creates a lot of problems (and controversy) is *lead*. The list of probable impacts of lead on medical and behavioral problems in humans includes aggression, violence, impulsive behavior, and attention deficit behavior, to impaired cognitive function, hypertension, cardiovascular disease, impaired renal function, and dental caries (Needleman, 1990; Bellinger et al., 1992; Bellinger and Dietrich, 1994; Schwartz, 1995; Freet, 1996; Hu et al., 1996; Kim et al., 1996; Tuthill, 1996; Payton et al., 1998; Moss et al., 1999). Lead in gasoline is fortunately being phased out in western nations. However, lead persists in dust (Mielke, 1999), and in the USA another exposure to lead from drinking water is getting increased attention. New research indicates that fluoride compounds (the silicofluorides fluosilic acid (H_2SiF_6) and sodium silicofluoride (Na_2SiF_6)) used in drinking water leads to increased uptake of lead (Masters, 2003; Masters et al., 1998; Masters and Coplan, 1999a,b). This will be most prevalent in poor people, particularly of African descent, with insufficient diet (e.g. low in calcium) that leads to increased uptake of lead from the water (Masters et al., 1997; Bruening et al., 1999).

An increasing number of other health problems are due to environmental changes, but the traditional focus on cancer in toxicology is too narrow (Colborn et al., 1996), and we have not evolved the defenses necessary to cope with some of these changes (Nesse and Williams, 1994). Insights into disciplines like toxicology and environmental medicine have to be integrated into future analyses of medical and social problems. In the long run, human impact on the environments will speed up the ongoing process of interdisciplinary integration, making various disciplines compatible, consistent or "consilient" with each other.

Interdisciplinary Consistency and Synthesis

Making various natural sciences consistent with each other and natural and social sciences compatible with each other is a process described as "vertical integration" (Barkow, 1989), "conceptual integration" (Cosmides et al., 1992; Tooby and Cosmides, 1992) and "consilience" (Wilson, 1998a). Evolutionary perspectives are increasingly central to the integration process between natural sciences (biology) and social sciences, necessary to address our environmental problems (Penn, 2003).

Conceptual integration refers to the principle that the various disciplines within the behavioral and social sciences should make themselves mutually consistent, and consistent with the natural sciences. The natural sciences are already mutually consistent: the laws of chemistry are compatible with the laws of physics, even though they are not reducible to them. Similarly, the theory of natural selection cannot be expressed solely in terms of the laws of physics and chemistry, yet it is compatible with those laws. A conceptually integrated theory is one framed so that it is compatible with data and theory from other relevant fields. Compatibility is taken for granted in the natural sciences, but not in the behavioral and social sciences. Training in any of the fields evolutionary biology, psychology, psychiatry, anthropology, sociology, history and economics does not regularly entail a shared understanding of the fundamentals of the others (Cosmides et al., 1992).

As mentioned in our introductory chapter, there is an ongoing process of integrating evolutionary biology with most medical, behavioral and social sciences, and the papers in this reader are examples where evolutionary biology are made consistent with behavioral and social sciences in an environmental context. Stated in another way, the papers represent important efforts of researchers that for several years have attempted to improve understanding of human behavior in an environmental context by building bridges between behavioral and social sciences. And as we have suggested in this chapter, much can be achieved by further integration within the natural sciences (e.g., evolutionary biology with medicine, and medicine with various proximate fields of relevance in an environmental context). We suspect that this is just the tip of the iceberg. Social and behavioral sciences will become better informed by integrating results from environmental sciences, toxicology, biochemistry and neuroscience. As Edward O. Wilson (2001b) points out, we live in the Age of Synthesis, "in which research will turn more and more from nearly pure reductionism to the reassembly into wholes of the complex systems whose parts and basic processes have been discovered by reductionism." It is our hope that this reader will contribute to this process. Integrating evolutionary perspectives into environmental sciences will be more attractive if they can resolve basic questions about human nature.

Individual Level Selection Versus Multilevel Selection

One of the most important problems in the environmental sciences is that the different political camps and disciplines hold very different views and assumptions about human nature, and especially about altruism, and these assumptions are generally unstated. The papers in this reader emphasize how cooperation and altruism in humans is the result of selection among individuals in the past. The context for this is a consensus in evolutionary biology that emerged in the 1960s when "naive" models of group selection (sensu Wynne-Edwards, 1962) were replaced with models emphasizing the individual level, and the importance of the level of the genes ("the gene's eye view of nature") (Hamilton, 1964; Williams, 1966; Trivers, 1971). It was argued that individual selection will be a much stronger evolutionary force than group selection so that in practice, we do not have to bother thinking of group selection as an evolutionary force for explaining individual adaptations. This argument has turned out to be consistent with what is observed in nature among most animal species (see Sober and Wilson, 1998 for a review of this debate). However, improved models of group selection are compatible with the gene's eye view of evolution (Sober and Wilson, 1998). Instead of focusing on individual level selection versus group level selection, this new approach applies to different levels in all hierarchies, individual versus group selection being just one example of a general process. The proponents of multilevel selection argue that humans may be an exception among mammals in that group selection has been important in our evolutionary history (Sober and Wilson, 1998). Everyone agrees that individual selection has been operating throughout human history. The bone of contention is whether humans are a group-selected species and how important group selection may have been. This debate is not settled, as a multi-leveled framework is emphasized by some (Maynard Smith and Szathmáry, 1995; Boehm, 1999; Keller, 1999), but dismissed by other evolutionary researchers as a sterile, semantic debate (Maynard Smith, 1998; Nunney, 1998).

The outcome of the group selection debate is of relevance—maybe crucial importance—for understanding human behavior and our environmental and social problems (Mysterud, 2000). Several researchers—also within the individual selectionist paradigm—have argued that other humans may have been our most important selective factor, i.e., that the main obstacle to reproductive success in the past has been hostile humans, and not predators, disease or lack of food (e.g., Alexander, 1975; 1990). If group selection has been operating on humans in the past, the outcome would have been group adaptations, i.e., behavioral strategies that benefit one's own group at the expense of other groups. Such adaptations would have a tendency to be activated in certain situations—conflicts with other groups (e.g., conflicts over access to some kind of resource—such as territory, water, shelter, a source of natural resources or women). As an example of a possible group-selected behavioral tendency (adaptation), humans have an inclination to develop moral codes that applies to members of own group (the in-group) at the expense of members of other groups (the out-groups). Thus morality can be viewed as a coin with one nice side, cooperation and moral rules for the benefit of one's own group, but also an ugly side, other moral rules and behavior strategies directed against other groups. The effect is cooperation with own group members to better compete with members of other groups (i.e., us versus them). Modern accounts of morality, as in the Bible (Old Testament), may have been a morality for the in-group (Hartung, 1995). For Moses, promoting the survival and reproduction of the Jews required social norms that led individuals to cooperate within their group to compete with other groups (Arnhart, 1998, p. 259). Darwinian theorists have therefore explained the Mosaic Law as promoting the reproductive interests of the Jews (MacDonald, 1994, pp. 35-55, Hartung, 1995). There is no reason to expect that Judaism is unusual in this respect. Modern social psychological literature abounds with articles discussing our tendency to distinguish between in-groups and out-groups (see Devine, 1995; Krebs and Denton, 1997). It is feasible that group selection acting on *human cultures* has lead to the evolution of human groupishness (Boyd and Richerson, 1990; Pagel and Mace, 2004).

The group selection debate is relevant to understanding why conflicts over resources and the environment are often more pronounced among rather than within groups (Porter and Brown, 1996). This debate is also important in today's context of a rapidly expanding megapopulation with individuals and various subgroups competing for all kinds of resources in order to survive and reproduce. War and violent conflict over resources are not new phenomena of course (Keeley, 1996), but they can be more disastrous, considering modern technology and weapons, the large global population, and human impacts on the environments. For future years, a lot of questions need to be answered (Mysterud, 2000), both in general and in an environmental context: How can human nature best be characterized and understood from an evolutionary perspective? What adaptations exist, and in which environment were they selected to contribute to reproductive success, and how do they function in our novel environment? How similar or dissimilar is the contemporary environment to the environments in which the relevant traits were selected (cf. Irons, 1998)? And how can one use this information for the political purposes of solving environmental and social problems including that of developing a morality for a sustainable future (Penn, 2003)?

If we have adaptations to benefit the group, such as a tendency to classify humans into in-groups and out-groups and to develop moralities favoring one's in-group, it may be problematic to extend our morality outside our own group, especially in environmental situations where resources become scarce. In an environmental context, it is important to clarify whether we need an "outer enemy" in order to be able to cooperate effectively with all of humanity, or if pressures and threats from our common global environmental problems can be enough (Mysterud, 2000). And if our society's existing morality is the result of negotiation between the members (Alexander, 1987) or coercion by some members on others (Irons, 1991), it is crucial to focus on the conditions under which it will be possible to develop a new morality for all humans (Mysterud, 2000). Both David Hume and Charles Darwin explained human morality as emerging from the complex cooperation within groups competing with other groups, and thus only gradually and with great difficulty does human moral concern expand to include those outside one's own group (Arnhart, 1998, p. 75). Herein lies the challenge.

The discussion of selection on the individual level versus group level has usually focused on *costs of cooperative behavior that benefits the group* (costs of "altruism"). A recent trend is instead to emphasize *benefits of cooperation* (Tooby and Cosmides, 1996; Smuts, 1999; Thompson, 1999). However, this is what Alexander (e.g. 1975, 1990) has been concerned with for a long time, although it has been done within a pure individual selectionist paradigm (see Wilson, 1999 for critique). Increased emphasis of multilevel selection—at the expense of individual selection—and a concern with benefits from cooperation—instead of costs from altruism—is a trend that not only applies to humans, but in nature in general (e.g., Corning, 1983, 1995, 1996, 1997, 1998a,b, 2003; Keller, 1999; Margulis, 1993, 1998; Margulis and Fester, 1991; Maynard Smith and Szathmáry, 1995, 1999). Corning (1998a) actually calls it a new evolutionary paradigm whose time seems to have come. This remains to be seen. It is clear that for too long, our species' dark sides (selfishness, greed, and immorality) have been attributed to nature and biology, whereas our bright sides (benevolence, altruism, and morality) have been credited to nurture and culture. It is time to recognize that our bright sides are also products of natural selection (de Waal, 1996, 2001). And theories about the evolution of higher level cooperation, either in humans or the rest of nature (such as the Gaia hypothesis), can be consistent with a Darwinian framework (Lenton, 1998; Maynard Smith and Szathmáry, 1995, 1999).

Whatever the outcome of the debate about whether humans are a group-selected species, most recognize the importance of explaining the evolution of human cooperation. Cooperation often creates synergistic effects, and this has been important in human history for millennia (Corning, 1983, 2003; Wright, 2000). For the coming years, it is a real challenge to see if humans may be able to cope with and solve environmental problems by creating synergistic outcomes through cooperative behavior.

Commitment and Reciprocity Models

There are two new developments in evolutionary social sciences with important relevance for environmental problems, but they have not yet been given due consideration in an environmental setting. Since the 1960s and 70s the two main models to explain cooperative behavior has been kin selection (Hamilton, 1964) and reciprocity (reciprocal altruism) (Trivers, 1971). They have both been influential, and they have been emphasized and utilized by several authors in this reader; however, several decades of research indicate that they are insufficient. Rational choice theory in economics is similarly proving incomplete. Even though rational choice models explain and predict some human behavior quite well, especially in an anonymous market setting, it is outdated as a general model of human behavior. Two new approaches from evolutionary social science expand kin selection and reciprocity as models of human behavior and contribute to greater consistency with economics in general and rational choice theory in particular. Both approaches come from evolutionary biologists and economists actively seeking to improve existing models of human behavior by linking natural and social sciences.

Robert Frank (1988) presented such improved models of human cooperation more than 15 years ago. His approach emphasizes the strategic role of emotions in social interactions. He showed how our seemingly irrational emotions actually can serve our rational self-interests in the long run. Frank's commitment approach integrates emotions into economic models, and this has turned out to be successful in its ability to explain more than traditional rational choice theory. A commitment approach solves puzzles raised by irrational behavior when viewed from rational choice theory. Many seemingly irrational behaviors are governed by strong emotions, and Frank's point is that these emotions can actually serve our self-interest in the long term (i.e. they can be rational), even though they may be costly in the short term (i.e. irrational) — at least in the evolutionary past. The emotions themselves have supposedly become a part of the human psyche during an evolutionary process in the past. Such a commitment approach is gaining acceptance in the community of evolutionary researchers (e.g., Nesse, 2001), and it will hopefully contribute to better communication and integration between traditional social scientists and evolutionary scientists, and eventually an improved understanding of human behavior.

In the field of experimental economics, evolutionary researchers have also developed better models than those in rational choice theory and traditional reciprocity theory (e.g., Fehr et al., 2002; Gintis, 2000a,b). Rational choice theory premises are incorrect outside an anonymous market setting. Actors in economic settings are cooperative and prosocial in ways not predicted by rational choice theory. Economic actors in many circumstances behave more like "strong reciprocators" who come to strategic interactions with a propensity to cooperate, respond to cooperative behavior by maintaining or increasing cooperation, and respond to noncooperative free-riders by retaliating against the "offenders". This is done even at a personal cost, and even when there is no reasonable expectation that future personal gains will flow from such retaliation. Humans seem to be equipped with the behavioral mechanisms suggested by the strong reciprocity model, and the theory of multilevel selection may explain how such behavioral tendencies may have become part of human nature. Finding effective solutions to address environmental problems depend critically upon holding an accurate model of human nature. Commitment and reciprocity models greatly improve our understanding of human nature. The challenge is to determine how these models might be utilized to address our environmental problems.

Finding Practical Ways to Go Forward

Environmental thinkers often erroneously assume that the existence of human nature is only an obstacle to change. Several papers in the reader have used evolutionary approaches to human nature as a point of departure for discussing *what can be done to change human behavior* to lessen the impacts on the environment. For example, Heinen and Low (1992) discuss several traditional approaches and incentives that can be utilized for social change, and how these can be utilized more efficiently by using evolutionary theory. They emphasize information, social and economic incentives, coalitions and negotiations, and government regulations. The environmental psychologists

Gardner and Stern (1996) have produced a good synthesis (even though they did not emphasize or integrate most of the papers in this reader). Their book, *Environmental Problems and Human Behavior,* is actually the first book to our knowledge that emphasizes human behavior as the cause of environmental problems, and argues that social sciences–not only the environmental sciences—should be consulted when preparing to understand and solve such problems. It contains one chapter that specifically addresses evolutionary approaches, and some evolutionary perspectives are integrated in other parts of the book. Gardner and Stern's discussion of interventions and incentives for changing human behavior in an environmental setting is exemplarily, and it should add to the consilient process of integrating social and natural sciences of importance for both understanding and handling human impact on the environments for the coming years, especially when it is seen together with the full range of papers in this reader and various topics we have addressed in this concluding chapter. Let us highlight a few wise suggestions by Gardner and Stern that environmental thinkers should keep in mind.

After a thorough analysis of various approaches, Gardner and Stern (1996) emphasize that it will *never be possible to write a "cookbook" for behavior change that gives specific instructions for intervention.* This is true in part because so many situations are unique: the barriers to behavior change vary, even for a single behavior, from place to place, time to time, and individual to individual. However, a deeper reason is that humans are continually responsive to interventions in their lives—even to the point of evading, blocking, or organizing to repeal the ones they find most objectionable. Treating people as inert objects to be moved is a good way to encourage resistance to any sort of effort to change behavior. Because of the inevitable interplay between change agents and those whose behavior they intend to change, some of the most important principles of intervention concern the relationships between these groups. Gardner and Stern (1996) have summarized several principles for environmental intervention (Table 1). They emphasize that a *combination of strategies* is one of the main messages from successful environmental programs. Gardner and Stern recommend combining solution strategies as a way to deal with the multiplicity of barriers, situations, and individuals, and to help overcome the difficulty of knowing which is the optimal intervention type for any particular situation. In addition, people themselves know best where the shoe pinches, and what is, and what is not, in their own interests. Many of the principles that are listed in the table may seem like common sense, yet they are commonly ignored by those who devise environmental policies and programs (Gardner and Stern, 1996).

We would add that the most powerful approaches to elicit change will be based on *individual incentives* (and simple heuristics), though individual interests must be recognized in a broader evolutionary sphere to include one's family, friends, and tribe (and not just narrow economic terms) (Penn, 2003).

Table 1
Principles for intervening to change environmentally destructive behavior (after Gardner & Stern, 1996)

1. Use multiple intervention types to address the factors limiting behavior change? Limiting factors are numerous (e.g., technology, attitudes, knowledge, money, convenience, trust)
   - Limiting factors vary with actor and situation, and over time
   - Limiting factors affect each other (interactive principle)
2. Understand the situation from the actor's perspective
   - Conduct surveys or experiments
   - Participatory approach to program design
3. When limiting factors are psychological, apply understanding of human choice processes
   - Get the actors' attention; make limited cognitive demands
   - Apply principles of community management (credibility, commitment, face-to-face communication, etc.)
4. Address conditions beyond the individual that constrain proenvironmental choice
5. Set realistic expectations about outcomes
6. Continually monitor responses and adjust programs accordingly
7. Stay within the bounds of the actors' tolerance for intervention
8. Use participatory methods of decision making

As we quoted E.O. Wilson (1984) in the Introduction: "The only way to make a conservation ethic work is to ground it in ultimately selfish reasoning... An essential component of this formula is the principle that people will conserve land and species fiercely if they foresee a material gain for themselves, their kin, and their tribe." We might call this Wilson's First Principle for Intervention. This means educating people how environmental destruction adversely affects health and costs money, using indices of economic growth that include the costs of environmental destruction, and finding ways to provide negative feedbacks. This view is consistent with ecological economics, but what is missing is the importance of social pressure. Moral exhortations, appeals to the common good, and other forms of social pressure used by environmentalist activists may not be as powerful as individual incentives, but they are not a waste of time. Human cooperation is to some extent based on direct and indirect reciprocity. Ecological economics is incomplete as long as it fails to consider social pressure and norms as ways to provide individual incentives. Such a synthesis is the goal of an evolutionary ecological economics (or Darwinian ecology) (Penn, 2003).

Gardner and Stern (1996, p. 256) suggest that there is an almost universal tendency among environmental researchers to overlook important inputs from disciplines other than their own (and researchers have tribal behaviors too). This is a trap we should all try to avoid. Also, we need to face that conflicts of interest arise over access to resources, and when trying to influence such access or consumption patterns in general, we are involved in political processes. This means that participants should not pretend to be completely unbiased scientists, but acknowledge that they are political actors. We hope this reader shows how evolutionary approaches offer fundamental insights into population growth, resource use, and our other environmental problems, without implying that such an approach is sufficient. This chapter has highlighted how social and mechanistic approaches are relevant as well, and undoubtedly, other perspectives are also needed. With this caveat in mind, we would agree that evolutionary biology provides the foundation necessary to synthesize and integrate these various perspectives. Cross-disciplinary cooperation and consistency between social sciences and biology will become increasingly important in the coming years, both to get better scientific understanding of human resource use and environmental problems and to get improved insight into humans themselves.

References

Acheson, J. 1987. The lobster fiefs revisited: Economic and ecological effects of territoriality in Maine lobster fishing. In *The Questions of the Commons: The Culture and Ecology of Communal Resources* (McCay, B. & Acheson, J., eds.), pp. 37-65. Tucson: University of Arizona Press.

—. 1988. *The Lobster Gangs of Maine*. Hanover, NH: University Press of New England.

Acheson, J. & Wilson, J. 1996. Order out of chaos. The case for parametric fisheries management. *Am Anthropol* 98(3): 579-594.

Adams, P. F., Schoenborn, C. A., Moss, A. J., Warren, C. W. & Kann, L. 1995. Health-risk behaviors among our nation's youth: United States, 1992. Bethesda, Maryland: *Vital Health Statistics* series 10, no. 192. National Center for Health Statistics.

Adams, W. M., Brockington, D., Dyson, J. & Vira, B. 2003. Managing tragedies: Understanding conflict over common pool resources. *Science* 302: 1915-1916.

Agrawal, A. 1999. *Greener Pastures: Politics, Markets, and Community Among a Migrant Pastoral People*. Durham, NC: Duke University Press.

—. 2003. Sustainable governance of common-pool resources: Context, methods, and politics. *Annu Rev Anthropol* 32: 243-262.

Ajzen, I., Brown, T. C. & Rosenthal, L. H. 1996. Information bias in contingent valuation: Effects of personal relevance, quality of information, and motivational orientation. *J Environ Econ Manage* 30: 43-57.

Alcock, J. 1984. *Animal Behavior: An Evolutionary Approach*. 3rd ed. MA: Sinauer Associates, Sunderland.

Alcorn, J. B. 1989. The traditional agricultural ideology of Bora and Huastec resource management and its implications for research. In *Resource Management in Amazonia: Indigenous and Folk Strategies* (Posey, D. A. & Balee, W., eds.), pp. 63-77. New York: The New York Botanical Garden.

—. 1991. Ethics, economies and conservation. In *Biodiversity: Culture, conservation, and ecodevelopment* (Oldefield, M. L. & Alcorn, J. B., eds.), pp. 317-349. Boulder, CO: Westview Press.

—. 1993. Indigenous people and conservation. *Conserv Biol* 7: 424-426.

—. 1995. Commentary. *Curr Anthropol* 36: 789-818.

Alexander, R. D. 1971. The search for an evolutionary philosophy of man. *Proc Roy Soc Victoria (Melbourne)* 84: 99-120.

—. 1974. The evolution of social behaviour. *Ann Rev Ecology and Systematics* 5: 325-83.

—. 1975. The search for a general theory of behavior. *Behavioral Science* 20: 77-100.

—. 1979. *Darwinism and human affairs*. Seattle: University of Washington Press.

—. 1987. *The Biology of Moral Systems*. New York: Aldine de Gruyter.

—. 1990. *How did Humans Evolve? Reflections on the Uniquely Unique Species*. Ann Arbor: Museum of Zoology, The University of Michigan.

Alvard, M. 1993. Testing the "ecologically noble savage" hypothesis: Interspecific prey choice by Piro hunters of Amazonian Peru. *Hum Ecol* 21(4): 355-387.

Alroy. J. 2001. A multispecies overkill simulation of the end-Pleistocene megafaunal mass extiction. *Science* 292: 1893-1896.

—. 1994a. Conservation by native peoples: Prey choice in depleted habitats. *Hum Nat* 5: 127-154.

—. 1994b. Prey choice in a depleted habitat. *Hum Nat* 5: 127-154.

—. 1995a. Intraspecific prey choice by Amazonian hunters. *Curr Anthropol* 36(5): 789-818.

—. 1995b. Shotguns and sustainable hunting in the neotropics. *Oryx* 29: 58-66.

—. 1996. *Exploring the transition from hunting to animal husbandry: An evolutionary approach.* Presented at the 95th Annual meeting of the American Anthropological Association, San Francisco, CA., November 20-24, 1996.

—. 1997. *Humans as predators: Contexts that favor the conservation of animal resources.* Presented at the annual meeting of the Ecology Society of America, Albuquerque, NM, August 11-14, 1997.

—. 1998a. Indigenous hunting in the neotropics: conservation or optimal foraging. In *Behavioral Ecology and Conservation Biology* (Caro, T., ed.), pp. 474-500. New York: Oxford University Press.

—. 1998b. Evolutionary ecology and resource conservation. *Evol Anthropol* 7: 62-64.

—. 2000. The impact of traditional subsistence hunting and trapping on prey populations: Data from the Wana of upland Central Sulawesi, Indonesia. In *Sustainability of Hunting in Tropical Forests* (Robinson, J. & Bennett, E., eds.), pp. 214-230. New York: Columbia Press.

Alvard, M. & Kaplan, H. 1990. *Juvenile nutritional condition and kin composition among the Piro of lowland tropical Peru.* Presented at the second annual meeting of the Human Behavior and Evolution Society, University of California, Los Angeles.

Alvard, M. & Kaplan, H. 1991. Procurement technology and prey mortality among indigenous neotropical hunters. In *Human Predators and Prey Mortality* (Stiner, M. C., ed.), pp. 79-104. Boulder, CO: Westview Press.

Alvard, M., Robinson, J., Redford, K. & Kaplan, H. 1997. The sustainability of subsistence hunting in the neotropics. *Conserv Biol* 11: 977-982.

Andamari, R., Subagiyo & Talaohu, S. H. 1991. Panen trochus (*Trochus niloticus*) di desa Nolloth, P. Saparua dengan sistem "sasi." *Jurnal Penelitian Perikanan Laut* 60: 31-36.

Anderson, A. 1984. The Extinction of the Moa in Southern New Zealand. In *Quaternary Extinctions: A Prehistoric Revolution* (Martin, P. S. & Klein, R. G., eds.), pp. 728-740. Tucson: University of Arizona Press.

—. 1989. Mechanics of overkill in the extinction of New Zealand moas. *J Archaeol Sci* 16: 137-151.

Anderson, T.L. & Leal, D. 1991. *Free Market Environmentalism.* Boulder, CO: Westview Press.

Anderson, T. L. & Simmons, R. T. (eds.). 1993. *The Political Economy of Customs and Culture: Informal Solutions to the Commons Problem.* Boston: Rowan and Littlefield Publishers.

Andersson, M. 1994. *Sexual Selection.* Princeton, NJ: Princeton University Press.

Appleton, J. 1975. *The Experience of Landscape.* New York: John Wiley & Sons.

—. 1990. *The Symbolism of Habitat: An Interpretation of Landscape in the Arts.* Seattle: University of Washington Press.

Arnhart, L. 1998. *Darwinian natural right: The biological ethics of human nature.* Albany: State University of New York Press.

Arnhem, K. 1976. Fishing and hunting among the Makana. *Arseryck (Annals)*: 22-47. Goteberg, Sweden.

Arnold, J.E.M. 1998. *Managing forests as common property.* FAO Forestry Paper 136, Rome.

Aspelin, L. 1975. *External articulation and domestic production.* Latin American Studies Dissertation Series, No. 58. Cornell University.

Atran, S. 1993. Itza Maya Tropical Agro-Forestry. *Curr Anthropol* 34: 633-689.

Auger, J., Kunstmann, J. M., Czyglik, F. & Jouannet, P. 1995. Decline in semen quality among fertile men in Paris during the past 20 years. *New Engl J Med* 332: 281-285.

Aunger, R. 1992. The nutritional consequences of rejecting food in the Ituri forest of Zaire. *Hum Ecol* 20: 263-291.

—. 1994. Are food avoidances maladaptive in the Ituri forest of Zaire? *J Anthropol Res* 50: 277-310.

Axelrod, R. 1984. *The Evolution of Cooperation.* New York: Basic Books.

—. 1986. An evolutionary approach to norms. *Am Pol Sci Rev* 80: 1095-1111.

Axelrod, R. & Dion, D. 1988. The further evolution of cooperation. *Science* 242: 13851390.

Axelrod, R. & Hamilton, W. D. 1981. The evolution of cooperation. *Science* 211: 1390-1396.

Baier, S. & Lovejoy, P. E. 1977. The Tuareg of the Central Sudan: Gradations in servility at the desert edge (Niger and Nigeria). In *Slavery in Africa: Historical and Anthropological Perspectives* (Miers, S. & Kopytoff, I., eds.), pp. 391-411. Madison: University of Wisconsin Press.

Bailey, C. & Zerner, C. 1992. Community based fisheries management institutions in Indonesia. *Maritime Anthropological Studies* 5(1): 1-17.

Baines, G. 1982. Traditional conservation practices and environmental management: The international scene. In *Traditional Conservation in Papua New Guineas: Implications for Today* (Morauta, L., Pernetta, J. & Heaney, W., eds.), pp. 45-58. Boroko, Papua New Guinea: Institute of Applied Social and Economic Research.

Baksh, M. 1985. Faunal food as a "limiting factor" on Amazonian cultural behavior: A Machiguenga example. *Res Econ Anthropol* 7: 145-175.

Baland, J. M. & Platteau, J. P. 1996. *Halting Degradation of Natural Resources: Is There a Role for Rural Communities?* Oxford: Clarendon.

Ball, S. B. & Eckel, C. C. 1996. Buying status: experimental evidence on status in negotiation. *Psychol Market* 13: 381-405.

—. 1998. The economic value of status. *J Socio-Econ* 27: 495-514.

Barkan, C.P.L. 1990. A field test of risk-sensitive foraging in black-capped chickadees (*Parus atricapillus*). *Ecology* 71: 391-400.

Barkow, J. H. 1989. *Darwin, Sex, and Status: Biological Approaches to Mind and Culture.* Toronto: University of Toronto Press.

Barkow, J. H., Cosmides, L. & Tooby, J. (eds.). 1992. *The Adapted Mind. Evolutionary Psychology and the Generation of Culture*, pp. 277-306. New York: Oxford University Press.

Barnett, C., Dunbar, R. & Lycett, S. 2002. *Human Evolutionary Psychology.* Princeton, NJ: Princeton University Press.

Barnett, H. G. 1938. The Nature of the Potlatch. *Am Anthropol* 40: 349-358.

Baron, J. & Greene, J. 1996. Determinants of insensitivity to quantity in valuation of public goods: contribution, warm glow, budget constraints, availability, and prominence. *J Exp Psychol Appl* 2: 107-125.

Basalla, G. 1988. *The Evolution of Technology.* Cambridge: Cambridge University Press.

Bateman, A. J. 1948. Intrasexual selection in Drosophila. *Heredity* 2: 349-368.

Bates, D. G. & Lees, S. H. 1979. The myth of population regulation. In *Evolutionary Biology and Human Social Behavior: An Anthropological Perspective* (Chagnon, N. & Irons, W., eds.), pp. 273-289. Belmont, CA: Wadesworth.

Bateson, M. & Kacelnik, A. 1996. Rate currencies and the foraging starling: The fallacy of the averages revisited. *Behav Ecol* 7: 341-352.

Beauchamp, G. 1994. The functional analysis of human fertility decisions. *Ethol Sociobiol* 15: 31-53.

Becker, G. 1976. Altruism, egoism and genetic fitness: economics and sociobiology. *J Econ Lit* 14: 817-826.

—. 1981. *A Treatise on the Family*. Cambridge, MA: Harvard University Press.

Becker, G. S. & Barro, R. J. 1988. A reformulation of the economic theory of fertility. *Quart J Econ* 103: 1-25.

Beckerman, S. 1983. Carpe diem: An optimal foraging approach to Bari fishing and hunting. In *Adaptive Responses of Native Amazonians* (Hames, R. & Vickers, W., eds.), pp. 269-325. New York: Academic Press.

Beckerman, S. & Valentine P. 1996. On native American conservation and the tragedy of the commons. *Curr Anthropol* 37: 659-66.

Bell, N. J. & Bell, R. W. 1993. *Adolescent Risk Taking*. Newbury Park, CA: Sage.

Bellinger, D. C. & Dietrich, K. N. 1994. Low-level lead exposure and cognitive function in children. *Pediatric Annals* 23: 600-605.

Bellinger, D. C., Stiles, K. M. & Needleman, H. L. 1992. Low-level lead exposure, intelligence and academic achievements: A long-term follow-up study. *Pediatrics* 90: 855-861.

Belovsky, G. 1988. An optimal foraging-based model of hunter-gatherer population dynamics. *J anthrop Archaeol* 7: 329-372.

Ben-Ner, A. & Putterman, L. 1998. Values and institutions in economic analysis. In *Economics, values and organizations* (BenNer, A. & Putterman, L., eds.), pp. 3-69. New York: Cambridge University Press.

Benson, K. & Stephens, D. W. 1996. Interruptions, tradeoffs and temporal discounting. *Am Zool* 36: 506-517.

Bergman, R. 1980. *Amazon Economics. The Simplicity of Shipibo Indian Wealth*. Ann Arbor, MI: University Microfilms International.

Bergstrom, T. C. 1995. On the evolution of altruistic ethical rules for siblings. *Am Econ Rev* 85: 58-81.

Bergstrom, T. C. & Bagnoli, M. 1993. Courtship as a waiting game. *J Polit Econ* 101: 185-202.

Berkes, F. 1987. Common-property resource management and Cree Indian fisheries in subartic Canada. In *The Questions of the Common: The Culture and Ecology of Communal Resources* (McCay, B. & Acheson, J., eds.), pp. 66-91. Tucson: University of Arizona Press.

—. 1989a. *Common Property Resources: Ecology and Community-Based Sustainable Development*. London: Belhaven Press.

—. 1989b. Cooperation from the perspective of human ecology. In *Common Property Resources: Ecology and Community-Based Sustainable Development* (Berkes, F., ed.), pp. 70-88. London: Belhaven Press.

Berkes, F. & Folke, C. (eds.). 1998. *Linking Social and Ecological Systems: Management Practices and Social Mechanisms for Building Resilience*. New York: Cambridge University Press.

Berkes, F., Feeny, D., McCay, B. J. & Acheson, J. M. 1989. The benefits of the commons. *Nature* 340: 91-93.

Betzig, L. 1986. *Despotism and Differential Reproduction: A Darwinian View of History*. New York, Chicago: Aldine de Gruyter.

—. 1997. *Human Nature: A Critical Reader*. New York: Oxford University Press.

Binger, B. R., Copple, R. & Hoffman, E. 1995. Contingent valuation methodology in the natural resource damage regulatory process: Choice theory and the embedding phenomenon. *Nat Res J* 35: 443-459.

Binmore, K. G., Samuelson, L. &Vaughn, R. 1995. Musical chairs: Modeling noisy evolution. *Games Econ Behav* 1: 135.

Bird, M. I. 1995. Fire, prehistoric humanity, and the environment. *Interdiscipl Sci Rev* 20: 141-154.

Birdsell, J. 1958. On population structure in generalized hunting and collecting populations. *Evolution* 12: 189-205.

—. 1968. Some predictions for the Pleistocence based on equilibruim systems among recent hunter-gatherers. In *Man the Hunter* (Lee, R. & Devore, I., eds.), pp. 229-240. Chicago: Aldine.

Bjorkbom, L. 1988. Resolution of environmental problems: The use of diplomacy. In *International Environmental Diplomacy* (Carroll, J. E., ed.), pp. 123-140. Cambridge: Cambridge University Press.

Bliege Bird, R. L., Smith, E. A. & Bird, D. W. 2001. The hunting handicap: Costly signaling in human foraging strategies. *Behav Ecol Sociobiol* 50: 9-19.

Blomquist, W. 1992. *Dividing the Waters: Governing Groundwater in Southern California*. San Francisco: ICS Press.

Blume, M. E. & Friend, I. 1978. *The Changing Role of the Individual Investor*. New York: John Wiley.

Blumenshine, R. J. 1987. Characteristics of an early hominid scavenging niche. *Curr Anthropol* 28: 383-407.

Blurton-Jones, N. G. 1987. Bushman birth spacing: Direct tests of some simple predictions. *Ethol Sociobiol* 8: 183-203.

Boaz, N. T. 2002. *Evolving Health: The Origins of Illness and How the Modern World is Making Us Sick*. New York: John Wiley.

Bock, J. 2000. Evolutionary approaches to population: implications for research and policy. *Popul Environ* 21: 193-222.

Bodley, J. H. 1990. *Victims of Progress*. 3rd ed. Mountain View, CA: Mayfield Publishing.

Bodmer, R. 1995. Susceptibility of mammals to overhunting in Amazonia. In *Integrating People and Wildlife for a Sustainable Future* (Bissonette, J. & Krausman, P., eds.), pp. 292-295. Bethesda: The Wildlife Society.

Boehm, C. 1999. *Hierarchy in the Forest: The Evolution of Egalitarian Behavior*. Cambridge, MA: Harvard University Press.

Bohm, P. 1994. CVM spells responses to *hypothetical* questions. *Nat Res J* 34:37-50.

Boone, J. L. 1986. Parental investment and elite family structure in preindustrial states: A case study of late medieval-early modern Portuguese genealogies. *Am Anthropol* 88: 859-878.

—. 1988. Parental investment, social subordination and population processes among the 15th and 16th century Portuguese nobility. In *Human Reproductive Behavior: A Darwinian perspective* (Betzig, L. L., Borgerhoff Mulder, M. & Turke, P., eds.), pp. 201-219. Cambridge: Cambridge University Press.

—. 1992. Competition, conflict, and the development of social hierarchies. In *Evolutionary Ecology and Human Behavior* (Smith, E. A. & Winterhalder, B., eds.), pp. 301-337. New York: Aldine de Gruyter.

—. 1992b. Conflict, cooperation and the formation of social hierarchies. In *Evolutionary Ecology and Human Behavior* (Smith, E. A. & Winterhalder, B., eds.), pp. 301-337. New York: Aldine de Gruyter.

—. 1998. The evolution of magnanimity: When is it better to give than to receive? *Hum Nat* 9(1): 1-21.

Boone, J. L. & Kessler, K. L. 1999. More status or more children? Social status, fertility reduction, and long-term fitness. *Evol Hum Behav* 20: 257-277.

Borgerhoff Mulder, M. 1989. Early maturing Kipsigis women have higher reproductive success than late maturing women and cost more to marry. *Behav Ecol Sociobiol* 24: 145-153.

—. 1991. Datoga pastoralists of Tanzania. *National Geographic Research and Exploration* 7: 166-187.

—. 1991b. Human behavioural ecology. In *Behavioural Ecology: An Evolutionary Approach* (Krebs, J. R. & Davies, N. B., eds.), pp. 69-98. Oxford: Blackwell Scientific.

—. 1998. The demographic transition: Are we any closer to an evolutionary explanation? *Trends Ecol Evol* 13: 266-270.

—. 2000. Optimizing offspring: The quantity-quality tradeoff in agropastoral Kipsigis. *Evol Hum Behav* 21: 391-410.

Borgerhoff Mulder, M. & Ruttan, L. M. 2000. Grassland conservation and the pastoralist commons. In *Behaviour and Conservation* (Gosling, L. M. & Sutherland, W. J., eds.), pp. 34-50. Cambridge: Cambridge University Press.

Borgerhoff Mulder, M. & Sellen, D. W. 1994. Pastoralist decision making: A behavioral ecological perspective. In *African Pastoralist Systems: An Integrated Approach* (Fratkin, E., Galvin, K. A. & Roth, E A., eds.), pp. 205-229. Boulder, CO: Lynne Reinner Publications.

Borgerhoff Mulder, M., Sieff, D. & Merus, M. 1989. Datoga history in Ngorongoro Crater. *Swara* 12: 32-35.

Boswell, J. 1988. *The Kindness of Strangers: The Abandonment of Children in Western Europe from Late Antiquity to the Renaissance*. New York: Random House.

Boulding, K. E. 1977. Commons and community: The idea of a public. In *Managing the Commons* (Hardin, G. & Baden, Jo, eds.), pp. 280-294. San Francisco, CA: Freeman.

Bowles, S., Boyd, R., Fehr E. & Gintis, H. 1997. *Homo reciprocans*: A research initiative on the origins, dimensions, and policy implications of reciprocal fairness. Working paper, University of Massachusetts.

Boyce, J. 1994. Inequality as a cause of environmental degradation. *Ecol Econ* 11: 169-178.

Boyce, M. S. & Perrins, C. M. 1987. Optimizing great tit clutch size in a fluctuating environment. *Ecology* 68: 142-153.

Boyce, R. R., Brown, T. C., McClelland, G. H., Peterson, G. L. & Schulze, W. D. 1992. An experimental examination of intrinsic values as a source of the WTAWTP disparity. *Am Econ Rev* 82:1366-1373.

Boyd, R. & Lorberbaum, J. 1989. No pure strategy is evolutionarily stable in the repeated Prisoner's Dilemma game. *Nature* 327: 58-59.

Boyd, R. & Richerson, P. J. 1985. *Culture and the Evolutionary Process*. Chicago: Chicago University Press.

—. 1990. Group selection among alternative evolutionary stable strategies. *J Theor Biol* 145: 331-342.

—. 1992. Punishment allows the evolution of cooperation (or anything else) in sizable groups. *Ethol Sociobiol* 13: 171-195.

Bradburd, D. 1982. Volatility of animal wealth among Southwest Asian pastoralists. *Hum Ecol* 10: 85-106.

Brewer, G. D. 1980. On the theory and practice of innovation. *Technol Soc* 2: 337-363.

Broecker, W. S. 1997. Thermohaline circulation, the Achilles heel of our climate system: Will man-made CO_2 upset the current balance? *Science* 278: 1582-1588.

Bromley, D. W. & Cernea, M. M. 1989. The management of common property natural resources: some conceptual and operational fallacies. World Bank Discussion Papers 57. The World Bank, Washington DC.

Bromley D. W. (ed.). 1992. *Making the Commons Work: Theory, Practice, and Policy.* San Francisco: ICS Press.

Brown, D. E. 1991. *Human Universals.* New York: McGraw-Hill.

Brown, I. D. & Groeger, J. A. 1988. Risk perception and decision taking during the transition between novice and experienced driver status. *Ergonomics* 31: 585-597.

Brown, J. L. 1964. The evolution of diversity in avian territorial systems. *Wilson Bull* 76: 160-169.

Brown, L. R. 1990. The illusion of progress. In *State of the World 1990* (Brown, L. R., ed.) (A Worldwatch Institute Report on Progress for a Sustainable Society), pp. 316. New York: W.W. Norton & Co.

—. 1991. The new world order. In *State of the World 1990* (Brown, L. R., ed.) (A Worldwatch Institute Report on Progress for a Sustainable Society), pp. 3-20. New York: W.W. Norton & Co.

Brown, L. R. & Young, J. F. 1990. Feeding the world in the nineties. In *State of the World 1990* (Brown, L. R., Flavin, C., Postel, S., Starke, I., Durning, A. B., Jacobson, J., Renner, M., French, H. F., Lowe, M. D & Young, J. E., eds.), pp. 59-78. New York: W.W. Norton & Co.

Bruening, K., Kemp, F. W., Simone, N., Holding, Y., Louria, D. B. & Bogden, J. D. 1999. Dietary calcium intakes of urban children at risk of lead poisoning. *Environ Health Persp* 107: 431-435.

Brunt, P. A. 1971. *Italian Manpower 225 BC-AD 14.* Oxford: Clarendon Press.

Buchanan J. & Tullock, G. 1962. *The Calculus of Consent.* Ann Arbor: University of Michigan Press.

Buck, S. 1998. *The Global Commons: An Introduction.* Washington, DC: Island.

Budiansky, S. 1995. *Nature's Keepers.* New York: Free Press.

Bulmer, R.N.H. 1982. Traditional conservation practices in Papua New Guinea. In *Traditional Conservation in Papua New Guineas: Implications for Today* (Morauta, L., Pernetta, J. & Heaney, W., eds.), pp. 59-78. Boroko, Papua New Guinea: Institute of Applied Social and Economic Research.

Bunyard, P. 1989a. Guardians of the Amazon. *New Sci* 17: 38-41.

—. *The Colombian Amazon: Policies for the Protection of Its Indigenous Peoples and Their Environment.* Cornwall, UK: Ecological Press.

Burger, J. 1998. Effects of motorboats and personal watercraft on flight behavior over a colony of common terns. *Condor* 100: 528-534.

—. 2001. In *Protecting the Commons: A Framework for Resource Management in the Americas* (Burger, J., Ostrom, E., Norgaard, R. B., Policansky, D. & Goldstein, B., eds.). Washington DC: Island.

Burke, E. 1757 (1958). *Philosophical Enquiry into the Origin of Ideas of the Sublime and Beautiful.* London: Routledge and Paul.

Buss, D. 1994. *The Evolution of Desire.* New York: Basic Books.

Buss, D. M. 1999. *Evolutionary Psychology: The New Science of the Mind.* Boston: Allyn and Bacon.

Buss, D. M. & Shackelford, T. K. 1997. Human aggression in evolutionary psychological perspective. *Clin Psychol Rev* 17: 605-619.

Butt-Colson, A. 1973. Inter-tribal trade in the Guiana Highlands. *Anthropologica* 34: 1-70.

Byelich, J., DeCapita, M. E., Irvine, G. W., Radtke, R. E., Johnson, N. I., Jones, W. R., Mayfield, H. & Mahalak, W. J. 1985. The Kirtland's Warbler Recovery Plan. Bethesda, MD: US Fish and Wildlife Service Reference Service.

Byrne, R. 1997. Machiavellian intelligence. *Evol Anthropol* 5: 172-180.

Byrne, R. & Whiten, A. 1988. *Machiavellian Intelligence: Social Expertise and the Evolution of Intellect in Monkeys, Apes and Humans*. Oxford: Clarendon Press.

Caldwell, L. K. 1988. Beyond environmental diplomacy: The changing institutional structure of international cooperation. In *International Environmental Diplomacy* (Carroll, J. E. ed.), pp. 13-28. Cambridge: Cambridge University Press.

Callicott, J. B. 1989. *In Defense of the Land Ethic*. Albany: State University of New York Press.

Campbell, B. 1985. *Human Evolution*. 3rd ed. New York: Aldine Press.

Caraco, T. & Lima, S. L. 1985. Foraging juncos: Interaction of reward mean and variability. *Anim Behav* 33: 216-224.

Caraco, T., Martindale, S. & Whittam, T. S. 1980. An empirical demonstration of risk-sensitive foraging preferences. *Anim Behav* 28: 820-830.

Carlisle, T. R. & Zahavi, A. 1986. Helping at the nest, allofeeding, and social status in immature Arabian Babblers. *Behav Ecol Sociobiol* 18: 339-351.

Carlsen, E., Giwercman, A., Keiding, N. & Skakkebæk, N. E. 1992. Evidence for decreasing quality of semen during past 50 years. *Br Med J* 305: 609-613.

Carlsson, G. 1966. The decline of fertility: innovation or adjustment process. *Popul Stud* 20: 149-174.

Carrier, J. 1982. Conservation and conceptions of the environment: A Manus case study. In *Traditional Conservation in Papua New Guineas: Implications for Today* (Morauta, L., Pernetta, J. & Heaney, W., eds.), pp. 39-44. Boroko, Papua New Guinea: Institute of Applied Social and Economic Research.

—. 1987. Marine tenure and conservation in Papua New Guinea: Problems in interpretation. In *The Questions of the Commons: The Culture and Ecology of Communal Resources* (McCay, B. & Acheson, J., eds.), pp. 142-170. Tucson: University of Arizona Press.

Carroll, J. E. 1988. Conclusion. In *International Environmental Diplomacy* (Carroll, J. E., ed.), pp. 275-279. Cambridge: Cambridge University Press.

Carson, R. T., Flores, N. E., Martin, K. M. & Wright, J. L. 1996. Contingent valuation and revealed preference methodologies: Comparing the estimates for quasi public goods. *Land Econ* 72: 80-99.

Carson, R. T. & Mitchell, R. C. 1993. The value of clean water: The public's willingness to pay for boatable, fishable, and swimmable quality water. *Water Resources Res* 29: 2445-2454.

Cashdan, E. 1983. Territoriality among human foragers: Ecological models and an application to four Bushman groups. *Curr Anthropol* 24: 47-66.

—. 1992. Spatial organization and habitat use. In *Evolutionary Ecology and Human Behavior* (Smith, E. A. & Winterhalder, B., eds.), pp. 237-266. New York: Aldine de Gruyter.

Cassels, R. 1984 Faunal extinctions and prehistoric man in New Zealand and the Pacific Islands. In *Quaternary Extinctions: A Prehistoric Revolution* (Martin, P. S. & Klein, R. G., eds.), pp. 768-782. Tucson: University of Arizona Press.

Caughley, G. 1977. *Analysis of Vertebrate Populations*. London: John Wiley and Sons.

CavalliSforza, L. L. & Feldman, M. W. 1981. *Cultural Transmission and Evolution*. Princeton, NJ: Princeton University Press.

Chagnon, N. 1982. Sociodemographic attributes of nepotism in tribal populations: Man the rule-breaker. In *Current Problems in Sociobiology* (Kings' College Sociobiology Group, eds.), pp. 291-318. Cambridge: Cambridge University Press.

—. 1988. Life histories, blood revenge, and warfare in a Tribal Society. *Science* 239: 985-992.

Chagnon, N. & Hames, R. 1979. Protein deficiency and tribal warfare in Amazonia: New data. *Science* 203: 910-913.

Chapman, G. B. 1996. Temporal discounting and utility for health and money. *J Exp Psychol Learn Memory Cognit* 22: 771-791.

Charnov, E. 1976a. Optimal foraging: The marginal value theorem. *Theor Popul Biol* 9: 129-136.

—. 1976b. Optimal foraging: The attack strategy of a Mantid. *Am Nat* 109: 343-352.

Charnov, E. & Berrigan, D. 1993. Why do female primates have such long lifespans and so few babies? Or life in the slow lane. *Evol Anthropol* 1: 191-194.

Charnov, E. & Orians, G. 1973. *Optimal Foraging: Some Theoretical Explorations.* Salt Lake City: University of Utah, Department of Biology.

Charnov, E., Hyatt, G. & Orians, G. 1976. Ecological implications of resource depression. *Am Nat* 110: 247-59.

Cheung, S. 1973. The fable of the bees: An economic investigation. *J Law Econ* 16: 11-33.

Chiras, D. D. 1988. *Environmental Science: A Framework for Decisionmaking.* 2nd ed. Menlo Park, CA: Benjamin Cummings.

Christakis, D. A., Zimmerman, F. J., DiGiuseppe, D. L. & McCarty, C. A. 2004. Early television exposure and subsequent attentional problems in children. *Pediatrics* 113: 708-713.

Christy, F. T., Jr. 1969. Fisheries goals and the rights of property. Transactions of the *American Fisheries Society* 2: 369-378.

Ciriacy-Wantrup, S. V. & Bishop, R. C. 1975. "Common property" as a concept in natural resources policy. *Nat Resour J* 15: 713-727.

Clark, C. 1973. The economics of overexploitation. *Science* 181: 630-634.

Clark, C. W. 1990. *The Optimal Management of Renewable Resources.* 2nd ed. New York: Wiley-Interscience.

—. 1991. Economic biases against sustainable development. In *Ecological Economics: The Science and Management of Sustainability* (Costanza, R., ed.), pp. 319-343. New York: Columbia University Press.

Clark, C. W. & Mangel, M. 1984. Foraging and flocking strategies: Information in an uncertain environment. *Am Nat* 123: 626-641.

Clark, C. W. & Munro, G. R. 1994. Renewable resources as natural capital: The fishery. In *Investing in Natural Capital: The Ecological Economics Approach to Sustainability* (Jansson, A. M., Hammer, M., Folke, C. & Costanza, R., eds.), pp. 343-361. Washington, DC: Island Press.

Clarke, A. L. 1993. *Behavioral ecology of human dispersal in 19th Century Sweden.* Ph.D. Dissertation, University of Michigan.

Clay, J. W. 1988. *Indigenous Peoples and Tropical Forests.* Cambridge, MA: Cultural Survival.

Clinton, W. L. & LeBoeuf, B. J. 1993. Sexual selection's effects on male life history and the pattern of male mortality. *Ecology* 74: 1884-1892.

Clutton-Brock, T. H. (ed.). 1988. *Reproductive Success: Studies of Individual Variation in Contrasting Breeding Systems.* Chicago: University of Chicago Press.

—. 1991. *The Evolution of Parental Care.* Princeton, NJ: Princeton University Press.

Clutton-Brock, T. H., Albon, S. D. & Harvey, P. H. 1980. Antlers, body size and breeding group size in the Cervidae. *Nature* 285: 565-567.

Clutton-Brock, T. & Parker, G. 1995. Punishment in animal societies. *Nature* 373: 209-216.

Coale, A. J. & Treadway, R. 1986. A summary of the changing distribution of overall fertility, marital fertility, and the proportion married in the provinces of Europe. In

The Decline of Fertility in Europe (Coale, A. J. & Watkins, S. C., eds.), pp. 31-181. Princeton, NJ: Princeton University Press.

Colborn, T., Dumanoski, D. & Myers, J. P. 1996. *Our Stolen Future: Are We Threatening Our Fertility, Intelligence, and Survival? A Scientific Detective Story*. New York: Dutton.

Colwell, R. K. 1974. Predictability, constancy, and contingency of periodic phenomena. *Ecology* 55(5): 1148-1153.

Common, M. 1995. *Sustainability and Policy: Limits to Economics*. Cambridge: Cambridge University Press.

Connor, R. C. 1986. Pseudo-reciprocity: Investing in mutualism. *Anim Behav* 34: 1562-1566.

—. 1995. The benefits of mutualism: A conceptual framework. *Biol Rev* 70: 427-457.

Conrad, C., Lechner, M. & Werner, W. 1996. East German fertility after unification: Crisis or adaptation. *Popul Dev Rev* 22: 331-358.

Cook, E. W., Hodes, R. L. & Lang, P. J. 1986. Preparedness and phobia: Effects of stimulus content on human visceral conditioning. *J Abnorm Psychol* 95: 195-207.

Cook, M. & Mineka, S. 1989. Observational conditioning of fears to fear-relevant versus fear-irrelevant stimuli in rhesus monkeys. *J Abnorm Psychol* 98: 448-459.

—. 1990. Selective associations in the observational conditioning of fear in rhesus monkeys. *J Exp Psychol* 16: 372-389.

Cordain, L. 1999. Cereal grains: Humanity's double-edged sword. *World Rev Nutr Diet* 84: 19-73.

Corning, P. A. 1983. *The Synergism Hypothesis: A Theory of Progressive Evolution*. New York: McGraw-Hill.

—. 1995. Synergy and self-organization in the evolution of complex systems. *Systems Research* 12: 89-121.

—. 1996. The co-operative gene: On the role of synergy in evolution. *Evol Theor* 11: 183-207.

—. 1997. Holistic Darwinism: "Synergistic selection" and the evolutionary process. *J Soc Evol Syst* 20: 363-400.

—. 1998a. "The synergism hypothesis": On the concept of synergy and its role in the evolution of complex systems. *J Soc Evol Syst* 21: 133-172.

—. 1998b. Synergy: Another idea whose time has come? *J Soc Evol Syst* 21: 1-6.

—. 2003. *Nature's Magic: Synergy in Evolution and the Fate of Humankind*. Cambridge: Cambridge University Press.

Cosgrove, D. 1986. Critiques and queries. In *Landscape Meanings and Values* (Penning-Rowsell E. C. & Lowenthal, D., eds.). London: Allen and Unwin.

Cosmides, L. 1989. The logic of social exchange: Has natural selection shaped how humans reason? Studies with the Wason selection task. *Cognition* 31: 187-276.

Cosmides, L. & Tooby, J. 1989. Evolutionary psychology and the generation of culture, Part II. Case study: A computational theory of social exchange. *Ethol Sociobiol* 10: 51-97.

—. 1992. Cognitive adaptations for social exchange. In *The Adapted Mind: Evolutionary Psychology and the Generation of Culture* (Barkow, J. H., Tooby, J. & Cosmides, L., eds.), pp. 163-228. New York: Oxford University Press.

—. 1994. Better than rational: Evolutionary psychology and the invisible hand. *Am Econ Rev* 84: 327-332.

—. 1996. Are humans good intuitive statisticians after all? Rethinking some conclusions from the literature on judgment under uncertainty. *Cognition* 58: 1-73.

Cosmides, L., Tooby, J. & Barkow, J. H. 1992. Introduction: Evolutionary psychology and conceptual integration. In *The Adapted Mind: Evolutionary Psychology and the*

Generation of Culture (Barkow, J. H., Cosmides, L. & Tooby, J., eds.), pp. 3-15. New York: Oxford University Press.

Costanza, R. 1987. Social traps and environmental policy. *BioScience* 37: 407-412.

—. 1991. *Ecological Economics: The Science and Management of Sustainability.* New York: Columbia University Press.

Costanza R. et al. 1998. Principles for sustainable governance of the oceans. *Science* 281: 198-199.

Cowhey, P.F. 1985. *The Problems of Plenty: Energy Policy and International Politics.* Berkeley: University of California Press.

Crawford, E. M. 1989. *Famine: The Irish Experience, 900-1900.* Edinburgh: J. Donald.

Cronk, L. 1991. Wealth, status and reproductive success among the Mukogodo. *Am Anthropol* 93: 345-360.

Cronk, L. 1991. Human behavioural ecology. *Annu Rev Anthropol* 20: 25-53.

Cropper, M., Aydede, S. & Portney, P. 1992. Rates of time preference for saving lives. *Am Econ Rev* 82: 469-472.

Crosby, A. W. 1986. *Ecological Imperialism: The Biological Expansion of Europe, 900-1900.* Cambridge: Cambridge University Press.

Cummings, R. G., Brookshire, D. S. & Schulze, W. D. 1986. *Valuing Public Goods: The Contingent Valuation Method.* Totowa, NJ: Rowman & Allenheld.

Cummings, R. G. & Harrison, G. W. 1994. Was the Ohio court well informed in its assessment of the accuracy of the contingent valuation method? *Nat Res J* 34: 136.

Cziko, G. 1995. *Without Miracles: Universal Selection Theory and the Second Darwinian Revolution.* Cambridge, MA: MIT Press.

Daly, H. E. & Cobb, J. B., Jr. 1989. *For the Common Good. Redirecting the Economy Toward Community, the Environment, and a Sustainable Future.* Boston, MA: Beacon Press.

Daly, M. & Wilson, M. 1983. *Sex, Evolution, and Behavior.* 2nd ed. Boston, MA: Willard Grant Press.

—. 1988. *Homicide.* Hawthorne, NY: Aldine de Gruyter.

—. 1990. Killing the competition: Female/female and male/male homicide. *Hum Nat* 1: 81-107.

—. 1994. Comment on "can selfishness save the environment?" *Hum Ecol Rev* 1: 42-45.

—.1997. Crime and conflict: Homicide in evolutionary perspective. *Crime Justice* 22: 251-300.

Darwin, C. 1859. *On the Origin of Species by Means of Natural Selection.* (Facsimile of the first edition, with an introduction by Ernst Mayr, published 1987.) Cambridge, MA: Harvard University Press.

—. 1871. *The Descent of Man, and Selection in Relation to Sex.* Vol I: 475 pp.; Vol. II: 423 pp. New York: Appleton.

Dasmann, R. 1976. Life-styles and nature conservation. *Oryx* 13: 281-286.

Davis, D. L. & Bradlow, H. L. 1995. Can environmental estrogens cause breast cancer? *Sci Am* 273(4): 144-149.

Davis, J. N. & Daly, M. 1997. Evolutionary theory and the human family. *Q Rev Biol* 72: 407-435.

Dawes, R. 1980. Social dilemmas. *Ann Rev Psych* 31: 169-193.

—. 1988. *Rational Choice in an Uncertain World.* New York: Harcourt Brace Jovanovich.

Dawkins, R. 1976. *The Selfish Gene.* Oxford: Oxford University Press.

—. 1982. *The Extended Phenotype: The Gene as the Unit of Selection.* Oxford: W. H. Freeman.

—. 1986. *The Blind Watchmaker.* New York: W.W. Norton & Co.

—. 1989. *The Selfish Gene*. New ed. Oxford: Oxford University Press.

de Graaf, J., Wann, D. & Naylor, T. H. 2001. *Affluenza: The All Consuming Epidemic*. San Francisco: Berrett-Koehler.

de Waal, F.B.M. 1996. *Good Natured: The Origins of Right and Wrong in Humans and Other Animals*. Cambridge, MA: Harvard University Press.

de Waal, F. 2001. *The Ape and the Sushi Master: Cultural Reflections by a Primatologist*. New York: Basic Books.

Deevey, F. S., Rice, D. S., Rice, P. M., Vaughan, H. H., Brenner, M. & Flannery, M. S. 1979. Mayan urbanism: Impact on a tropical Karst environment. *Science* 206: 298-305.

Demsetz, H. 1967. Toward a theory of property rights. *Am. Econ. Rev.* 57, 2 (Papers & Proc.): 347-359.

Denevan, W. 1992. The pristine myth: The landscape of the Americas in 1492. *Ann Assoc Am Geogr* 82: 369-385.

Denevan, W. M. (ed.). 1976. *The Native Population of the Americas in 1492*. Madison: University of Wisconsin Press.

Deshmukh, I. 1989. On the limited role of biologists in biological conservation. *Conserv Biol* 3(3): 321.

Devine, P. G. 1995. Prejudice and out-group perception. In *Advanced social psychology* (Tesser, A., ed.), pp. 467-524. New York: McGraw-Hill.

Dewar, R. 1984. Extinctions in Madagascar: The loss of the subfossil fauna. In *Quaternary Extinctions* (Martin, P. & Klein, R., eds.), pp. 574-593. Tucson: University of Arizona Press.

Diamond, J. M. 1984. Historic extinction: A Rosetta Stone for understanding prehistoric extinctions. In *Quaternary Extinctions: A Prehistoric Revolution* (Martin, P. & Klein, R., eds.), pp. 824-864. Tucson: University of Arizona Press.

—. 1992. *The Third Chimpanzee*. New York: Harper Collins.

—. 1995. Easter's end. *Discover*, August: 63-69.

—. 1998. *Guns, Germs, and Steel: the Fates of Human Societies*. New York: W.W. Norton and Company.

Diamond, N., Bartholomew, K., McCarthy, S. & Farmer, N. 1987 Update One. *Federal Fisheries Management: A Guidebook to the Magnuson Fishery Conservation and Management Act*. Eugene, OR: Ocean & Coastal Law Center, University of Oregon Law School.

Dibb, S. 1995. Swimming in a sea of estrogens: Chemical hormone disrupters. *The Ecologist* 25: 27-31.

Dietz, T., Ostrom, E. & Stern, P.C. 2003. The struggle to govern the commons. *Science* 302: 1907-1912.

Dixon, J. A. & Sherman, P.B. 1990. *Economics of Protected Areas: A New Look at Benefits and Costs*. Washington, DC: Island Press.

Dixon, S. 1988. *The Roman Mother*. Norman: University of Oklahoma Press.

—. 1992. *The Roman Family*. Baltimore: Johns Hopkins University Press.

Dove, M. R. 1983. Theories of swidden agriculture, and the political economy of ignorance. *Agroforest Syst* 1: 85-99.

Downing, D. 2000. Mercury again. *J Nut Environ Med* 10: 267-269.

Drucker, P. & Heizer, R.F. 1967. *To Make My Name Good: A Reexamination of the Southern Kwakiutl Potlatch*. Berkeley: University of California Press.

Duff, A. M. 1958. *Freedman in the Early Empire*. Cambridge: Heffer & Sons.

Dugatkin, L. A. 1997. The evolution of cooperation: Four paths to the evolution and maintenance of cooperative behavior. *BioScience* 47(6): 355-362.

—. 1997b. *Cooperation among Animals: An Evolutionary Perspective*. Oxford: Oxford University Press.

Dugatkin, L. A., Mesterton-Gibbons, M. & Houston, A. I. 1992. Beyond the prisoner's dilemma: Toward models to discriminate among mechanisms of cooperation in nature. *Trends Ecol Evol* 7(6): 202-205.

Dunbar, R.I.M. 1987. *Primate Social Systems.* Ithaca, NY: Comstock, Cornell University Press.

Dunlap, T. R. 1988. *Saving America's Wildlife.* Princeton, NJ: Princeton University Press.

Dupont, D. 1996. Contingent valuation study of recreational opportunities in the Hamilton Harbour ecosystem. In *Proceedings of the 3rd Annual EcoWise Workshop,* McMaster University.

Durham, W. 1981. Overview: Optimal foraging analysis in human ecology. In *Hunter-Gatherer Foraging Strategies* (Winterhalder, B. & Smith, E. A., eds.), pp. 218-31. Chicago: University of Chicago Press.

—. 1991. *Coevolution: Genes, Culture, and Human Diversity.* Stanford: Stanford University Press.

Durning, A. 1992. *How Much is Enough? The Consumer Society and the Future of the Earth.* New York: W.W. Norton and Co.

—. 1993. Guardians of the land: Indigenous peoples and the health of the earth. Washington, DC: Worldwatch Paper 12.

Dyer, O. 1995. Studies highlight chemical threats to reproduction. *Br Med J* 311: 347-347.

Dyson, T. 1999. Prospects for feeding the world. *Br Med J* 319: 988-991.

Dyson-Hudson, R. & Dyson-Hudson, N. 1969. Subsistence herding in Uganda. *Sci Am* 220: 76-89.

Dyson-Hudson, R. & Smith E. A. 1978. Human territoriality: An ecological reassessment. *Am Anthropol* 80: 21-41.

Eaton, S. B., Eaton III, S. B. & Konner, M. J. 1997. Paleolithic nutrition revisited: A twelve-year retrospective on its nature and implications. *Eur J Clin Nutr* 51: 207-216.

Eaton, S. B., Konner, M. & Shostak, M. 1988. Stone agers in the fast lane: Chronic degenerative diseases in evolutionary perspective. *Am J Med* 84: 739-749.

Eaton, S. B., Strassmann, B. I., Nesse, R. M., Neel, J. V., Ewald, P. W., Williams, G. C., Weder, A. B., Eaton III, S. B., Lindeberg, S., Konner, M. J., Mysterud, I. & Cordain, L. 2002. Evolutionary health promotion. *Prev Med* 34: 109-118.

Eckel, C. C. & Grossman, P. J. 1996. The relative price of fairness: Gender differences in a punishment game. *J Econ Behav Org* 30: 143-158.

Edgerton, R. 1992. *Sick Societies: Challenging the Myth of Primitive Harmony.* New York: Free Press.

Eggan, F. 1966. *The American Indian: Perspectives for the Study of Social Change.* Chicago: Aldine.

Eggertson, T. 1993. Analyzing institutional successes and failures: A millennium of common mountain pastures in Iceland. In *The Political Economy of Customs and Culture: Informal Solutions to the Commons Problem* (Anderson, T. L. & Simmons, R. T., eds.), pp. 109-126. Boston: Rowan and Littlefield Publishers.

Ehrlich, P. R. & Ehrlich, A. H. 1981. *Extinction: The Causes and Consequences of the Disappearance of Species.* New York: Random House.

—. 1990. *The Population Explosion.* New York: Simon & Schuster.

Ehrlich, P. R. & Holdren, J. P. 1971. Impact of population growth. *Science* 171: 1212-1217.

Eibl-Eibesfeldt, I. 1989. *Human Ethology.* New York: Aldine de Gruyter.

Ellickson, R. 1991. *Order Without Law: How Neighbors Settle Disputes.* Cambridge: Harvard University Press.

Ellis, J. E. & Swift, D. M. 1988. Stability of African pastoral ecosystems: Alternate paradigms and implications for development. *J Range Manage* 41: 450-459.

Ellison, P. T. et al. 1993. Population variation in ovarian function. *Lancet* 342: 433-434.

Elster, J. 1983. *Explaining Technical Change: A Case Study in the Philosophy of Science.* Cambridge: Cambridge University Press.

Enquist, M. & Leimar, O. 1990. The evolution of fatal fighting. *Anim Behav* 39: 19.

Ensminger, J. & Knight, J. 1997. Changing social norms: common property, bridewealth, and clan exogamy. *Curr Anthropol* 38: 1-24.

Essock-Vitale, S. 1984. The reproductive success of wealthy americans. *Ethol Sociobiol* 5: 45-49.

Ewald, P. W. 1994. *Evolution of infectious disease.* Oxford: Oxford University Press.

Feeny, D., Berkes, F., McCay, B. J. & Acheson, J. M. 1990. The tragedy of the commons: Twenty-two years after. *Hum Ecol* 18: 1-19.

Feeny, D., Hannah, S. & McEvoy, A. 1996. Questioning the assumptions of the "tragedy of the commons" model of fisheries. *Land Econ* 72(2): 187-205.

Fehr, E., Fischbacher, U. & Gächter, S. 2002. Strong reciprocity, human cooperation, and the enforcement of social norms. *Hum Nat* 13: 1-25.

Feit, H. 1973. The Ethno-Ecology of the Waswanipi Cree; or, how hunters can handle their resources. In *Cultural Ecology* (Cox, B., ed.), pp. 115-125. Toronto: McClelland and Stewart.

Fershtman, C., Murphy, K. & Weiss, Y. 1996. Social Status, Education, and Growth. *J Polit Econ* 104: 108-133.

Finlayson, A. C. & McCay, B. J. 1998. Crossing the threshold of ecosystem resilience: The commercial extinction of northern cod. In *Linking Social and Ecological Systems: Management Practices and Social Mechanisms for Building Resilience* (Berkes, F. & Folke, C., eds.), pp. 311-337. New York: Cambridge University Press.

Finn, R. & Bragg, B.W.E. 1986. Perception of the risk of an accident by young and older drivers. *Accident Anal Prev* 18: 289-298.

Firth, R. 1936. *We, The Tikopia: A Sociological Study of Kinship in Primitive Polynesia.* New York: American Book Company.

—. 1957. *We, the Tikopia.* Stanford, CA: Stanford University Press.

—. 1959. *Social Change in Tikopia: Re-study of a Polynesian Community after a Generation.* London: George Allen and Unwin, Ltd.

—. 1965. *Primitive Polynesian Economy.* London: Routledge and Kegan Paul.

Fischoff, B. 1991. Value elicitation. Is there anything in there? *Am Psychol* 46: 835-847.

Fisher, I. 1930. *The Theory of Interest.* New York: Kelley and Millman.

Fisher, R. A. 1930. *The genetical theory of natural selection.* Oxford, Clarendon Press.

Fisher, R. 1958. *The Genetical Theory of Natural Selection.* 2nd ed. New York: Dover.

Fitzgibbon, C. 1990. Why do hunting cheetahs prefer male gazelles? *Anim Behav* 40: 837-845.

Fitzgibbon, C., Mogaka, H. & Fanshawe, J. 1995. Subsistence hunting in Arabuko-Sokoke forest, Kenya, and its effects in mammal populations. *Conservation Biol* 9: 1116-1126.

Flannery, T. 2001. *The Eternal Frontier: An Ecological History of North America and Its Peoples.* New York: Atlantic Monthly Press.

Flynn, J., Slovic, P. & Mertz, C. K. 1994. Gender, race, and perception of environmental health risks. *Risk Anal* 14: 1101-1108.

Foley, R. 1995. The adaptive legacy of human evolution: A search for the environment of evolutionary adaptedness. *Evol Anthropol* 4: 194-203.

Francis, C. & Cooper-Marcus, C. 1991. Places people take their problems. In *Proceedings of the 22nd Annual Conference of the Environmental Design Research Association* (Urbina-Soria, J., Ortega-Andeane, P. & Bechtel, R., eds.). Oklahoma City: EDRA.

Frank, R. H. 1985. *Choosing the Right Pond: Human Behavior and the Quest for Status.* New York: Oxford University Press.

—. 1988. *Passions within Reason.* New York: W.W. Norton and Co.

—. 1999. *Luxury Fever: Money and Happiness in an Era of Excess.* Princeton, NJ: Princeton University Press.

Fratkin, E. 1989. Household production and gender inequality in Ariaal Rendille pastoral production. *Am Anthropol* 91: 45-55.

—. 1997. Pastoralism: Governance and development issues. *Annu Rev Anthropol* 26: 235-261.

Fratkin, E. & Roth, E.A. 1990. Drought and economic differentiation among Ariaal pastoralists of Kenya. *Hum Ecol* 18: 385-402.

Fratkin, E. & Smith, K. 1994. Labor, livestock, and land: The organization of pastoral production. In *African Pastoralist Systems: An Integrated Approach* (Fratkin, E., Galvin, K. A. & Roth, E. A., eds.), pp. 91-112. Boulder, CO: Lynne Reinner Publications.

Freeman, D. 1983. *Margaret Mead and Samoa: The Making and Unmaking of an Anthropological Myth.* Cambridge, MA: Harvard University Press.

—. 1998. *The Fateful Hoaxing of Margaret Mead: A Historical Analysis of Her Samoan Research.* Boulder, CO: Westview.

Freet, J. C. 1996. Bone lead as a risk factor for hypertension in men. *Nutr Rev* 54: 180-182.

Frumkin, H. 2001. Beyond toxicity: Human health and the natural environment. *Am J Prev Med* 20: 234-240.

Gadgil, M. & Berkes, F. 1991. Traditional resource management systems. *Resour Manage Opt* 8: 127-141.

Galaty, J. G. 1994. Rangeland tenure and pastoralism in Africa. In *African Pastoralist Systems: An Integrated Approach* (Fratkin, E., Galvin, K. A. & Roth, E. A., eds.), pp. 185-204. Boulder, CO: Lynne Reinner Publications.

Gardner, G. T. & Stern, P. C. 1996. *Environmental Problems and Human Behavior.* Boston: Allyn and Bacon.

Gardner, W. 1993. A life-span rational-choice theory of risk taking. In *Adolescent risk taking* (Bell, N. J. & Bell, R. W., eds.), pp. 66-83. Newbury Park, CA: Sage.

Gibson, C. 1999. *Politicians and Poachers: The Political Economy of Wildlife Policy in Africa.* New York: Cambridge University Press.

Gibson, C. C. & Marks, S. A. 1995. Transforming rural hunters into conservationists: An assessment of community-based wildlife management programs in Africa. *World Dev* 23: 941-956.

Gifford, R. 1997. *Environmental Psychology.* Boston: Allyn & Bacon.

Gigerenzer, G., Hell, W. & Blank, H. 1988. Presentation and content: The use of base rates as a continuous variable. *J Exp Psychol Hum Percept Perform* 14: 513-525.

Gigerenzer, G. & Hoffrage, U. 1995. How to improve Bayesian reasoning without instruction: Frequency formats. *Psychol Rev* 102: 684-704.

Gigerenzer, G., Hoffrage, U. & Kleinbölting, H. 1991. Probabilistic mental models: A Brunswikian theory of confidence. *Psychol Rev* 98: 506-528.

Gillespie, J. H. 1977. Natural selection for variances in offspring numbers: A new evolutionary principle. *Am Nat* 111: 1010-1014.

Gintis, H. 2000a. *Game Theory Evolving: A Problem-Centered Introduction to Modeling Strategic Interaction.* Princeton, NJ: Princeton University Press.

—. 2000b. *Beyond Homo economicus*: Evidence from experimental economics. *Ecol Econ* 35: 311-322.

Glass, D. C. & Singer, J. E. 1972. *Urban Stress: Experiments on Noise and Social Stressors*. New York: Academic Press.

Gleditsch, N. P. (ed.). 1997. *Conflict and the Environment*. Dordrecht: Kluwer.

Glick, D., Carr, M. & Harting, B. (eds.). 1991. *Environmental Profile of the Greater Yellowstone Ecosystem*. Bozeman, MT: Greater Yellowstone Coalition.

Global Governance. 1997. *Drawing Insights from the Environmental Experience*. Cambridge, MA: MIT Press.

Goodall, J. 1986. *The Chimpanzees of Gombe: Patterns of Behavior*. Cambridge, MA: Belknap Press of Harvard University Press.

Goodin, R. 1982. Discounting discounting. *J Pub. Policy* 2: 53-72.

Goodwin, B. K., Offenbach, L. A., Cable, T. T. & Cook, P. S. 1993. Discrete/continuous contingent valuation of private hunting access in Kansas. *J Environ Manage* 39: 1-12.

Gordon, H. S. 1954. The economic theory of a common-property resource: The fishery. *J Polit Econ* 62: 124-142.

Gorsline, J. & House, L. 1974 Future primitive. *Planet Forum* 3: 1-13.

Gove, W. R. 1985. The effect of age and gender on deviant behavior: A biopsychological perspective. In *Gender and the life course* (Rossi, A. S., ed.), pp. 115-144. New York: Aldine.

Grabowski, R. 1988. Theory of induced institutional innovation: A critique. *World Dev* 16: 385-394.

Grafen, A. 1990a. Biological signals as handicaps. *J Theor Biol* 144: 517-546.

—. 1990b. Sexual selection unhandicapped by the Fisher process. *J Theor Biol* 144: 473-516.

Gramlich, E. M. 1990. *A Guide to BenefitCost Analysis*. 2nd ed. Englewood Cliffs, NJ: Prentice Hall.

Gray, L. E., Wolf, C., Lambright, C., Mann, P., Price, M., Cooper, R. L. & Ostby, J. 1999. Administration of potentially antiandrogenic pesticides (procymidone, linuron, iprodione, chlozolinate, p,p'-DDE, and ketoconazole) and toxic substances (dibutyl- and diethylhexyl phthalate, PCB 169, and ethane dimethane sulphonate) during sexual differentiation produces diverse profiles of reproductive malformations in the male rat. *Toxicol Ind Health* 15: 94-118.

Green, L. & Myerson, J. 1996. Exponential versus hyperbolic discounting of delayed outcomes: Risk and waiting time. *Am Zool* 36: 496-505.

Gregory, R., Lichstenstein, S., Brown, T.C., Peterson, C. L. & Slovic, P. 1995. How precise are monetary representations of environmental improvements? *Land Econ* 71: 462-473.

Gross, D. 1975. Protein capture and cultural development in the Amazon Basin. *Am Anthropol* 77: 526-49.

—. 1983. Village movement in relation to resources in Amazonia. *In Adaptive Responses of Native Amazonians* (Hames, R. & Vickers, W., eds.), pp. 429-50. New York: Academic Press.

Guagnano, G .A., Dietz, T. & Stern, P. C. 1994. Willingness to pay for public goods: A test of the contribution model. *Psychol Sci* 5: 411-415.

Guilford, T. & Dawkins, M. S. 1995. What are conventional signals? *Anim Behav* 49: 1689-1695.

Haas, P., Keohane, R. & Levy, M. 1993. *Institutions for the Earth: Sources of Effective Environmental Protection*. Cambridge, MA: MIT Press.

Hackworth, D. H. 1989. *About Face: The Odyssey of an American Warrior*. With Julie Sherman. New York: Simon & Schuster.

Haines, F. 1970. *The Buffalo: The Story of American Bison and Their Hunters from Prehistoric Times to the Present.* New York: Thomas Y. Crowell.

Hames, R. 1979. A comparison of the shotgun and the bow in neotropical forest hunting. *Hum Ecol* 7: 219-252.

——. 1980. Game depletion and hunting zone rotation among the Ye'kwana and Yanomamo of Amazonas, Venezuela. *In Studies of Hunting and Fishing in the Neotropics* (Hames, R., ed.) Working Papers on South American Indians, vol. 2. Bennington, VT: Bennington College.

——. 1983. The settlement pattern of a Yanomamo population bloc. In *Adaptive Responses of Native Amazonians* (Hames, R. & Vickers, W., eds.), pp. 393-427. New York: Academic Press.

——. 1987a. Garden labor exchange among the Ye'kwana. *Ethol Sociobiol* 8: 259-284.

——. 1987b. Game conservation or efficient hunting? In *The Questions of the Commons: The Culture and Ecology of Communal Resources* (McCay, B. & Acheson, J., eds.), pp. 97-102. Tucson: University of Arizona Press.

——. 1989. Time, efficiency, and fitness in the Amazonian protein quest. *Res Econ Anthropol* 11: 43-85.

——. 1990. Sharing among the Yanomamo: Part I, The effects of risk. In *Risk and Uncertainty in Tribal and Peasant Economies* (Cashdan, E., ed.), pp. 89-106. Boulder, CO: Westview Press.

——. 1991. Wildlife conservation in tribal societies. In *Biodiversity: Culture, Conservation, and Ecodevelopment* (Oldfield, M. L. & Alcorn, J. B., eds.), pp. 172-199. Boulder, CO: Westview Press.

Hames, R. & Vickers, W. 1982 Optimal diet breadth theory as a model to explain variability in Amazonian hunting. *Am Ethnol* 9: 357-378.

——. 1983 Optimal foraging theory as a model to explain variability in Amazonian hunting. *Am Ethnol* 9: 379-391.

Hamilton, W. 1964. The genetical evolution of social behavior I, II. *J Theoretical Biology* 7: 1-52.

——. 1964a. The genetical evolution of social behaviour I. *J Theor Biol* 7: 1-16.

Hamilton, W. D. 1975. Innate social attitudes of Man: An approach from evolutionary genetics. In *Biosocial Anthropology* (Fox, R., ed.), pp. 133-155. New York: John Wiley.

Hamilton, W. D. & Zuk, M. 1982. Heritable true fitness and bright birds: A role for parasites. *Science* 218: 384-387.

Hanna, S. S., Folke, C. & Maler, K. G. (eds.). 1996. *Rights to Nature: Ecological, Economic, Cultural, and Political Principles of Institutions for the Environment.* Washington, DC: Island Press.

Harcourt, A. H., Harvey, P. H., Larson, S. G. & Short, R. V. 1981. Testis weight, body weight, and breeding system in primates. *Nature* 293: 55-57.

Hardin, G. 1968. The tragedy of the commons. *Science* 162: 1243-1248.

——. 1991. The Tragedy of the Unmanaged Commons: Population and the Disguises of Providence. In *Commons without Tragedy* (R.V. Andelson, ed.). Savage, MD: Barnes and Noble.

——. 1993. *Living Within Limits.* Oxford: Oxford University Press.

——. 1998. Extensions of "the tragedy of the commons." *Science* 280 (5364): 682-683.

Hardin, G. & Baden, J. 1977. *Managing the Commons.* San Francisco, CA: W.H. Freeman & Co.

Hargrove, E. C. 1988. *Foundations of Environmental Ethics.* Englewood Cliffs, NJ: Prentice Hall.

Harpending, H. C. & Rogers, A. R. 1990. Fitness in stratified societies. *Ethol Sociobiol* 11: 497-509.

Harris, M. 1979. *Cultural Materialism: The Struggle for a Science of Culture.* New York: Random House.

Harris, M. & Ross, E. B. (eds.). 1967. *Food and Evolution.* Philadelphia: Temple University.

Hartig, T., Mang, M. & Evans, G. W. 1991. Restorative effects of natural environmental experiences. *Environ Behav* 23: 3-26.

Hartung, J. 1995. Love thy neighbor: The evolution of in-group morality. *Skeptic* 3(4): 86-99.

Harvey, P. H. & Bradbury, J. W. 1991. Sexual selection. In *Behavioural Ecology: An Evolutionary Approach* (Krebs, J. R. & Davies, N. B., eds.), pp. 203-233. Oxford: Blackwell Scientific.

Hassan, F. 1982. Demographic archaeology. In *Advances in Archaeological Method and Theory: Selections by Students from Volumes 1 through 4* (Schiffer, M. B., ed.), pp. 225-279. New York: Academic Press.

Hastjarjo, T., Silberberg, A. & Hursh, S. R. 1990. Risky choice as a function of amount and variance in food supply. *J Exp Anal Behav* 53: 155-161.

Hauert, C., De Monte, S., Hofbauer, J. & Sigmund K. 2002. Volunteering as Red Queen mechanism for cooperation in public goods games. *Science* 296: 1129-1132.

Hausfater, G. & Hrdy, S. B. 1984. *Infanticide: Comparative and Evolutionary Perspectives.* New York: Aldine.

Hawken, P., Lovins, A. B. & Lovins, L. H. 1999. *Natural Capitalism: The Next Industrial Revolution.* Boston, MA: Little, Brown.

Hawkes, K. 1992. Sharing and collective action. In *Evolutionary Ecology and Human Behavior* (Smith, E. A. & Winterhalder, B., eds.), pp. 269-300. New York: Aldine de Gruyter.

—. 1993. Why hunter-gatherers work: An ancient version of the problem of public goods. *Curr Anthropol* 34:341-361.

Hawkes, K., Hill, K. & O'Connell, J. 1983. Why hunters gather: Optimal foraging and the Ache of Eastern Paraguay. *Am Ethnol* 9: 379-391.

Hawkes, R. & Charnov, E. L. 1988. On human fertility: Individual or group benefit? *Curr Anthropol* 29: 469-471.

Hecht, S. & Cockburn, A. 1990. *The Fate of the Forest: Developers, Destroyers, and Defenders of the Amazon.* New York: Harper Collins.

Hechter, M., Nadel, L. & Michod, R. E. 1993. *The Origin of Values.* New York: Aldine de Gruyter.

Hechter, M., Opp, K.-D. & Wippler, R. 1990. Introduction. In *Social Institutions: Their Emergence, Maintenance and Effects* (Hechter, M., Opp, K.-D. & Wippler, R., eds.), pp. 1-9. New York: Aldine de Gruyter.

Heerwagen, J. H. & Orians, G. H. 1993. Humans, habitats, and aesthetics. In *The Biophilia Hypothesis* (Kellert, S. R. & Wilson, E. O., eds.), pp. 138-172. Washington, DC: Island Press.

Heinen, J. T. 1993. Park/people relations in Kosi Tappu Wildlife Reserve, Nepal: A socioeconomic analysis. *Environ Conserv* 20(1): 25-34.

—. 1994. Emerging, diverging and converging paradigms for sustainable development. *Int J Sust Dev World* 1(1): 22-33.

Heinen, J. T. & Kattel, B. 1992a. Parks, people, and conservation: A review of management issues in Nepal's protected area. *Popul Environ* 14(1): 49-84.

—. 1992b. A review of conservation legislation in Nepal: Past progress and future needs. *Environ Manage* 16(6): 723-733.

Heinen, J. T. & Leisure, B. 1993. A new look at the Himalayan fur trade. *Oryx* 27(4): 231-238.

Heinen, J. & Low, B. 1992. Human behavioral ecology and environmental conservation. *Environ Conserv* 19(2): 105-116.

Heinen, J. T. & Yonzon, P.B. 1994. A review of conservation issues and programs in Nepal: From a single-species focus toward biodiversity protection. *Mt Res Dev* 14(1): 61-76.

Heller, W. P. & Starrett, D. A. 1976. On the nature of externalities. In *Theory and Measurement of Economic Externalities* (Lin, S.A.Y., ed.), pp. 9-22. New York: Academic Press.

Henderson, N. & Bateman, I. 1995. Empirical and ppublic choice evidence for hyperbolic social discount rates and the implications for intergenerational discounting. *Environ Res Econ* 5: 413-423.

Henderson, N. & Sutherland, W. J. 1997. Discounting and conservation: Another final word. *Trends Ecol Evol* 12: 402.

Hepburn, R. W. 1976. Aesthetic appreciation in nature. In *Aesthetics in the Modern World* (Osborne, H., ed.). London: Thames and Hudson.

Hermy, C. 1983. Gunung Mulu National Park, Sarawak. *Oryx* 17: 6-14.

Herren, U. J. 1990. Socioeconomic stratification and small stock production in Mukogodo Division, Kenya. *Res Econ Anthropol* 12: 111-148.

Herrnstein, R. J. 1990. Rational choice theory: Necessary but not sufficient. *Am Psychol* 45: 356-367.

Hess, C. 1996. *Common Pool Resources and Collective Action: A Bibliography*, Vol. 3, and *Forest Resources and Institutions: A Bibliography*. Bloomington: Workshop in Political Theory and Policy Analysis, Indiana University. www.indiana.edu/~workshop/wsl/wsl.html.

Hewlett, B. S. 1988. Sexual selection and paternal investment among Aka pygmies. In *Human Reproductive Behaviour* (Betzig, L., Borgerhoff Mulder, M. & Turke, P., eds.). Cambridge, MA: Cambridge University Press.

Heyde, J. M. 1995. Is contingent valuation worth the trouble? *Univ Chicago Law* Rev 62: 331-362.

Hildebrand, G. 1991. *The Wright Space: Pattern and Meaning in Frank Lloyd Wright's Houses*. Seattle: University of Washington Press.

Hildebrandt, W. & Jones, T. 1992. Evolution of marine mammal hunting: A view from the California and Oregon coasts. *J Anthropol Archeol* 11: 360-401.

Hill, J. 1984. Prestige and reproductive success in man. *Ethol Sociobiol* 5: 77-95.

Hill, K. 1988. Macronutrient modifications of optimal foraging theory: An approach using indifference curves applied to some modern foragers. *Hum Ecol* 16: 157-197.

—. 1993 Life history and evolutionary anthropology. *Evol Anthropol* 2: 78-88.

—. 1995. Comment on Alvard M., "Intraspecific prey choice by Amazonian hunters." *Curr Anthropol* 36: 805-807.

—. 1995b. Commentary. *Curr Anthropol* 36: 789-818.

Hill, K. & Hawkes, K. 1983. Neotropical hunting among the Ache of Eastern Paraguay. In *Adaptive Responses of Native Amazonians* (Hames, R. & Vickers, W., eds.), pp. 139-188. New York: Academic Press.

Hill, K. & Hurtado, M. 1996. *Ache Life History: The Ecology and Demography of a Foraging People*. Hawthorne, New York: Aldine de Gruyter.

Hill, K. & Kaplan, H. 1988. Tradeoffs in male and female reproductive strategies among the Ache: Part I. In *Human Reproductive Behavior: A Darwinian Perspective* (Betzig, L., Borgerhoff Mulder, M. & Turke, P., eds.), pp. 227-289. Cambridge, UK: Cambridge University Press.

Hilton, R. M. 1992. Institutional incentives for resource mobilization: An analysis of irrigation schemes in Nepal. *J Theor Pol* 4: 283-308.

Hirsch, F. 1978. *Social Limits to Growth*. Cambridge, MA: Harvard University.

Hirshleifer, J. 1991. The paradox of power. *Econ Politics* 3: 177-200.

Hoch, S J. & Loewenstein, G. F. 1991. Timeinconsistent preferences and consumer selfcontrol. *J Consum Res* 17: 492-507.

Hockey, R. (ed.). 1983. *Stress and Fatigue in Human Performance*. New York: John Wiley.

Hodgson, G. (ed.). 1995. *Economics and Biology*. Aldershot: Edward Elgar.

Hoffman, E., McCabe, K. & Smith, V. L. 1996. Social distance and otherregarding behavior in dictator games. *Am Econ Rev* 86: 653-660.

Holdaway, R. N. & Jacomb, C. 2000. Rapid extinction of the moas (Aves: Dinornithiformes): Model, test, and implications. *Science* 287: 2250-2274.

Holdren, J. P. 1991. Population and the energy problem. *Popul Environ* 12: 231-255.

Holdren, J. P. & Ehrlich, P. R. 1974. Human population and the global environment. *Am Sci* 62: 282-292.

Holinger, P.C. 1987. *Violent Deaths in the United States*. New York: Guilford Press.

Holinger, P. C. & Klemen, E. 1982. Violent deaths in the United States 1900-1975: Relation-ships between suicide, homicide and accidental deaths. *Soc Sci Med* 16: 1929-1938.

Holling, C. 1959. Some characteristics of simple types of predation and parasitism. *Can Entomol* 91: 385-398.

Holmes, R. 1985. *Acts of War: The Behavior of Men in Battle*. New York: Free Press.

Homer-Dixon, T. F. 1991. On the threshold: Environmental changes as causes of acute conflict. *Int Security* 16: 76-116.

—. 1999. *Environment, Scarcity, and Violence*. Princeton, NJ: Princeton University Press.

Homewood, K. M. & Rodgers, W. A. 1991. *Maasailand Ecology*. Cambridge: Cambridge University Press.

Hooper, A. 1989. Tokelau fishing in traditional and modern contexts. In *Traditional Marine Resource Management in the Pacific Basin: An Anthology* (Ruddle, K. & Johannes, R. E., eds.), pp. 213-240. Jakarta, Indonesia: UNESCO/ROSTSEA.

Hopkins, K. 1965. Contraception in the Roman Empire. *Comp Stud Soc Hist* 8: 124-151.

—. 1983. *Death and Renewal*. Cambridge: Cambridge University Press.

Hough, J. L. & Sherpa, M. N. 1989. Bottom up vs. basic needs: Integrated conservation and development in the Annapurna and Michuru Mountain Conservation Areas of Nepal and Malawi. *Ambio* 18(8): 434-441.

Houston, D. 1979. The adaptation of scavengers. In *Serengeti* (Sinclair, A. & Norton-Griffiths, M., eds.), pp. 263-286. Chicago: University of Chicago Press.

Howarth, R. B. 1996. Status effects and environmental externalities. *Ecol Econ* 16: 25-34.

Howell, N. 1979. *The Demography of the Dobe! Kung*. New York: Academic Press.

—. 1986. Feedbacks and buffers in relation to scarcity and abundance: Studies of hunter-gatherer populations. In *The State of Population Theory: Forward from Malthus*. (Coleman, D. & Schofield, R. eds.), pp. 156-187. London: Basil Blackwell.

Hrdy, S. B. & Judge, D. S. 1993. Darwin and the puzzle of primogeniture. *Hum Nat* 4: 1-45.

Hsieh, C. C. & Pugh, M. D. 1993. Poverty, income inequality, and violent crime: A metaanalysis of recent aggregate data studies. *Crim Justice Rev* 18: 182-202.

Hu, H., Aro, A., Payton, M., Korrick, S., Sparrow, D., Weiss, S. T. & Rotnitzky, A. 1996. The relationship of bone and blood lead to hypertension. *J Amer Med Assoc* 275: 1171-1176.

Hugdahl, K. 1978. Electrodermal conditioning to potentially phobic stimuli: Effects of instructed extinction. *Behav Res Ther* 16: 315-321.

Hugdahl, K. & Karker, A. C. 1981. Biological vs. experiential factors in phobic conditioning. *Behav Res Ther* 19: 109-115.

Hughes, J. 1983. *American Indian Ecology*. El Paso: Texas Western Press.

Humphrey, C. & Sneath, D. (eds.). 1996. *Culture and Environment in Inner Asia*. Vol. 1. Cambridge: White Horse Press.

Humphrey, N. K. 1983. *Consciousness Regained: Chapters in the Development of Mind*. Oxford: Oxford University Press.

Hunn, E. 1982. Mobility as a factor limiting resource use in the Columbia Plateau of North America. In *Resource Managers: North American and Australian Hunter-Gatherers* (Williams, N. & Hunn, E., eds.), pp. 17-43. Boulder, CO: Westview Press.

Hviding, E. 1989. Keeping the sea: Aspects of marine tenure in Marovo lagoon, Solomon Islands. In *Traditional Marine Resource Management in the Pacific Basin: An Anthology* (Ruddle, K. & Johannes, R. E., eds.), pp. 7-44. Jakarta, Indonesia: UNESCO/ROSTSEA.

IIED. 1994. Whose Eden? *An overview of community approaches to wildlife management*. London: International Institute for Environment and Development.

IWIGIA. 1992. Declaration by the indigenous peoples. *IWIGIA Yearbook*: 157-163.

Irons, W. G. 1979. Cultural and biological success. In *Evolutionary Biology and Human Social Behavior: An Anthropological Perspective* (Chagnon, N. A. & Irons, W., eds.), pp. 257-272. North Scituate, MA: Duxbury.

—. 1983. Human female reproductive strategies. In *Social Behavior of Female Vertebrates* (Wasser, S.K., ed.), pp. 169-213. New York Academic Press.

—. 1991. How did morality evolve? *Zygon* 26: 49-89.

—. 1998. Adaptively relevant environments versus the environment of evolutionary adaptedness. *Evol Anthropol* 6: 194-204.

Irwin, C.E. 1993. Adolescence and risk taking. In *Adolescent risk taking* (Bell, N. J. & Bell, R. W., eds.), pp. 17-28. Newbury Park, CA: Sage.

Irwin, J. R., Slovic, P., Lichtenstein, S. & McClelland, G. H. 1993. Preference reversals and the measurement of environmental values. *J Risk Uncert* 6: 5-18.

Isen, A. M. 1990. The influence of positive and negative affect on cognitive organization: Some implications for development. In *Psychological and Biological Approaches to Emotion* (Stern, N. L., Leventhal, B. & Trabasso, T., eds.). Hillsdale, NJ: Lawrence Erlbaum Associates.

IUCN. 1990. *United Nations List of National Parks and Protected Areas*. Cambridge: IUCN Publication Services.

Ives, J. D. & Messerli, B. 1989. *The Himalayan Dilemma: Reconciling Development and Conservation*. New York: Routledge.

Jensen, T. K., Carlsen, E., Jorgensen, N., Berthelsen, J. G., Keiding, N., Christensen, K., Petersen, J. H., Knudsen, L. B. & Skakkebaek, N. E. 2002. Poor semen quality may contribute to recent decline in fertility rates. *Hum Reprod* 17: 1437-1440.

Johannes, R. E. 1978. Traditional marine conservation methods in Oceania and their demise. *Ann Rev Ecol Syst* 9: 349-364.

—. 1981. *Words of the Lagoon: Fishing and Marine Lore in the Palau District of Micronesia*. Berkeley: University of California Press.

—. 1982. Traditional conservation methods and protected marine areas in Oceania. *Ambio* 11: 258-261.

—.1989. *Traditional Ecological Knowledge: A Collection of Essays*. Gland, Switzerland and Cambridge: IUCN.

Johannes, R. E. & MacFarlane, J. W. 1989. Assessing traditional fishing rights systems in the context of marine resource management: A Torres Strait example. In *Traditional Marine Resource Management in the Pacific Basin: An Anthology* (Ruddle, K. & Johannes, R. E., eds.), pp. 241-262. Jakarta, Indonesia: UNESCO/ROSTSEA.

Johansson, S. R. 1987. Status anxiety and demographic contraction of privileged populations. *Popul Dev Rev* 13: 349-470.

Johnson, A. 1982. Reductionism in cultural wcology: The Amazonian case. *Curr Anthropol* 23: 413-428.

Johnson, A. 1989. How the Machiguenga manage resources: Conservation or exploitation of nature? In *Resource Management in Amazonia: Indigenous and Folk Strategies* (Posey, D. A. & Balée, W., eds.). Advances in Economic Botany 7. New York: The New York Botanical Society.

Johnstone, R. 1997. The evolution of animal signals. In *Behavioral Ecology: An Evolutionary Approach* (Krebs J. R. & Davies, N. B., eds.), pp. 155-178. London: Blackwell Scientific.

Jonah, B. A. 1986. Accident risk and risktaking behaviour among young drivers. *Accident Anal Prev* 18: 255-271.

Jones, D. 1996. Witchcraft, sorcery, magic and social order among the Ibibio of Nigeria. *Africa* 66: 321-333.

Kacelnik, A. 1997. Normative and descriptive models of decision making: Time discounting and risk sensitivity. In *Characterizing human psychological adaptations* (Bock, G. & Cardew, G., eds.). Ciba Foundation Symposium 208, pp. 51-70. London: John Wiley.

Kacelnik, A. & Bateson, M. 1996. Risky theories—the effects of variance on foraging decisions. *Am Zool* 36: 402-434.

Kagel, J. H., Green, L. & Caraco, T. 1986. When foragers discount the future: Constraint or adaptation? *Anim Behav* 34: 271-283.

Kagel, J. H. & Roth, A. E. (eds.). 1995. *The Handbook of Experimental Economics*. Princeton, NJ: Princeton University Press.

Kahneman, D., 1986. Comments on the contingent valuation method. In *Valuing Environmental Goods: An Assessment of the Contingent Valuation Method* (Cummings, R. G., Brookshire, D. S. & Schulze, W. D., eds.), pp. 185-193. Totowa, NJ: Rowman & Allanheld.

Kahneman, D. & Knetsch, J. L. 1992. Valuing public goods: The purchase of moral satisfaction. *J Environ Econ Manage* 22: 57-70.

Kahneman, D., Ritov, I., Jacowitz, K. E. & Grant, P. 1993. Stated willingness to pay for public goods: A psychological perspective. *Psychol Sci* 4: 310-315.

Kahneman, D., Slovic, P. & Tversky, A. 1982. *Judgement under Uncertainty*. New York: Cambridge University Press.

Kahneman, D. & Tversky, A. 1979. Prospect theory: An analysis of decision under risk. *Econometrika* 47: 263-291.

—. 1984. Choices, values, and frames. *Am Psychol* 39: 341-350.

Kaplan, H. 1996. A theory of fertility and parental investment in traditional and modern human societies. *Yearb Phys Anthropol* 39: 91-135.

Kaplan, H. & Hill, K. 1985a. Hunting ability and reproductive success among male Ache foragers: preliminary results. *Curr Anthropol* 26: 131-133.

—. 1985b. Food sharing among Ache foragers: Tests of explanatory hypotheses. *Curr Anthropol* 26: 223-246.

—. 1986. Sexual strategies and social-class differences in fitness in modern societies. *Behav Brain Sci* 9: 198-201.

—. 1992. The evolutionary ecology of food acquisition. In *Evolutionary Ecology and Human Behavior* (Smith, E. A. & Winterhalder B., eds.), pp. 167-201. New York: Aldine de Gruyter.

Kaplan, H. & Kopishke, K. 1992. Resource use, traditional technology, and change among native peoples of lowland South America. In *Conservation of neotropical forests* (Redford, K. H. & Padoch, C., eds.), pp. 83-107. New York: Columbia University Press.

Kaplan, H. S. & Lancaster, J. B. 2000a. The evolutionary economics and psychology of the demographic transition to low fertility. In *Adaptation and Human Behavior: An Anthropological Perspective* (Cronk, L., Chagnon N. A. & Irons, W., eds.), pp. 283-322. New York: Aldine de Gruyter.

—. 2000b. The life histories of men in Albuquerque: An evolutionary-economic analysis of parental investment and fertility in modern society. In *Adaptation and Human Behavior: An Anthropological Perspective* (Cronk, L., Chagnon, N. A. & Irons, W., eds.). Hawthorne, NY: Aldine de Gruyter.

Kaplan, H. S. et al. 1995. Fertility and fitness among Albuquerque men: A competitive labour market theory. In *Human Reproductive Decisions* (Dunbar, R.I.M., ed.), pp. 96-136. St. Martin's Press.

Kaplan, S. & Talbot, J. F. 1983. Psychological benefits of a wilderness experience. In *Behavior in the Natural Environment* (Altman, I. & Wohlwill, J. F., eds.). New York: Plenum.

Kaplan, S. 1987. Aesthetics, affect, and cognition: environmental preferences from an evolutionary perspective. *Environ Behav* 19: 3-32.

—. 1992. Environmental preferences in a knowledge-seeking, knowledge-using organism. In *The Adapted Mind: Evolutionary Psychology and the Generation of Culture* (Barkow, J. H., Tooby, J. & Cosmides, L., eds.), pp. 581-598. Oxford: Oxford University Press.

Kauffman, W. 1995. *No Turning Back: Dismantling the Fantasies of Environmental Thinking.* New York: Basic Books.

Kay, C. 1990. *Yellowstone's northern elk herd: A critical evaluation of the "natural regulation" Paradigm,* Ph.D. Dissertation, Utah State University.

—. 1994. Aboriginal overkill: The role of American Indians in structuring western ecosystems. *Hum Nat* 5: 359-398.

—. 1995. Aboriginal overkill and native burning: Implications for modern ecosystem management. *Western J Appl Forestry* 10: 121-126.

Kay, J. 1985. Native Americans in the fur trade and wildlife depletion. *Environmental Review* 9: 118-130.

Keckler, C.N.W. 1997. Catastrophic mortality in simulations of forager age-at-death: Where did all the humans go? In *Integrating Archaeological Demography: Multidisciplinary Approaches to Prehistoric Populations* (Paine, R., ed.). Center for Archaeological Investigations Occasional Papers No. 24. pp. 205-228. Carbondale: Southern Illinois University Press.

Keeley, L. H. 1996. *War before Civilization: The Myth of the Peaceful Savage.* New York & Oxford: Oxford University Press.

Keller, L. (ed.). 1999. *Levels of Selection in Evolution.* Princeton, NJ: Princeton University Press.

Kellert, S. J. & Wilson, E. O. (eds.). 1993. *The Biophilia Hypothesis.* Washington, DC: Island Press, Shearwater Books.

Kendler, K. S., Neale, M. C., Kessler, R. C., Heath, A. C. & Eaves, L. J. 1992. The genetic epidemiology of phobias in women. *Arch Gen Psychiat* 49: 273-281.

Kensinger, K. & Kracke, W. (eds.). 1981. *Food Taboos in Lowland South America.* Working Papers on South American Indians, No. 3. Bennington, VT: Bennington College.

Keohane, R. 1984. *After Hegemony*. Princeton, NJ: Princeton University Press.

Keohane, R. & Ostrom, E. (eds.). 1995. *Local Commons and Global Interdependence*: *Heterogeneity and Cooperation in Two Domains*. London: Sage.

Keohane, R., Ostrom, E. & McGinnis, M. 1993. Linking local and global commons: Monitoring, sanctioning, and theories of self-organization in common pool resources and international regimes. In *Proceedings of a Conference on Linking Local and Global Commons* (Keohane, R., McGinnis, M. & Ostrom, E., eds.), pp. 1-15. Bloomington: Indiana University Workshop in Political Analysis and Theory.

Keyfitz, N. 1984. Introduction: Biology and demography. In *Population and Biology* (Keyfitz, N., ed.), pp. 1-7. Ordina Editions.

Kiltie, R. 1980. More on Amazonian cultural ecology. *Cultural Anthropology* 23: 541-544.

Kim, R., Rotnitzky, A., Sparrow, D., Weiss, S. T., Wager, C. & Hu, H. 1996. A longitudinal study of low-level lead exposure and impairment of renal function. *J Amer Med Assoc* 275: 1177-1181.

King, B. 1994. *Information Continuum: Evolution of Social Information Transfer in Monkeys, Apes and Hominids*. Santa Fe: School of American Research.

Kingsland, S.E. 1985. *Modeling Nature. Episodes in the History of Population Ecology*. Chicago: University of Chicago Press.

Kirby, K. N. & Herrnstein, R. J. 1995. Preference reversals due to myopic discounting of delayed reward. *Psychol Sci* 6: 83-89.

Kirch, P. V. & Yen, D. 1982. Tikopia: The prehistory and ecology of a Polynesian outlier. Bernice P. Bishop Museum Bulletin 238.

Klein, R. 1995. Anatomy, behavior and modern human origins. *J World Prehist* 9: 167-198.

Klepper, S. & Nagin, D. 1989. Tax compliance and perceptions of the risks of detection and criminal prosecution. *Law Soc Rev* 23: 209-240.

Klima, G. 1964. Jural relations between the sexes among the Barabaig. *Africa* 34: 9-19.

—. 1965. *Kinship, property and jural relations among the Barabaig*. Ph.D. dissertation, University of California.

Kluger, M. J. 1991. The adaptive value of fever. In *Fever: Basic Mechanisms and Management* (Mackowiak, P. A., ed.), pp. 105-124. New York: Raven Press.

Kluger, M. J., Kozak, W., Conn, C. A., Leon, L. R. & Soszynski, D. 1996. The adaptive value of fever. *Infect Dis Clin N Am* 10: 1-20.

Knetsch, J. L. 1995. Asymmetric valuation of gains and losses and preference order assumptions. *Econ Inq* 33: 134-141.

Knodel, J. et al. 1982. *Fertility in Thailand: Trends, Differentials, and Proximate Determinants*. National Academy of Sciences.

Knopf, R. C. 1987. Human behavior, cognition, and affect in the natural environment. In *Handbook of Environmental Psychology* (Stokols, D. & Altman, I., eds.). New York: John Wiley.

Kollock, P. 1998. Social dilemmas: The anatomy of cooperation. *Annu Rev Anthropol* 24: 183-214.

Krebs, D. L. & Denton, K. 1997. Social illusions and self-deception: The evolution of biases in person perception. In *Evolutionary Social Psychology* (Simpson, J. A. & Kenrick, D. T., eds.), pp. 21-47. Mahwah, NJ: Lawrence Erlbaum.

Krebs, J. 1978. Optimal foraging: Decision rules for predators. In *Behavioural Ecology: An Evolutionary Approach* (Krebs, J. & Davies, N., eds.), pp. 23-63. Oxford: Blackwell Scientific Publications.

Krebs, J. R. & Davies, N. B. 1991a. *Behavioural Ecology: An Evolutionary Approach*, 3rd ed. Oxford: Blackwell Scientific.

—. 1991b. *An Introduction to Behavioral Ecology*. MA: Sinauer Associates, Sunderland.

Krech, S. I. 1999. *The Ecological Indian: Myth and History*. New York: W.W. Norton.

Kreps, D. M., Milgrom, P., Roberts, J. & Wilson, R. 1982. Rational cooperation in the finitely repeated prisoner's dilemma. *J Econ Theory* 27: 245-252.

Krier, J. E. 1992. The tragedy of the commons, part two. *Harvard J Law Pub Policy* 15: 325-347.

Krupnik, I. 1993. Prehistoric Eskimo whaling in the arctic: Slaughter of calves or fortuitous ecology? *Arctic* 30: 1-12.

Kurland, J. & Beckerman, S. 1985. Optimal foraging and hominid evolution: Labor and reciprocity. *Am Anthropol* 87: 73-93.

Lack, D. 1947. The significance of clutch size. *Ibis* 89: 302-352.

Lahr, M. & Foley, R. 1992. Multiple dispersals and modern human origins. *Evol Anthropol* 3: 48-60.

Lam, W. F. 1996. Improving the performance of small-scale irrigation systems: The effects of technological investments and governance structure on irrigation performance in Nepal. *World Dev* 24(8): 1301-1315.

—. 1998. *Governing Irrigation Systems in Nepal: Institutions, Infrastructure, and Collective Action*. Oakland, CA: ICS Press.

Lamprey, H. F. 1983. Pastoralism yesterday and today: the overgrazing problem. In *Tropical savannas* (Bourliere, F., ed.), pp. 643-666. Amsterdam: Elsevier.

Lande, R., Engen, S. & Saether, B. 1994. Optimal harvesting, economic discounting and extinction risk in fluctuating populations. *Nature* 372(3): 88-90.

Lane, C. 1990. *Barabaig natural resource management: Sustainable land use under threat of destruction*. United Nations Research Institute for Social Development, Discussion Paper 12, June 1990.

—. 1991. *Alienation of Barabaig pasture land: Policy implications for pastoral development in Tanzania*. Ph.D. Dissertation, University of Sussex.

—. 1996. *Pastures Lost: Barabaig Economy, Resource Tenure, and the Alienation of Their Land in Tanzania*. Nairobi, Kenya: Initiatives Publishers.

Lang, K. & Ruud, P. 1986. Returns to schooling, implicit discount rates, and black-white wage differentials. *Rev Econ Sta* 68: 41-47.

LaPointe, J. 1970. *Residential Patterns and Wayana Social Organization*. Ph.D. Dissertation, Columbia University.

Lappé, M. 1994. *Evolutionary Medicine: Rethinking the Origins of Disease*. San Francisco, CA: Sierra Club Books.

Latane, B. & Rodin, J. 1969. A lady in distress: Inhibiting effects of friends and strangers on bystander intervention. *J Experimental Social Psychology* 5: 189-202.

Leacock, E. 1954. The Montagnais "hunting territory" and the fur trade. *American Anthropological Association Memoir* 78, 56(5): pt. 2: 1-59.

Lenton, T. M. 1998. Gaia and natural selection. *Nature* 394: 439-437.

Lesthaeghe, R. & Wilson, C. 1986. Modes of production, secularization, and the pace of fertility decline in Western Europe, 1870-1930. In *The Decline of Fertility in Europe* (Coale, A. J. & Watkins, S. C., eds.), pp. 261-292. Princeton, NJ: Princeton University Press.

Levy, J. 1992. *Orayvi Revisited: Social Stratification in an "Egalitarian" Society*. Santa Fe: School of American Research Press.

Lewis, H. T. 1982. Fire technology and resource management in aboriginal North America and Australia. In *Resource Managers: North American and Australian Hunter-Gatherers* (Williams, N. & Hunn, E., eds.), pp. 45-67. Boulder, CO: Westview Press.

Lewontin, R. C. 1970. The units of selection. *Annu Rev Ecol Syst* 1: 1-18.

Libecap, G. D. 1989. Distributional issues in contracting for property rights. *J Instl Theor Econ* 145: 6-24.

Lind, R. 1990. Reassessing the government's discount rate policy in light of new theory and data in a world economy with a high degree of capital mobility. *J Environ Econ Manage* 18: S8-S28.

Liverman, M. 1990. The (endangered) Endangered Species Act: Political economy of the Northern Spotted Owl. *Endangered Species* UPDATE 7(10): 14.

Livi-Bacci, M. 1979. Social group forerunners of fertility control in Europe. In *The Decline of Fertility in Europe* (Coale, A. J. & Watkins, S. C., eds.), pp. 182-200. Princeton, NJ: Princeton University Press.

Lloyd, W. F. 1833. *Lectures on population, value, poor laws and rent* [Facs. ed. (1968)]. Augustus M. Kelley.

Loewenstein, G. & Adler, D. 1995. A bias in the prediction of tastes. *Econ J* 105: 929-937.

Loewenstein, G. & Thaler, R.H. 1989. Intertemporal choice. *J Econ Perspect* 3: 181-193.

Logue, A. 1988. Research on self-control: An integrating framework. *Behav Brain Sci* 11: 665-710.

Loomis, J., Lockwood, M. & DeLacy, T. 1993. Some empirical evidence on embedding effects in contingent valuation of forest protection. *J Environ Econ Manage* 24: 45-55.

Lopes, L. L. 1987. Between hope and fear: The psychology of risk. *Adv Exp Soc Psychol* 2: 255-295.

—. 1993. Reasons and resources: The human side of risk taking. In *Adolescent risk taking* (Bell, N. J. & Bell, R. W., eds.), pp. 29-54. Newbury Park, CA: Sage.

Lovejoy, C.O. 1981. The origin of man. *Science* 211: 341-350.

Lovelock, J. 1979. *Gaia: A New Look at Life on Earth*. Oxford: Oxford University Press.

—. 1988. *The Ages of Gaia*. New York: W. W. Norton & Co.

Lovelock, J. & Margulis, L. 1974. Biological modulation of the earth's atmosphere. *Icarus* 21: 471-489.

Low, B.S. 1978. Environmental uncertainty and the parental strategies of marsupials and placentals. *Am Nat* 112: 197-213.

—. 1989a. Crosscultural patterns in the training of children: An evolutionary perspective. *J Comp Psychol* 103: 311-319.

—. 1989b. Occupational status and reproductive behavior in 19th century Sweden: Locknevi parish. *Soc Biol* 36: 82-101.

—. 1989c. Human responses to environmental extremeness and uncertainty. A cross-cultural perspective. In *Risk and Uncertainty in Tribal and Peasant Societies* (Cashdan, E., ed.), pp. 229-255. Boulder, CO: Westview Press.

—. 1990a. Sex, power, and resources: Ecological and social correlates of sex differences. *J Contemp Sociol* 27: 45-71.

—. 1990b. Land ownership, occupational status, and reproductive behavior in 19th century Sweden: Tuna parish. *Amer Anthropologist* 92(2): 457-468.

—. 1992. Sex, coalitions, and politics in preindustrial societies. *Polit Life Sci* 11(1): 63-80.

—. 1993. Ecological demography: A synthetic focus in evolutionary anthropology. *Evol Anthropol* 1(5): 177-187.

—. 1993b. An evolutionary perspective on war. In *Behavior, Culture, and Conflict in World Politics* (Zimmerman, W. & Jacobson, H., eds.), pp. 13-55. Ann Arbor: University of Michigan Press.

—. 1994. Men in demographic transition. *Hum Nat* 5: 223-254.

—. 1995. We're not environmental altruists—But we can solve environmental problems. Guest Essay. In *Living in the Environment* (Miller, G. T.), pp. 189-190. Belmont, CA: Wadsworth.

—. 1996. Behavioral ecology of conservation in traditional societies. *Hum Nat* 7: 353-379.

—. 2000a. Sex, wealth, and fertility: Old rules, new environments. In *Adaptation and human behavior: An anthropological perspective* (Cronk, L., Chagnon, N., & Irons, W., eds.), pp. 323-344. New York: Aldine de Gruyter.

—. 2000b. *Why sex matters: A Darwinian look at human behavior.* Princeton, NJ: Princeton University Press.

Low, B. S. & Heinen, J. T. 1993. Population, resources, and environment: Implications of human behavioural ecology for conservation. *Popul Environ* 15(1): 7-41.

Low, B. S. & Ridley, M. 1994. Why we're not environmentalist altruists—and what we can do about it. *Hum Ecol Rev* 1: 107-124.

Lowdermilk, W. C. 1953. *Conquest of the Land through 7000 Years.* Washington, DC: U.S. Department of Agriculture.

Ludwig, D., Hilborn, R. & Walters, C. 1993. Uncertainty, resource exploitation, and conservation: Lessons from history. *Science* 260: 17, 36.

Lumsden, C. J. & Wilson, E. O. 1981. *Genes, Mind, and Culture: The Coevolutionary Process.* Cambridge, MA: Harvard University Press.

Luttbeg, B., Borgerhoff Mulder, M. & Mangel, M. 2000. To marry again or not? A dynamic model of marriage behavior and demographic transition. In *Human Behavior and Adaptation: an Anthropological Perspective* (Cronk, L., Chagnon, N. A. & Irons, W., eds.), pp. 345-368. Hawthorne, NY: Aldine de Gruyter.

Lutz, W. & Qiang, R. 2002. Determinants of human population growth. *Phil Trans R Soc Lond* 357: 1197-1210.

Lyman, R. 1995. On the evolution of marine mammal hunting on the west coast of North America. *J Anthropol Archeol* 14: 45-77.

Lyng, S. 1990. Edgework: A social psychological analysis of voluntary risk taking. *Am J Sociol* 95: 851-856.

—. 1993. Dysfunctional risk taking: criminal behavior as edgework. In *Adolescent risk taking* (Bell, N. J. & Bell, R. W., eds.), pp. 107-130. Newbury Park, CA: Sage.

MacArthur, R. H. & Wilson, E. O. 1967. *The Theory of Island Biogeography.* Vol. 1 of *Monographs in Population Biology.* Princeton, NJ: Princeton University Press.

MacDonald, H. & McKenney, D. W. 1996. Varying levels of information and the embedding problem in contingent valuation: The case of Canadian wilderness. *Can J Forest Res* 26: 1295-1303.

MacDonald, K. 1994. *A People that Shall Dwell Alone: Judaism as a Group Evolutionary Strategy.* Westport, CT & London: Praeger.

Mace, R. 1996. When to have another baby: A dynamic model of reproductive decision-making and evidence from Gabbra pastoralists. *Ethol Sociobiol* 17: 263-273.

—. 2000a. Evolutionary ecology of human life history. *Anim Behav* 59: 1-10.

—. 2000b. An adaptive model of human reproductive rate: why people have small families. In *Adaptation and Human Behavior: an Anthropological Perspective* (Cronk, L., Chagnon N. A. & Irons, W., eds.), pp. 261-282. Hawthorne, NY: Aldine de Gruyter.

MacKinnon, J., MacKinnon, K., Child, G. & Thorsell, J. 1985. *Managing Protected Areas in the Tropics.* xvi + 295 pp., illustr. Cambridge: IUCN Publications.

Mangel, M. & Clark, C. W. 1988. *Dynamic Modeling in Behavioral Ecology.* Princeton, NJ: Princeton University Press.

Mangel, M. & Tier, C. 1993. Dynamics of metapopulations with demographic stochasticity and environmental catastrophes. *Theor Popul Biol* 44: 1-31.

Manson, J. & Wrangham, R. 1991. The evolution of hominid intergroup aggression. *Curr Anthropol* 32: 369-390.

Marcus, G. B. 1986. Stability and change in political attitudes: observe, recall, and "explain." *Polit Behav* 8: 21-44.

Margulis, L. 1993. *Symbiosis in cell evolution.* 2nd ed. New York: W.H. Freeman.

—. 1998. *Symbiotic Planet: A New View of Evolution.* New York: Basic Books.

Margulis, L. & Fester, R. (eds.). 1991. *Symbiosis as a Source of Evolutionary Innovation: Speciation and Morphogenesis.* Cambridge, MA: MIT Press.

Marler, P. 1961. The filtering of external stimuli during instinctive behavior. In *Current Problems in Animal Behavior* (Thorpeand, W. H. & Zangwill, O. L., eds.). Cambridge University Press.

Martin, C. 1978. *Keepers of the Game.* Berkeley: University of California Press.

Martin, L. L. 1993. Common dilemmas: Research programs in common-pool resources and international cooperation. In *Proceedings of a Conference on Linking Local and Global Commons* (Keohane, R., McGinnis, M. & Ostrom, E., eds.), pp. 147-170. Bloomington: Indiana University Workshop in Political Analysis and Theory.

Martin, P. S. 1984. Prehistoric overkill: The global model. In *Quarternary Extinctions: A Prehistoric Revolution* (Martin, P. S. & Klein, R. G., eds.), pp. 354-402. Tucson: University of Arizona Press.

Martin, P. S. & Klein, R. G. 1984. *Quaternary Extinctions.* Tucson: University of Arizona Press.

Masters, R. D. 2003. The social implications of evolutionary psychology: Linking brain biochemistry, toxins, and violent crime. In *Evolutionary Psychology and Violence: A Primer for Policymakers and Public Policy Advocates* (Bloom, R. W. & Dess, N., eds.), pp. 23-56. Westport, CT: Praeger.

Masters, R. D. & Coplan, M. J. 1999a. A dynamic, multifactorial model of alcohol, drug abuse, and crime: Linking neuroscience and behavior to toxicology. *Soc Sc Inform* 38: 591-624.

—. 1999b. Water treatment with silicofluorides and lead toxicity. *Int J Environ Stud* 56: 435-449.

Masters, R. D., Grelotti, D. J., Hone, B. T., Gonzalez, D. & Jones, D. 1997. Brain biochemistry and social status: The neurotoxicity hypothesis. In *Intelligence, Political Inequality, and Public Policy* (White, E., ed.), pp. 141-183. Westport, CT: Praeger.

Masters, R. D., Hone, B. & Doshi, A. 1998. Environmental pollution, neurotoxicity, and criminal violence. In *Environmental Toxicology: Current Developments* (Rose, J., ed.), pp. 13-48. Amsterdam: Gordon and Breach Science Publishers.

Matthews, M. L. & Moran, A. R. 1986. Age differences in male drivers' perception of accident risk: The role of perceived driving ability. *Accident Anal Prev* 18: 299-313.

Maynard Smith, J. 1964. Group selection and kin selection. *Nature* 201: 1145-1147.

—. 1982. *Evolution and the Theory of Games.* Cambridge: Cambridge University Press.

—. 1991. Honest signalling: The Philip Sidney game. *Anim Behav* 42: 1034-1035.

—. 1994. Must reliable signals always be costly? *Anim Behav* 47: 1115-1120.

—. 1998. The origin of altruism. *Nature* 393: 639-640.

Maynard Smith, J. & Szathmáry, E. 1995. *The Major Transitions in Evolution.* Oxford: Freeman.

—. 1999. *The Origins of Life: From the Birth of Life to the Origin of Language.* Oxford: Oxford University Press.

McAlpin, M. B. 1983. *Subject to Famine: Food Crises and Economic Change in Western India, 1860-1920.* Princeton, NJ: Princeton University Press.

McCabe, J. T. 1990. Turkana pastoralism: A case against the tragedy of the commons. *Hum Ecol* 18: 81-103.

McCabe, K. A., Rassenti, S. J. & Smith, V. L. Smart computer-assisted markets. *Science* 254: 534.

McCartney, A. & Savalle, J. 1985. Thule Eskimo whaling in the Central Canadian Arctic. *Arctic Anthropol* 22: 37-58.

McCay, B. J. 1980. A fishermen's cooperative, limited: Indigenous resource management in a complex society. *Anthropol Quart* 53: 29-38.

—. 1995. Social and ecological implications of ITQs: An overview. *Coastal Ocean Manage* 28: 3-22.

McCay, B. J. & Acheson, J. M. (eds.) 1987. *The Questions of the Commons: The Culture and Ecology of Communal Resources*. Tucson: University of Arizona Press.

McDonald, D. 1977. Food taboos: A primitive environmental protection agency (South America). *Anthropos* 72: 734-748.

McGinley, M. A. & Charnov, E. L. 1988. Multiple resources and the optimal balance between size and number of offspring. *Evol Ecol* 2: 77-84.

McGinnis, M. & Ostrom, E. 1996. In *The International Political Economy and International Institutions* (Young, O. R., ed.), vol. 2, pp. 465-493. Cheltenham, UK: Elgar.

McInish, T. H. 1982. Individual investors and risk-taking. *J Econ Psychol* 2: 125-136.

McIntosh, R. P. 1985. *The Background of Ecology: Concept and Theory*. Cambridge: Cambridge University Press.

McKean, M. A. 1992. Success on the commons: a comparative examination of institutions for common property resource management. *J Theor Pol* 4: 243-281.

McMichael, A. J. 1993. *Planetary Overload: Global Environmental Change and the Health of the Human Species*. Cambridge: Cambridge University Press.

McNally, R. J. 1987. Preparedness and phobias: A review. *Psychol Bull* 101: 283-303.

McNamara, J. M. 1996. Riskprone behaviour under rules which have evolved in a changing environment. *Am Zool* 36: 484-495.

McNamara, J. M. & Houston, A. I. 1992. State-dependent life-history theory and its implications for optimal clutch size. *Evol Ecol* 6: 170-185.

—. 1996. State-dependent life histories. *Nature* 380: 215-221.

McNeely, J. A. 1988. *Economics and Biological Diversity: Developing and Using Economic Incentives to Conserve Biological Resources*. Gland, Switzerland: IUCN Publication Services.

—. 1993. People and protected areas: partners in prosperity. In *Law of the Mother: Protecting Indigenous Peoples in Protected Areas* (Kemf, E., ed.), pp. 249-257. San Francisco: Sierra Club Books.

McNeely, J. A., Miller, K. R., Reid, W. V., Mittermeier, R. A. & Werner, T. B. 1990. *Conserving the World's Biological Diversity*. Gland, Switzerland: IUCN Publication Services.

McWhirter, N. & Greenberg, S. 1979. *Guinness Book of Records*. Edition 26. London: Guinness Superlatives Ltd.

Mealey, L. 1985. The relationship between social status and biological success: A case study of the Mormon religious hierarchy. *Ethol Sociobiol* 6: 249-257.

Mellor, M. 1993. *Breaking the Boundaries: Towards a Feminist Green Socialism*. London: Virgo Press.

Meltzer, D. 1993. Pleistocene peopling of the Americas. *Evol Anthropol* 1: 157-169.

Merchant, C. 1982. *The Death of Nature: Women, Ecology and the Scientific Revolution*. London: Wildwood House.

Mesterton-Gibbons, M. & Dugatkin, L. A. 1992. Cooperation among unrelated individuals: evolutionary factors. *Q Rev Biol* 67: 267-281.

Metcalfe, S. 1994. The Zimbabwe communal areas management programme for indigenous resources (CAMPFIRE). In *Natural Connections: Perspectives in Community-based Conservation* (Western, D. & Wright, M., eds.), pp. 161-192. Washington, DC: Island Press.

Michelman, F. 1990. In *Liberty, Property, and the Future of Constitutional Development* (Paul, E. F. & Dickman, H., eds.), pp. 127-171. Albany: State University of New York Press.

Mielke, H. W. 1999. Lead in the inner cities. *Am Sci* 87: 62-73.

Milinski, M., Semmann, D. & Krambeck, H. 2001. Reputation helps solve the "tragedy of the commons." *Nature* 415: 424-426.

Miller, C. H., Magee, J. W., Johnson, B. J., Fogel, M. L., Spooner, N. A., McCulloch, M. T. & Ayliffe, L. K. 1999. Pleistocene extinction of *Genyornis newtoni*: Human impact on Australian megafauna. *Science* 283: 205-208.

Miller, G. 1999. Waste is good. *Prospect* Feb.: 18-23.

Millstein, S. G. 1989. Adolescent health. Challenges for behavioral scientists. *Am Psycho* 44: 837-842.

—. 1993. Perceptual, attributional, and affective processes in perceptions of vulnerability through the life span. In *Adolescent Risk Taking* (Bell, N. J. & Bell, R. W., eds.), pp. 55-65. Newbury Park, CA: Sage.

Mineka, S., Davidson, M., Cook, M. & Keir, R. 1984. Observational conditioning of snake fears in Rhesus Monkeys. *J Abnorm Psychol* 93: 355-372.

Minnis, P. E. 1985. *Social Adaptation to Food Stress. A Prehistoric Southwestern Example*. Prehistoric Archaeology and Ecology Series. Chicago: University of Chicago Press.

Mohai, P. 1992. Men, women, and the environment: an examination of the gender gap in environmental concern and activism. *Society Nat Res* 5: 1-19.

Møller, A. P. 1988. Ejaculate quality, testes size and sperm competition in primates. *J Hum Evol* 17: 479-488.

Molte, K. von 1988. International commissions and implementation of international environmental law. In *International Environmental Diplomacy*, pp. 87-94. Cambridge: Cambridge University Press.

Monbiot, G. 1994. The tragedy of enclosure. *Sci Am* January 140.

Monod, J. 1941. *Chance and Necessity: An Essay on the Natural Philosophy of Modern Biology*. New York: Alfred A. Knopf.

Montagu, A. 1968. *Man and Aggression*. Oxford: Oxford University Press.

Moore, M. & Viscusi, W. 1990. Discounting environmental health risks: New evidence and policy implications. *J Environ Econ Manage* 18: 551-562.

Moore, O. 1957. Divination—a new perspective. *Am Anthropol* 59: 69-74.

Moran, C. & Andrews, G. 1985. The familial occurrence of agoraphobia. *Brit J Psychiat* 146: 262-267.

Moss, C. 1988. *Elephant Memories*. New York: Fawcett Columbine.

Moss, M. E., Lanphear, B. P. & Auinger, P. 1999. Association of dental caries and blood lead levels. *J Amer Med Assoc* 281: 2294-2298.

Mulligan, C. B. 1997. *Parental Priorities and Economic Inequality*. Chicago: University of Chicago Press.

Murdock, G. P. 1967. *Ethnographic Atlas*. Pittsburgh: University of Pittsburgh Press.

—. 1981. *Atlas of World Cultures*. Pittsburgh: University of Pittsburgh Press.

Murdock, G. P. & Morrow, D. O. 1970 Subsistence economy and supportive practices: Cross-cultural codes I. *Ethnology* 9: 302-330.

Murdock, G. P. & White, D. 1969. Standard cross-cultural sample. *Ethnology* 8: 329-369.

Myers, D. G. 1990. *Social Psychology*. New York: McGraw Hill.

Myers, N. 1984. *The Primary Source*. London: W.W. Norton and Co.

—. 1990. Tropical forests and life on earth. In *Lessons from the Rainforest* (Head, S. & Heinzman, R., eds.), pp. 13-24. San Francisco: Sierra Club Books.

—. 2000. Sustainable consumption. *Science* 287: 2419-2419.

Mysterud, I. 2000. Group selection, morality, and environmental problems. *J Consciousness Stud* 7: 225-227.

—. 2004. One name for the evolutionary baby? A preliminary guide for everyone confused by the chaos of names. *Soc Sc Inform* 43: 95-114.

National Research Council. 1999. *Sustaining Marine Fisheries*. Washington, DC: National Academy Press.

Needleman, H. L. 1990. Low-level lead exposure and the IQ of children. *J Amer Med Assoc* 263: 673-678.

Neel, J. V. 1999. When some fine old genes meet a "new" environment. *World Rev Nutr Diet* 84: 1-18.

Neiman, F. D. 1997. Conspicuous consumption as wasteful advertising: A Darwinian perspective on spatial patterns in classic Maya terminal monument dates. In *Rediscovering Darwin: Evolutionary Theory in Archaeological Explanation* (Clarke, G. & Barton, M., ed.), pp. 267-290. Archaeological Papers of the American Anthropological Association, 7.

Nelson, R. 1979. Athabaskan subsistence sdaptation in Alaska. In *Alaska Native Cultures and History*, Senri Ethnological Studies No. 4 (Kotani, Y. & Workman, M., eds.), pp. 38-54. Osaka: National Museum of Ethnology.

—. 1982. A conservation ethic and environment: The Koyukon of Alaska. In *Resource Managers: North American and Australian Hunter-Gatherers* (Williams, N. & Hunn, E., eds.), pp. 211-238. Boulde, CO: Westview Press.

Nesse, R. 1990. Evolutionary explanations of emotions. *Hum Nat* 1: 261-289.

—. (ed.). 2001. *Evolution and the Capacity for Commitment*. New York: Russell Sage Foundation.

Nesse, R. M. & Williams, G. C. 1994. *Why We Get Sick: The New Science of Darwinian Medicine*. New York: Vintage Books.

—. 1997. Evolutionary biology in the medical curriculum—what every physician should know. *BioScience* 47: 664-666.

—. 1998. Evolution and the origins of disease. *Sci Am* 279(5): 58-65.

Netting, R. M. 1976. What alpine peasants have in common: Observations on communal tenure in a Swiss village. *Hum Ecol* 4: 135-146.

Newman, P. C. 1989. *Empire of the Bay: An Illustrated History of the Hudson's Bay Company*. New York: Viking.

Ng, Y. K. & Wang, J. 1993. Relative income, aspiration, environmental quality, individual and political myopia: Why may the ratrace for material growth be welfare reducing? *Math Soc Sci* 26: 3-23.

Nisbett, R. E. & Ross, L. 1980. *Human Inference: Strategies and Shortcomings of Social Judgment*. Englewood Cliffs, NJ: Prentice Hall.

Nisbett, R. E. & Wilson, T. 1977. Telling more than we can know: verbal reports on mental processes. *Psychol Rev* 84: 231-259.

Noë, R., van Schaik, C. P. & van Hoof, J.A.R.A.M. 1991. The market effect: An explanation for payoff asymmetries among collaborating animals. *Ethology* 87: 97-118.

Norgaard, R. B. 1995. Intergenerational commons, globalization, economism, and unsustainable development. *Adv Hum Ecol* 4: 141-171.

Norgaard, R. B. & Howarth, R. B. 1991. Sustainability and discounting the future. In *Ecological Economics: the Science and Management of Sustainability* (Costanza, R., ed.), pp. 88-101. New York: Columbia University Press.

North, D. C. 1994. Economic performance through time. *Am Econ Rev* 84(3): 359-368.

Nowak, M. A. & Sigmund, K. 1992. Tit for tat in heterogeneous populations. *Nature* 355: 250-252.

Nowak, M. & Sigmund, K. 1993. A strategy of win-stay, lose-shift that outperforms tit-for-tat in the prisoner's dilemma game. *Nature* 364: 56-58.

Nunney, L. 1998. Are we selfish, are we nice, or are we nice because we are selfish? *Science* 281: 1619-1620.

Nylander, M., Friberg, L. & Lind, B. 1987. Mercury concentrations in the human brain and kidneys in relation to exposure from dental amalgam fillings. *Swed Dent J* 11: 179-187.

Oelschlaeger, M. 1991. *The Idea of Wilderness: Prehistory to the Age of Ecology*. New Haven, CT: Yale University Press.

Öhman, A. 1986. Face the beast and fear the face: animal and social fears as prototypes for evolutionary analyses of emotion. *Psychophysiology* 21: 123-145.

Öhman, A. & Dimberg, U. 1984. An evolutionary perspective on human social behavior. In *Sociophysiology* (Waid, W. M., ed.). New York: Springer-Verlag.

Öhman, A. & Soares, J.J.F. 1993. On the automaticity of phobic fear: Conditioned responses to masked phobic stimuli. *J Abnorm Psychol* 102: 121-132.

—. 1994. Unconscious anxiety: Phobic responses to masked stimuli. *J Abnorm Psychol* 103: 231-240.

Olsen, M.E., Lodwick, D.G. & Dunlop, R.E. 1992. *Viewing the world ecologically*. Boulder CO: Westview Press.

Olson, M. 1965. *The Logic of Collective Action: Public Goods and the Theory of Groups*. Cambridge: Harvard University Press.

—.1967. *The Logic of Collective Action: Public Goods and the Theory of Groups*. Boston, MA: Harvard University Press.

—. 1971. *The Logic of Collective Action: Public Goods and the Theory of Groups* (first published in 1965). New York: Schocken Books.

Olson, S. L. & James, H. F. 1984. The role of Polynesians in the extinction of Avifauna of the Hawaiian Islands. In *Quaternary Extinctions: A Prehistoric Revolution* (Martin, P. S. & Klein, R. G., eds.), pp. 768-783. Tucson: University of Arizona Press.

Orbell, J. M., van de Kragt, A. & Dawes, R. M. 1988. *J Personality Soc Psych* 54: 811.

Oreskes, N., Shrader-Frechette, K. & Belitz, K. 1994. Verification, validation, and confirmation of numerical models in the Earth sciences. *Science* 263: 641-646.

Organisation for Economic Cooperation and Development (OECD). 1997. *Towards Sustainable Fisheries: Economic Aspects of the Management of Living Marine Resources*. Paris: OECD.

Orians, G. H. 1980. Habitat selection: General theory and applications to human behavior. In *The Evolution of Human Social Behavior* (Lockard, J. S., ed.). New York: Elsevier.

—. 1986. An ecological and evolutionary approach to landscape aesthetics. In *Landscape Meanings and Values* (Penning-Rowsell, E. C. & Lowenthal, D., eds.). London: Allen and Unwin.

—. 1998. Human behavioral ecology: 140 years without Darwin is too long. *Bulletin of the Ecological Society of America* 79: 15-28.

Orians, G. H. & Heerwagen, J. H. 1992. Evolved responses to landscapes. In *The Adapted Mind: Evolutionary Psychology and the Generation of Culture* (Barkow, J., Cosmides, L. & Tooby, J., eds.), pp. 138-172; pp. 555-580. New York: Oxford University Press New York.

Ornstein, R. & Ehrlich, P. 1989. *New World, New Mind: Moving Toward Conscious Evolution*. New York: Touchstone.

Ostrom, E. 1990. *Governing the Commons: The Evolution of Institutions for Collective Action*. Cambridge: Cambridge University Press.

—. 1998: A Behavioral Approach to the Rational Choice Theory of Collective Action. *Am Pol Sci Rev* 92: 1-22.

—. 2001. Reformulating the commons. In *Protecting the Commons: A Framework for Resource Management in the Americas* (Burger, J., Ostrom, E., Norgaard, R. B., Policansky, D. & Goldstein, B., eds.), pp. 17-41. Washington, DC: Island Press.

Ostrom, E. & Walker, J. M. 1997. Neither markets nor states: Linking transformation processes in collective action arenas. In *Perspectives on Public Choice: A Handbook* (Mueller, D. C., ed.), pp. 35-72. New York: Cambridge Univ. Press.

Ostrom, E., Walker, J. & Gardner, R. 1994. *Rules, games and common pool resources*. Ann Arbor: University of Michigan Press.

Ostrom, V. 1990. Courts and collectivities. *Brigham Young Univ Law Rev* 3: 857-872.

OTA. 1987. *Technologies to Maintain Biological Diversity*. Washington, DC: U.S. Government Printing Office, Office of Technological Assessment, U.S. Congress.

Oye, K. A. & Maxwell, J. H. 1995. Self-interest and environmental management. In *Local Commons and Global Interdependence* (Keohane, R. & Ostrom, E., eds.), pp. 191-221. London: Sage.

Packer, C. & Pusey, A. 1982. Cooperation and competition within coalitions of male lions: Kin selection or game theory? *Nature* 296: 740-742.

Packer, C. & Ruttan, L. M. 1988. The evolution of cooperative hunting. *Am Nat* 132: 159-198.

Pagel, M. & Mace, R. 2004. The cultural wealth of nations. *Nature* 428: 275-278.

Palmer, C. T. 1991. Kin-selection, reciprocal altruism, and information sharing among Maine lobstermen. *Ethol Sociobiol* 12: 221-235.

Parkin, T. G. 1992. *Demography and Roman Society*. Baltimore: Johns Hopkins University Press.

Payton, M., Riggs, K. M., Spiro, A., Weiss, S. T. & Hu, H. 1998. Relations of bone and blood lead to cognitive function: The VA normative aging study. *Neurotoxicol Teratol* 20: 19-27.

Pearce, D. & Turner, R. 1990. *Economics of Natural Resources and the Environment*. Harvester Wheatsheaf.

Pearce, F. 1992. First aid for the Amazon. *New Sci* 3: 42-46.

Penn, D. 1999. Explaining the human demographic transition. *Trends Ecol Evol* 14: 32.

—. 2003. The evolutionary roots of our environmental problems: Towards a Darwinian ecology. *Q Rev Biol* 78: 275-301.

Pennington, R. & Harpending, H. 1988. Fitness and fertility among Kalahari !Kung. *Am J Phys Anthropol* 77: 202-319.

Peoples, J. G. 1982. Individual or group advantage? A reinterpretation of the Maring ritual cycle. *Curr Anthropol* 23(3): 291-309.

Peres, C. 1999. Effects of subsistence hunting and forest types on the structure of Amazonian primate communities. In *Primate Communities* (Fleagle, J., Janson, C. & Reed, K., eds.). Cambridge: Cambridge University Press.

Peres, C. 2000. Effects of subsistence hunting on vertebrate community structure in Amazonian forests: A large-scale cross-site comparison. *Conservation Biol*, 14(1): 240-253.

Perusse, D. 1993. Cultural and reproductive success in industrial societies: Testing relationship at the proximate and ultimate levels. *Behav Brain Sci* 16: 267-322.

Pinker, S. 2002. *The Blank Slate: The Modern Denial of Human Nature*. New York: Penguin Putnam.

Pinkerton, E. (ed.). 1989. *Cooperative Management of Local Fisheries: New Directions for Improved Management and Community Development*. Vancouver, Canada: University of British Columbia Press.

Platt, J. 1973. Social traps. *Amer Psychol* August: 641-651.

Pleva, J. 1992. Mercury from dental amalgams: exposure and effects. *International Journal of Risk & Safety in Medicine* 3: 1-22.

—. 1994. Dental amalgam—a public health hazard. *Rev Environ Health* 10: 1-27.

Plott, C. R. 1986. Laboratory experiment in economics: The implications of posted-price institutions. *Science* 232: 732-738.

Pollard, J. H. 1973. *Mathematical Models for the Growth of Human Populations*. New York: Cambridge University Press.

Polunin, N. 1984. Do traditional marine "reserves" conserve? A view of Indonesian and New Guinean evidence. In *Maritime Institutions in the Western Pacific*. Senri Ethnological Studies No. 17 (Ruddle, K. & Akimichi, J., eds.), pp. 267-283. Osaka: National Museum of Ethnology.

Porter, G. & Brown, J. W. 1996. *Global Environmental Politics*. Boulder, CO: Westview Press.

Posey, D. 1985. Native and indigenous guidelines for new Amazonian development strategies: Understanding biodiversity through ethnoecology. In *Change in the Amazon* (Hemming, J., ed.), pp. 156-181. Manchester: Manchester University Press.

Posey, D. A. & Balee, W. (eds.). 1989. *Resource Management in Amazonia: Indigenous and Folk Strategies*. The Bronx: New York Botanical Garden.

Profet, M. 1991. The function of allergy: Immunological defense against toxins. *Q Rev Biol* 66: 23-62.

—. 1992. Pregnancy sickness as adaptation: A deterrent to maternal ingestion of teratogens. In *The Adapted Mind: Evolutionary Psychology and the Generation of Culture* (Barkow, J., Cosmides, L. & Tooby, J., eds.), pp. 327-365. New York: Oxford University Press.

Pugh, G. E. 1977. *The Biological Origin of Human Values*. New York: Basic Books.

Pulliam, H. R. & Dunford, C. 1980. *Programmed to Learn. An Essay on the Evolution of Culture*. New York: Columbia University Press.

Puri, R. K. 1995. Commentary. *Curr Anthropol* 36: 789-818.

Putnam, R. D. 1988. Diplomacy and domestic politics: The logic of twolevel games. *Int Organ* 42(3): 427-460.

—. 1993. Democracy, development, and the civic community: Evidence from an Italian experiment. In *Proceedings of a Conference on Linking Local and Global Commons* (Keohane, R., McGinnis, M. & Ostrom, E., eds.), pp. 95-146. Bloomington: Indiana University Workshop in Political Analysis and Theory.

Pyke, G., Pulliam, R. & Charnov, E. 1976. Optimal foraging: A selective review of theories and tests. *Q Rev Biol* 52: 137-154.

Rachlin, H. 1991. *Introduction to Modern Behaviorism*. 3rd ed. Freeman.

Rappaport, R. 1969. Ritual regulation of environmental relations among a New Guinea people. In *Environment and Cultural Behavior: Ecological Studies in Cultural Anthropology* (Vayda, A. P., ed.), pp. 181-201. Garden City, NY: The Natural History Press.

Real, L. 1987. Objective benefit versus subjective perception in the theory of risksensitive foraging. *Am Nat* 130: 399-411.

Real, L. & Caraco, T. 1986. Risk and foraging in stochastic environments. *Annu Rev Ecol Syst* 17: 371-390.

Redclift, M. 1987. *Sustainable Development: Exploring the Contradictions*. New York: Routledge.

Redford, K. 1991a. The ecologically noble savage. *Cult Survival Q* 15(1): 46-48.

—. 1991b. The ecologically noble savage. *Orion* 9: 24-29.

Regelmann, K. & Curio, E. 1986. Why do great tit *(Parus major)* males defend their brood more than females do? *Anim Behav* 34: 1206-1214.

Reichel-Dolmatoff, G. 1976. Cosmology as ecological analysis: A view from the rain forest. *Man* 11: 307-318.

Remedial Action Plan 1992. *Hamilton Harbour Stage I Report: Environmental Conditions and Problem of Definition.*

Repetto, R. 1986. *Skimming the Water: Rent Seeking and the Performance of Public Irrigation Systems.* Research Report 4. Washington, DC: World Resources Institute.

Repton, H. 1907. *The Art of Landscape Gardening.* Boston, MA: Houghton Mifflin.

Revelle, P. & Revelle, C. 1988. *The Environment: Issues and Choices for Society.* 3rd ed. Boston, MA: Portola Publishers.

Reznick, D. & Travis, J. 1996. The empirical study of adaptation in natural populations. In *Adaptation* (Rose, M. & Lauder, G., eds.), pp. 243-289. New York: Academic Press.

Richerson, P. J. & Boyd, R. 1992. Cultural inheritance and evolutionary ecology. In *Evolutionary Ecology and Human Behavior* (Smith, E. A. & Winterhalder, B., eds.), pp. 61-92. New York: Aldine de Gruyter.

Richter, L. K. 1989. *The Politics of Tourism in Asia.* Honolulu: University of Hawaii Press.

Ridley, M. 1996. *The Origins of Virtue: Human Instincts and the Evolution of Cooperation.* New York: Penguin.

—. 2003. *Nature Via Nurture: Genes, Experience, and What Makes Us Human.* New York: HarperCollins.

Ridley, M. & Low, B. 1993. Can selfishness save the environment? *Atlantic Monthly* (September): 76-86.

—. 1994. Can selfishness save the environment? *Hum Ecol Rev* 1: 1-13.

Rightmire, P. 1993. *The Evolution of Homo erectus.* Cambridge: Cambridge University Press.

Rimland, B. 2000. The autism epidemic, vaccinations, and mercury. *J Nutr Envir Med* 10: 261-266.

Roberts, R. G., Flannery, T. F., Ayliffe, L. K., Yoshida, H., Olley, J. M., Prideaux, G. J., Laslett, G. M., Baynes, A., Smith, M. A., Jones, R. & Smith, B. L. 2001. New ages for the last Australian megafauna: Continent-wide extinction about 46,000 years ago. *Science* 292: 1888-1892.

Robinson, J. & Redford, K. 1986a. Intrinsic rate of natural increase in neotropical forest mammals: Relationship to phylogeny and diet. *Oecologia* 68: 516-520.

—. 1986b. Body size, diet, and population density of neotropical forest mammals. *Am Nat* 128: 665-680.

Robinson, J. G. & Redford, K. H. (eds.). 1991. *Neotropical Wildlife Use and Conservation.* Chicago: University of Chicago Press.

Robinson, W. C. 1997. The economic theory of fertility over three decades. *Popul Stud* 51: 63-74.

Roebroeks, W., Conrad, N. & van Kolfschoten, T. 1992. Dense forests, cold steppes, and the Paleolithic settlement of Northern Europe. *Curr Anthropol* 33: 551-586.

Roelofsma, P.H.M.P., 1996. Modelling intertemporal choices: An anomaly approach. *Acta Psychol* 93: 5-22.

Rogers, A. R. 1988. Does biology constrain culture? *Am Anthropol* 90: 819-831.

—. 1990a. Birth, death, and the evolution of initial time preference. Ms. in author's possession.

—. 1990b. The evolutionary economics of human reproduction. *Ethol Sociobiol* 11: 479-495.

—. 1991. Conserving resources for children. *Hum Nat* 2: 73-82.

—. 1992. Resources and population dynamics. In *Evolutionary Ecology and Human Behavior* (Smith, E.A. & Winterhalder, B., eds.), pp. 375-402. New York: Aldine de Gruyter.

—. 1994. Evolution of time preference by natural selection. *Am Econ Rev* 84: 460-481.

—. 1995. For love or money: The evolution of reproductive and material motivations. In *Human Reproductive Decisions* (Dunbar, R.I.M., ed.), pp. 76-95. London: St. Martin's Press.

—. 1997. The evolutionary theory of time preference. In *Characterizing Human Psychological Adaptations* (Bock, G. & Cardew, G., eds.), pp. 231-252. Ciba Foundation Symposium 208. London: John Wiley.

Rogers, A. R. & Hawkes, K. 1990. Conservation, discounting, and evolution. Ms. in author's possession.

Roitberg, B. D., Mangel, M., Lalonde, R. G., Roitberg C. A., van Alphen, J.J.M. & Vet, L. 1992. Seasonal dynamic shifts in patch exploitation by parasitic wasps. *Behav Ecol* 3: 156-165.

Romer, P. M. 1995. Preferences, promises, and the politics of entitlement. In *Individual and Social Responsibility* (Fuchs, V. R., ed.), pp. 195-220. Chicago: University of Chicago Press.

Roosevelt, A. 1980. *Parmana: Prehistoric Maize and Manoic Subsistence Along the Amazon and Orinoco*. New York: Academic Press.

Rose, C. M. 1994. *Property & Persuasion: Essays on the History, Theory, and Rhetoric of Ownership*. Boulder, CO: Westview Press.

Rose, M. C. 1976. Nature as aesthetic object: An essay in meta-aesthetics. *Brit J Aesthet* 16: 3-12.

Rose, M. R. 1997. Toward an evolutionary demography. In *Between Zeus and the Salmon: The Biodemography of Longevity* (Wachter, K. & Finch, C., eds.), pp. 96-107. Washington, DC: National Academy Press.

Ross, E. 1978. Food taboos, diet, and hunting strategy: The adaptation to animals in Amazonian cultural ecology. *Curr Anthropol* 19: 1-36.

—. 1980. Reply. *Curr Anthropol* 21: 544-546.

Rubin, P. H. & Paul, C. W. 1979. An evolutionary model of taste for risk. *Econ Inq* 17: 585-596.

Ruddle, K. & Akimichi, T. (eds.). 1984. *Senri Ethnological Studies* (No. 17). Osaka: National Museum of Ethnology.

Ruddle, K. & Johannes, R. E. (eds.). 1989. *Traditional Marine Resource Management in the Pacific Basin: An Anthology*. Jakarta, Indonesia: UNESCO/ROSTSEA.

Runge, C. F. 1981. Common property externalities: Isolation, assurance, and resource depletion in a traditional grazing context. *Am J Agr Econ* 63(4): 595-606.

Ruttan, L. 1998. Closing the commons: Cooperation for gain or restraint? *Hum Ecol* 26(1): 43-66.

Ruttan, L. & Borgerhoff-Mulder, M. 1999. Wealth asymmetries and grazing restraint among East African pastoralists. *Curr Anthropol* 40: 621-652.

Sachs, C. 1996. *Gendered Fields: Rural Women, Agriculture and Environment*. Boulder, CO: Westview Press.

—. 1997. *Resourceful Natures, Women, and Environment*. Washington, DC: Francis & Taylor.

Sacks, O. 1996. *The Island of the Colorblind*. New York: Vintage Books.

Saffirio, J. & Hames, R. 1983. The forest and the highway. In *The Impact of Contact* (Kensinger, K., ed.), pp. 4-39. Cambridge, MA: Joint Publication of Cultural Survival, Report #11, and Working Papers on South American Indians, vol 6.

Sahlins, M. 1968. Notes on the original affluent society. In *Man the Hunter* (Lee, R. & DeVore, I., eds.), pp. 85-89. Chicago: Aldine.

—. 1972. *Stone Age Economics*. Chicago: Aldine Press.

Safina, C. 1994. Where have all the fishes gone? *Issues Sc. Technol* 10(3): 37-43.

Salisbury, R. F. 1962. *From Stone to Steel: Economic Consequences of a Technological Change in New Guinea*. Cambridge: Cambridge University Press.

Samuelson, P. A. 1993. Altruism as a problem involving group versus individual selection in economics and biology. *Am Econ Rev* 83: 143-148.

Sandford, S. 1983. *Management of Pastoral Development in the Third World*. London: John Wiley and Sons.

Savelle, J. & McCartney, A. 1991. Thule Eskimo subsistence and bowhead whale procurement. In *Human Predators and Prey Mortality* (Stiner, M., ed.), pp. 201-216. Boulder, CO: Westview Press.

Schaller, G. B., Jinchu, H., Wenshi, P. & Jing, Z. 1985. *The Giant Pandas of Wolong*. Chicago: University of Chicago Press.

Schama, S. 1996. *Landscape and Memory*. New York: Vintage Books.

Schlager, E., Blomquist, W. & Tang, S. Y. 1994. Mobile flows, storage, and self-organized institutions for governing common-pool resources. *Land Econ* 70(3): 294-317.

Schmid, A. 1989. *Benefit-Cost Analysis*. Boulder, CO: Westview Press.

Schneider, S. H. 1989. *Global Warming: Are We Entering the Greenhouse Century?* New York: Vintage Books.

Schroeder, H. W. 1989. Environment, behavior, and design research on urban forests. In *Advances in Environment, Behavior, and Design* (Zube, E. H. & Moore, G. T., eds.). Vol. 2. New York: Plenum.

Schwartz, J. 1995. Lead, blood pressure, and cardiovascular disease in men. *Arch Environ Health* 50: 31-37.

Scott, A. 1955. The fishery: The objectives of sole ownership. *J Polit Econ* 63: 116-124.

Seeley, T. D. 1985. *Honeybee Ecology: A Study of Adaptation in Social Life*. Princeton, NJ: Princeton University Press.

Seligman, M.E.P. 1970. On the generality of the laws of learning. *Psychol Rev* 77: 406-418.

—. 1971. Phobias and preparedness. *Behav Ther* 2: 307-320.

Sen, A. 1981. *Poverty and Famines: An Essay on Entitlement and Deprivation*. Oxford: Clarendon Press.

Sessions, G. 1995. *Deep Ecology for the Twenty-First Century*. Boston: Shambhala.

Sethi, R. & Somanathan, E. 1996b. The evolution of social norms in common property resource use. *Am Econ Rev* 86(4): 766-788.

Shafir, E., 1993. Choosing versus rejecting: Why some options are both better and worse than others. *Memory Cognit* 21: 546-556.

Sharpe, R. M. & Skakkebaek, N. E. 1993. Are oestrogens involved in falling sperm counts and disorders of the male reproductive tract? *Lancet* 341: 1392-1395.

Shermer, M. 1997. The beautiful people myth: Why the grass is always greener in the other century. *Skeptic* 5: 72-79.

Sherpa, M. N. 1993. Grass roots in a Himalayan Kingdom. In *Law of the Mother: Protecting Indigenous Peoples in Protected Areas* (Kemf, E., ed.), pp. 45-51. San Francisco: Sierra Club Books.

Sieff, D. F. 1995. *The effects of resource availability on the subsistence strategies of Datoga pastoralists of north west Tanzania*. Ph.D., University of Oxford.

Simms, S.R. 1992. Wilderness as a human landscape. In *Wilderness Tapestry: An Eclectic Approach to Preservation* (Zeveloff, S. I., Vause, L. M. & McVaugh, W. H., eds.), pp. 183-201. Reno: University of Nevada Press.

Simon, H. 1993. Altruism and economics. *Am Econ Rev* 83: 156-161.

Simon, H.A. 1990. A mechanism for social selection and successful altruism. *Science* 250. pp. 1665-1668.

Simon, J. L. 1977. *The Economics of Population Growth*. Princeton, NJ: Princeton University Press.

Simpson, J. & Kenrick, D. 1997. *Evolutionary Social Psychology*. Englewood Cliffs, NJ: Lawrence Erlbaum Associates.

Slobodkin, L. 1968. How to be a predator. *Am Zool* 8: 43-51.

Smil, V. 2000. *Feeding the World: A Challenge for the Twenty-first Century*. Cambridge, MA: MIT Press.

Smith, C. C. & Fretwell, S. D. 1974. The optimal balance between size and number of offspring. *Am Nat* 108: 499-506.

Smith, D. M. 1982. *Where the Grass is Greener: Living in an Unequal World*. Harmondsworth: Penguin.

Smith, E. A. 1983. Anthropological applications of optimal foraging theory: A critical review. *Curr Anthropol* 24(5): 625-651.

—. 1991. *Inujjuamiut Foraging Strategies: Evolutionary Ecology of an Arctic Hunting Economy*. New York: Aldine de Gruyter.

—. 1992. Human behavioral ecology: I. *Evol Anthropol* 1: 15-25.

—. 1995. Comment on Alvard M. Intraspecific prey choice by Amazonian hunters. *Curr Anthropol* 36: 805-807.

—. 1995b. Commentary. *Curr Anthropol* 36: 789-818.

—. 2000. Three styles in the evolutionary analysis of human behavior. In *Adaptation and Human Behavior: An Anthropological Perspective* (Cronk, L., Chagnon N. A. & Irons, W., eds.). New York: Aldine de Gruyter.

Smith, E. A. & Bird, R.L.B. 2000. Turtle hunting and tombstone opening: Public generosity as costly signaling. *Evol Hum Behav* 21: 245-261.

Smith, E. A. & Winterhalder, B. 1992. Natural selection and decision making: Some fundamental principles. In *Evolutionary Ecology and Human Behavior* (Smith, E. A. & Winterhalder, B., eds.), pp. 25-60. New York: Aldine de Gruyter.

Smith, E. A. & Winterhalder, B. (eds.). 1992b. *Evolutionary Ecology and Human Behavior*. New York: Aldine de Gruyter.

Smith, E. A. & Wishnie, M. 2000. Conservation and subsistence in small-scale societies. *Annu Rev Anthropol* 29: 493-524.

Smith, V. K. 1994. Lightning rods, dart boards, and contingent valuation. *Nat Res J* 34: 121-152.

Smuts, B. 1999. Multilevel selection, cooperation, and altruism: Reflections on *Unto others: The evolution and psychology of unselfish behavior*. *Hum Nat* 10: 311-327.

Sneath, D. 1998. State policy and pasture degradation in Inner Asia. *Science* 281(5380): 1147-1148.

Sober, E. & Wilson, D. S. 1998. *Unto others: The evolution and psychology of unselfish behavior*. Cambridge, MA & London: Harvard University Press.

Soltis, J., Boyd, R. & Richerson, P. J. 1995. Can group-functional behaviors evolve by cultural group selection? *Curr Anthropol* 36(3): 473-494.

Sosis, R. 2000. Costly signaling and torch fishing on Ifaluk atoll. *Evol Hum Behav* 21: 223-244.

Spackman, M. 1991. *Discount Rates and Rates of Return in the Public Sector: Economic Issues*, Her Majesty's Treasury.

Speck, F. G. 1939. Aboriginal conservators. *Bird Lore* 40: 258-261.

Speck, F. G. & Hadlock, W. S. 1946. A report on tribal boundaries and hunting areas of the Malecite Indian of New Brunswick. *Am Anthropol* 48: 355-374.

Sperling, L. & Galaty, J. G. 1990. Cattle, culture, and economy. In *The World of Pastoralism: Herding Systems in Comparative Perspective* (Galaty, J. G. & Johnson, D. L., eds.), pp. 69-98. New York: Guildford Press.

Speth, J. D. 1983. *Bison Kills and Bone Counts: Decision Making by Ancient Hunters*. Chicago: University of Chicago Press.

Spillius, J. 1957. Natural disaster and political crisis in a Polynesian society. *Hum Relat* 10: 3-27.

Spoehr, A. 1956. Cultural differences in the interpretation of natural resources. In *Man's Role in Changing the Face of the Earth*. An International Symposium under the Co-Chairmanship of Sauer, C.O., Bates, M. & Mumford, L. (Thomas, Jr., W. L., ed.), pp. 93-102. Chicago: University of Chicago Press.

Steadman, D. W. 1995. Prehistoric extinctions of Pacific island birds: Biodiversity meets zooarcheology. *Science* 267: 1123-1131.

Steadman, D. W., Pregill, G. K. & Burley, D. V. 2002. Rapid prehistoric extinction of iguanas and birds in Polynesia. *Proc Natl Acad Sci USA* 99: 3673-3677.

Stearman, A. M. 1994. "Only slaves climb trees": Revisiting the myth of the ecologically noble savage in Amazonia. *Hum Nat* 5: 339-357.

—. 1995. Commentary. *Curr Anthropol* 36: 789-818.

Stearns, S. C. 1992. *The Evolution of Life Histories*. Oxford: Oxford University Press.

Stearns, S. C. (ed.). 1999. *Evolution in Health & Disease*. Oxford: Oxford University Press.

Stephens, D. W. & Krebs, J. 1986. *Foraging Theory*. Princeton, NJ: Princeton University Press.

Stevens, S. 1997. Consultation, co-management, and conflict in Sagarmatha (Mount Everest) National park, Nepal. In *Conservation through Cultural Survival: Indigenous Peoples and Protected Areas* (Stevens, S., ed.), pp. 63-97. Washington, DC: Island Press.

Stiner, M. 1991. An interspecific perspective on the emergence of the modern human predatory niche. In *Human Predators and Prey Mortality* (Stiner, M., ed.), pp. 149-185. Boulder, CO: Westview Press.

Stocks, A. 1983a. Cocamilla fishing: Patch modification and environmental buffering in the Amazon Varzea. In *Adaptive Responses of Native Amazonians* (Hames, R. & Vickers, W. eds.), pp. 239-68. New York: Academic Press.

—. 1983b. Native enclaves in the Upper Amazon: A case of regional non-integration. *Ethnohistory* 30(2): 77-92.

Stone, R. 1992. Swimming against the PCB tide. *Science* 225: 798-799.

Strassmann, B. I. & Dunbar, R.I.M. 1999. Human evolution and disease: Putting the stone age in perspective. In *Evolution in health & disease* (Stearns, S.C., ed.), pp. 91-101. Oxford: Oxford University Press.

Strassmann, B. I. & Gillespie, B. 2002. Life-history theory, fertility and reproductive success in humans. *Proc Roy Soc Lond B Bio* 269: 553-562.

Strong, D. H. 1988. *Dreamers and Defenders: American Conservationists*. Lincoln: University of Nebraska Press.

Strotz, R. 1956. *Rev Econ Stud* 23: 165-180.

Stuart-Macadam, P. & Kent, S. (eds.). 1992. *Diet, Demography, and Disease: Changing Perspectives on Anemia*. New York: Aldine de Gruyter.

Sunstein, C. R. 1994. Incommensurability and valuation in law. *Mich Law Rev* 92: 779-861.

Sutherland, W. J. 1985. Chance can produce a sex difference in variance in mating success and explain Bateman's data. *Anim Behav* 33: 1349-1352.

Suttles, W. 1960. Affinal ties, subsistence, and prestige among the Coast Salish. *Am Anthropol* 62: 269-305.

—. 1991. Streams of property, armor of wealth: The traditional Kwakiutl Potlatch. In *Chiefly Feasts: The Enduring Kwakiutl Potlatch* (Jonaitis, A., ed.), pp. 71-133. New York: American Museum of Natural History and Seattle: University of Washington Press.

Tang, S. Y. 1992. *Institutions and Collective Action: SelfGovernance in Irrigation*. San Francisco: ICS Press.

Taylor, K. 1981. Knowledge and praxis in Sanuma food prohibitions. In *Food Taboos in Lowland South America* (Kensinger, K. & Kracke, W., eds.). Working Papers on South American Indians No. 3. pp. 24-54. Bennington, VT: Bennington College.

Taylor, M. 1992. The economics and politics of property rights and common pool resources. *Nat Resour J* 32: 633-648.

Taylor, M. & Singleton, S. 1993. The communal resource: Transaction costs and the solution of collec-tive action problems. In *Proceedings of a Conference on Linking Local and Global Commons* (Keohane, R., McGinnis, M. & Ostrom, E., eds.), pp. 66-94. Bloomington: Indiana University Workshop in Political Analysis and Theory.

Temple, S. 1987. Do predators always capture substandard individuals disproportionately from prey populations? *Ecology* 68: 669-674.

Templeton, A. & Lawlor, L. 1981. The fallacy of averages in ecological optimization theory. *Am Nat* 117: 390-393.

The Economist. 1997. Salmon war on two fronts: Canada. June 28, 1997.

Thomas, K. 1983. *Man and the Natural World. A History of the Modern Sensibility*. New York: Pantheon Books.

Thompson, N. S. 1999. Group selection & the origins of evil. *Skeptic* 7(2): 70-73.

Tietenberg, T. 1996. *Environmental & Resource Economics*. 4th ed. New York: Scott-Foresman.

Tiger, L. & Fox, R. 1971. *The Imperial Animal*. New York: Holt, Rinehart and Winston.

Todd, J. 1986. *Earth dwelling: The Hopi environmental ethos and its architectural symbolism—A model for the deep ecology movement*. Ph.D. Dissertation, University of California at Santa Cruz.

Tomikawa, M. 1979. The migrations and inter-tribal relations of the pastoral Datoga. *Senri Ethnological Studies* 5:1-46.

Tooby, J. 2001. Is human nature hidden in the genome? *Nat Genet* 29: 363.

Tooby, J. & Cosmides, L. 1992. The psychological foundations of culture. In *The Adapted Mind* (Barkow, J., Tooby, J. & Cosmides, L., eds.), pp. 19-136. New York: Oxford University Press.

—. 1996. Friendship and the banker's paradox: Other pathways to the evolution of adaptations for altruism. In *Evolution of Social Behaviour Patterns in Primates and Man* (Runciman, W.G., Maynard Smith, J. & Dunbar, R.I.M., eds.), pp. 119-143. Oxford: Oxford University Press.

Trevathan, W. R., Smith, E.O. & McKenna, J.J. (eds). 1999. *Evolutionary Medicine*. Oxford: Oxford University Press.

Trimpop, R. M. 1994. *The Psychology of Risk Taking Behavior*. Amsterdam: North-Holland.

Trivers, R. L. 1971. The evolution of reciprocal altruism. *Q Rev Biol* 46: 35-57.

—. 1972. Parental investment and sexual selection. In *Sexual Selection and the Descent of Man, 1871-1971* (Campbell, B., ed.), pp. 136-179. Chicago: Aldine.

—. 1985. *Social Evolution*. Menlo Park, CA: Benjamin Cummings.

Trotter, M. M. & McCulloch, B. 1984. Moas, men, and Middeus. In *Quaternary Extinctions: A Prehistoric Revo-lution* (Martin, P. S. & Klein, R. G., eds.), pp. 708-726. Tucson: University of Arizona Press.

Tuan, Yi-Fu. 1968. Discrepancies between environmental attitude and behaviour: Examples from Europe and China. *Canadian Geography* 12: 176-191.

—. 1970. Our treatment of the environment in ideal and actuality. *Am Sci* 58: 244-249.

Tucker, W. T. n.d. Relationships between Access to Resources, Mean Fertility, and Offspring Quality in the National Survey of Families and Households Sample. Ms. in the author's possession.

Turelli, M., Gillespie, J. & Schoener, T. 1982. The fallacy of the fallacy of the averages in ecological optimization theory. *Am Nat* 119: 879-884.

Turke, P. W. 1989. Evolution and the demand for children. *Popul Dev Rev* 15: 61-90.

Turke, P. W. & Betzig, L. L. 1985. Those who can do: Wealth, status, and reproductive success on Ifaluk. *Ethol Sociobiol* 6: 79-87.

—. 1986. Food sharing on Ifaluk. *Curr Anthropol* 27: 397-400.

Turnbull, C. M. 1972. *The Mountain People*. New York: Simon & Schuster.

Turner, B. L. II. 1982. Pre-Colombian agriculture: Review of Maya subsistence. *Science* 217: 345-346.

Tuthill, R. W. 1996. Hair lead levels related to children's classroom attention-deficit behavior. *Arch Environ Health* 51: 214-220.

Tutor, B. M. 1962. Nesting studies of the boat-tailed grackle. *Auk* 79: 77-84.

Ulrich, R. S. 1984. View through a window may influence recovery from surgery. *Science* 224: 420-421.

Ulrich, R. S. & Addoms, D. 1981. Psychological and recreational benefits of a neighborhood park. *J Leisure Res* 13: 43-65.

Ulrich, R. S. & Lundén, O. 1990. Effects of nature and abstract pictures on patients recovering from open heart surgery. Paper presented at Internat. Congr. of Behavioral Medicine, Uppsala. June 27-30, 1990.

Ulrich, R. S., Dimberg, O. & Driver, B. L. 1991. Psychophysiological indicators of leisure benefits. In *Benefits of Leisure* (Driver, B. L., Brown, P. J. & Peters, G. L., eds.). PA: Ventura State College.

United Nations. 1953. *The Determinants and Consequences of Population Trends*. United Nations.

Vander Wall, S. 1990. *Food Hoarding in Animals*. Chicago: University of Chicago Press.

Veblen, T. 1899/1981. *The Theory of the Leisure Class*. New York: Penguin Books.

—. 1973. *The Theory of the Leisure Class* [originally published in 1899]. Boston: Houghton Mifflin.

Vehrencamp, S. L. 1983. A model for the evolution of despotic versus egalitarian societies. *Anim Behav* 31: 667-682.

Viazzo, P. P. 1990. *Upland Communities: Environment, Population, and Social Structure in the Alps since the Sixteenth Century*. Cambridge: Cambridge University Press.

Vickers, W. 1980. An analysis of Amazonian hunting yields as a function of settlement age. In *Studies of Hunting and Fishing in the Neotropics* (Hames, R., ed.). Working Papers on South American Indians, vol. 2. pp. 7-29. Bennington, VT: Bennington College.

—. 1988. Game depletion hypothesis of Amazonian adaptation: Data from a native community. *Science* 239: 1521-1522.

—. 1991. Hunting yields and game composition over ten years in an Amazon Indian Territory. In *Neotropical Wildlife Use and Conservation* (Robinson, J. & Redford, R., eds.), pp. 53-81. Chicago: Chicago University Press.

—. 1994. From opportunism to nascent conservation: The case of the Siona-Secoya. *Hum Nat* 5: 307-337.

Vining, D. R. 1986. Social versus reproductive success—the central theoretical problem of human sociobiology. *Behav Brain Sci* 9: 167-216.

Vogel, G. 2004. Behavioral evolution. The evolution of the golden rule. *Science* 303: 1128-1131.

Voland, E. 1998. Evolutionary ecology of human reproduction. *Annu Rev Anthropol* 27: 347-374.

Voland, E. & Dunbar, R. 1995. Resource competition and reproduction: The relationship between economic and parental strategies in the Krummhorn population (1720-1874). *Hum Nat* 6: 33-49.

von Schantz, T. 1984. Spacing strategies, kin selection, and population regulation in altricial vertebrates. *OIKOS* 42: 48-58.

Wade, R. 1994. *Village Republics: Economic Conditions for Collective Action in South India.* San Francisco, CA: ICS Press.

Wallace, A. R. 1869. *The Malay Archipelago: The Land of the Orang-utan and the Bird of Paradise.* Oxford, New York: Oxford University Press.

Wang, X. T. 1996. Domainspecific rationality in human choices: Violations of utility axioms and social contexts. *Cognition* 60: 3163.

Wasser, S. K. 1994. Psychosocial stress and infertility. *Hum Nat* 5: 293-306.

Watkins, S. C. 1990. From local to national communities: The transformation of demographic regimes in western Europe, 1870-1960. *Popul Dev Rev* 16: 241-272.

Wazir-Jahan, B. K. 1981. *Ma Betisek Concepts of Living Things.* London and Atlantic Highlands, NJ: Athlone Press and Humanities Press.

Weeks, H. 1993. Arboreal food caching by long-tailed weasels. *Prairie Nat* 25: 39-42.

Weick, K. E. 1984. Small wins: Redefining the scale of social problems. *Am Psychol* 39(1): 40-49.

Weinberg, E. D. 1984. Iron withholding: A defense against infection and neoplasia. *Physiological Review* 64: 65-102.

Western, D. 1975. Water availability and its influence on the structure and dynamics of a savannah mammal community. *East African Wildlife Journal* 13: 265-286.

Wharton, E. 1962. *The Age of Innocence.* New York: Signet Classics.

Whary, M. T. & Fox, J. G. 2004. Th1-mediated pathology in mouse models of human disease is ameliorated by concurrent Th2 responses to parasite antigens. *Curr Top Med Chem* 4(5): 531-538.

Whelan, T. (ed.). 1991. *Nature Tourism: Managing for the Environment.* Washington, DC: Island Press.

White, A. T. 1988. The effect of community-managed marine reserves in the Philippines on their associated coral reef fish populations. *Asian Fisheries Science* 2: 27-41.

White, L. 1967. The historical roots of our ecologic crisis. *Science* 155: 1203-1207.

White, T. A. & Runge, C. F. 1995. The emergence and evolution of collective action: Lessons from watershed management in Haiti. *World Dev* 23(10): 1683-1698.

Whiten, A. & Byrne, R. (eds.). 1997. *Machiavellian Intelligence II: Extensions and Evaluations.* Cambridge: Cambridge University Press.

Whiteman, L. 1997. Making waves. *National Parks* 71: 22-25.

Wiener, J. B. 1999a. Global environmental regulation: Instrument choice in legal context. *Yale Law J* 108(4): 677-800.

—. 1999b. On the political economy of global environmental regulation. *Georgetown Law J* 87: 749-794.

Williams, G. C. 1966. *Adaptation and Natural Selection. A Critique of Some Current Evolutionary Thought.* Princeton, NJ: Princeton University Press.

—. 1989. A sociobiological expansion of evolution and ethics. In *T.H. Huxley's Evolution and Ethics, with New Essays on its Victorian and Sociobiological Context* (Paradis, J. & Williams, G. C., eds.), pp. 179-214. Princeton, NJ: Princeton University Press.

—. 1992. Gaia, nature worship, and biocentric fallacies. *Q Rev Biol* 67(4): 479-486.

Williams, G. C. & Nesse, R. M. 1991. The dawn of Darwinian medicine. *Q Rev Biol* 66: 1-22.

Willis, K. G. & Garrod, G. D. 1993. Valuing landscape: A contingent valuation approach. *J Environ Manage* 37: 1-22.

Wilson, D. S. 1983. The group selection controversy: History and current status. *Annu Rev Ecol Syst* 14: 159-187.

—. 1998. Game theory and human behavior. In *Game Theory and Animal Behavior* (Dugatkin, L. & Reeve, H., eds.), pp. 261-282. Oxford: Oxford University Press.

—. 1999. A critique of R. D. Alexander's views on group selection. *Biol Philos* 14: 431-449.

Wilson, D. S. & Dugatkin, L. A. 1991. Nepotism vs. tit-for-tat, or why should you be nice to your rotten brother? *Evol Ecol* 5: 291-299.

Wilson, D. S. & Sober, E. 1994. Re-introducing group selection to the human behavioral sciences. *Behav Brain Sci* 17: 585-654.

Wilson, E. O. 1975. *Sociobiology: The New Synthesis*. Cambridge, MA: Harvard University Press.

—. 1984. *Biophilia: The Human Bond with Other Species*. Cambridge, MA: Harvard University Press.

—. (ed.). 1988. *Biodiversity*. Washington, DC: National Academy Press.

—. 1992. *The Diversity of Life*. Cambridge, MA: Belknap Press.

—. 1993. Biophilia and the conservation ethic. In *The Biophilia Hypothesis* (Kellert, S. R. & Wilson, E. O., eds.), pp. 31-41. Washington, DC: Island Press, Shearwater Books.

—. 1998a. *Consilience: The Unity of Knowledge*. New York: Alfred Knopf.

—. 1998b. Integrating science and the coming century of the environment. *Science* 279: 2048-2049.

—. 2001a. *The Future of Life*. New York: Alfred A. Knopf.

—. 2001b. Nature matters. *Am J Prev Med* 20: 241-242.

Wilson, J. 1977. A test of the tragedy of the commons. In *Managing the Commons* (Hardin, G & Braden, J., eds.), pp. 96-111. San Francisco, CA: Freeman.

Wilson, J. Q. & Herrnstein, R. 1985. *Crime and human nature*. New York: Simon & Schuster.

Wilson, M. & Daly, M. 1985. Competitiveness, risk-taking and violence: The young male syndrome. *Ethol Sociobiol* 6: 59-73.

—. 1993. Lethal confrontational violence among young men. In *Adolescent risk taking* (Bell, N. J. & Bell, R. W., eds.), pp. 84-106. Newbury Park, CA: Sage.

—. 1997. Life expectancy, economic inequality, homicide, and reproductive timing in Chicago neighbourhoods. *British Medical J* 314: 1271-1274.

Wilson, M., Daly, M. & Gordon, S. 1998. The evolved psychological apparatus of human decision-making is one source of environmental problems. In *Behavioral Ecology and Conservation Biology* (Caro, T., ed.), pp. 501-523. Oxford: Oxford University Press.

Wilson, M., Daly, M., Gordon, S. & Pratt, A. 1996. Sex differences in valuations of the environment? *Popul Environ* 18: 143-159.

Wilson, M. S. & Maizels, R. M. 2004. Regulation of allergy and autoimmunity in helminth infection. *Clin Rev Allergy Immunol* 26(1): 35-50.

Winston, G. & Woodbury, R. 1991. Myocopic discounting: Empirical evidence. In *Handbook of Behavioral Economics,* Vol. 2B (Kaish, S. & Gilad, B., eds.), pp. 325-342. Greenwich, CT: JAI Press.

Winterhalder, B. 1981. Foraging strategies in the Boreal Forest: An analysis of Cree hunting and gathering. In *Hunter-Gatherer Foraging Strategies* (Winterhalder, B. & Smith, E. A., eds.), pp. 66-98. Chicago: Chicago University Press.

—. 1983. Opportunity-cost foraging models for stationary and mobile predators. *Am Nat* 122: 73-84.

—. 1996. Social foraging and the behavioral ecology of intragroup resource transfers. *Evol Anthropol* 5: 46-57.

—. 1997. Gifts given, gifts taken: The behavioral ecology of nonmarket, intragroup exchange. *J Archaeol Res* 5: 121-168.

—. 1997b. *Foraging strategy adaptations of the boreal forest Cree: An evaluation of theory and models from evolutionary ecology.* Ph.D. dissertation, Cornell University.

Winterhalder, B., Baillargeon, W., Cappelletto, F., Daniel, I. & Presctt, C. 1988. The population ecology of hunter-gatherers and their prey. *J Anthrop Archaeol* 7: 289-328.

Winterhalder, B. & Leslie, P. Risk-sensitive fertility: The variance compensation hypothesis. Unpublished.

Winterhalder, B. & Lu, F. 1997. A forager-resource population ecology model and implications for indigenous conservation. *Conservation Biol* 11: 1354-1364.

Wolfson, R. 1991. *Nuclear Choices.* Cambridge, MA: MIT Press.

Woodwell, D. A. 1997. *National Ambulatory Medical Care Survey: 1995 Summary.* Advance Data, Number 286. Bethesda, Maryland: National Center for Health Statistics.

Woolfenden, G. E. & Fitzpatrick, J. W. 1984. *The Florida Scrub Jay: Demography of a Cooperativelybreeding Bird.* Princeton, NJ: Princeton University Press.

Wrangham, R. & Peterson, D. 1996. *Demonic Males: Apes and the Origins of Human Violence.* London: Bloomsbury.

Wright, R. 1994. *The Moral Animal: The New Science of Evolutionary Psychology.* New York: Vintage Books.

—. 2000. *Nonzero: The Logic of Human Destiny.* New York: Pantheon.

Wrigley, C. 1978. Fertility strategy for the individual and the group. In *Historical Studies of Changing Fertility* (Tilly, C., ed.), pp. 135-154. Princeton, NJ: Princeton University Press.

Wynne-Edwards, V. 1962. *Animal Dispersion in Relation to Social Behavior.* Edinburgh: Oliver & Boyd.

Yochim, M. J. 2001. Aboriginal overkill overstated: errors in Charles Kay's hypothesis. *Hum Nat* 12: 141-167.

Yost, J. & Kelley, P. 1983. Shotguns, blowguns and spears: The analysis of technological efficiency. In *Adaptive Responses of Native Amazonians* (Hames, R. & Vickers, W., eds.), pp. 189-223. New York: Academic Press.

Young, O. (ed.). 1999. *Science Plan for Institutional Dimensions of Global Environmental Change.* Bonn, Germany: International Human Dimensions Programme on Global Environmental Change.

Young, T. P. 1993. Natural die-offs of large mammals: implications for conservation. *Conserv Biol* 8: 410-418.

Zahavi, A. 1975. Mate selection: Selection for a handicap. *J Theor Biol* 53: 205-214.

—. 1977. Reliability in communications systems and the evolution of altruism. In *Evolutionary Ecology* (B. Stonehouse and C. Perrins, eds.), pp. 253-259. London: Macmillan.

—. 1990. Arabian Babblers: The quest for social status in a cooperative breeder. In *Cooperative Breeding in Birds* (Stacey, P. B. & Koenig, W. D., eds.), pp. 103-130. Cambridge: Cambridge University Press.

—. 1993. The fallacy of conventional signalling. *Philosophical Transactions of the Royal Society of London* 340: 227-230.

Zahavi, A. & Zahavi, A. 1997. *The Handicap Principle*: *A Missing Piece of Darwin's Puzzle*. Oxford: Oxford University Press.

Zala, S. M. & Penn, D. J. 2004. Abnormal behaviours induced by chemical pollution: a review of the evidence and new challenges. *Anim Behav* 68: 649-664.

Zann, L. P. 1989. Traditional management and conservation of fisheries in Kiribati and Tuvalu Atolls. In *Traditional Marine Resource Management in the Pacific Basin: An Anthology* (Ruddle, K. & Johannes, R. E., eds.), pp. 77-102. Jakarta, Indonesia: UNESCO/ROSTSEA.

Zerner, C. 1990. Community management of marine resources in the Maluku Islands. Unpublished manuscript prepared for the Fisheries Research and Development Project. Jakarta, Indonesia.

Zerner, C. 1994. Transforming customary law and coastal management practices in the Maluku Islands, Indonesia, 1870-1992. In *Natural Connections: Perspectives in Community-Based Conservation* (Western, D. & Wright, R. M., eds.; Strum, S. C., assoc. ed.), pp. 80-112. Washington, DC: Island Press.

Zuckerman, M. 1994. *Behavioral Expressions and Biosocial Bases of Sensation Seeking*. Cambridge: Cambridge University Press.

Contributors

Alvard, M. S.	Associate Professor of Anthropology Department of Anthropology 4352 TAMU Texas A&M University College Station, Texas 77843-4352 Office phone: 979-862-3492 Alvard@tamu.edu http://anthropology.tamu.edu/faculty/alvard/profile.htm
Boone, J. L.	Department of Anthropology University of New Mexico Albuquerque, New Mexico 87131 Office telephone: 505-277-6558 jboone@unm.edu http://www.unm.edu/~hebs/pubs_boone.html
Borgerhoff-Mulder, M.	Professor Department of Anthropology (Evolutionary Wing) University of California, Davis One Shields Avenue Davis, CA 95616 U.S.A. Phone: 530-752-0659 Fax: 530-752-8885 mborgerhoffmulder@ucdavis.edu http://www.anthro.ucdavis.edu/mbmulder
Daly, M.	Department of Psychology McMaster University 1280 Main Street West Hamilton, Ontario, Canada L8S 4K1 Fax: 905-529-6225 daly@mcmail.cis.mcmaster.ca http://psych.mcmaster.ca/dalywilson/pubs.html
† Hardin, G.	
Hames, R.	Department of Anthropology and Geography University of Nebraska 836 Oldfather House Lincoln, NE 68588 Office Phone 402-472-6240 rhames@unl.edu http://www.unl.edu/rhames/

Heinen, J. T.	Associate Professor and Chair Department of Environmental Studies ECS 340, Florida International University Miami, FL 33199 Phone: 305-348-3732 Fax: 305-348-6137 heinenj@fiu.edu http://www.fiu.edu/~envstud/faculty/heinen.html
Low, B. S.	Professor of Natural Resources and Environment *Evolutionary and Behavioral Ecology* Phone: 734-763-4518 Office: G142A Dana bobbilow@umich.edu http://www.snre.umich.edu/contact/faculty-detail.php?people_id=15
Mysterud, I.	Department of Biology, University of Oslo PO Box 1066 Blinern, NO-0316 Oslo, Norway Phone: ++4722858192 Fax: ++4722854605 mysterud@bio.uio.no
Orians, G. H.	Professor Emeritus of Zoology University of Washington Seattle, Washington blackbrd@u.washington.edu blackbrd@serv.net
Ostrom, E.	Arthur F. Bentley Professor of Political Science Co-Director, *Workshop in Political Theory and Policy Ananlysis* Professor of Public and Environmental Affairs, Part-Time Co-Director, *Center for the Study of Institutions, Population, and Envirnmental Change* (CIPEC) 408 N. Indiana Ave., Room 230A, Bloomington, IN 47408 Phone: 812-855-0441 Fax: 812-855-3150 ostrom@indiana.edu http://www.cogs.indiana.edu/people/homepages/ostrom.html

Penn, D. J.	Currently Director and Senior Scientist of the Konrad Lorenz Institute for Ethology of the Austrian Academy of Sciences SavoyenstraBe 1a 1160 Vienna, Austria Phone: ++43 1 51581 2723 Fax: ++43 1 51581 2800 d.penn@klivv.oeaw.ac.at http://www.oeaw.ac.at/klivv/en/persons/penn.html
Rogers, A.	Professor of Anthropology University of Utah Stewart Building Salt Lake City, UT 84112 Phone: 801-581-5529 Fax: 801-581-6252 rogers@anthro.utah.edu http://www.anthro.utah.edu/~rogers/
Ruttan, L.	Assistant Professor Department of Environmental Studies Emory University 400 Dowman Drive, Suite N510 Atlanta GA 30322, USA Department phone: 404-727-4216 Office phone: 404-727-4217 Fax: 404-727-4448 lruttan@emory.edu http://www.envs.emory.edu/Faculty/ruttan.html
Sutherland, W. J.	Centre for Ecology, Evolution and Conservation School of Biological Sciences, University of East Anglia, Norwich NR4 7TJ, United Kingdom Phone: +44 01603 592778 Fax: +44 01603 592250 w.sutherland@uea.ac.uk http://bioweb2.bio.uea.ac.uk/bioperson/facultyasp/ SutherlandW.aspx
Wilson, M.	Department of Psychology McMaster University 1280 Main Street West Hamilton, Ontario, Canada L8S 4K1 Fax: 905-529-6225 wilson@mcmaster.ca http://psych.mcmaster.ca/dalywilson/pubs.html

Wilson, E. O.

Professor Emeritus Pelegrino
University Research Professor
Harvard University 26 Oxford Street Cambridge
Massachusetts 02138-2902 USA
Phone: 617-495-2315
Fax: 617-495-1224
ewilson@oeb.harvard.edu
http://en.wikipedia.org/wiki/Edward_Osborne_
Wilson

Author Index

Acheson, J.M. (J.), 4, 91, 103, 109, 114, 139
Adams, P.F., 49
Adams, W.M., 4
Addoms, D., 272
Adler, D., 36, 37
Agrawal, A., 4, 139
Ajzen, I., 36
Akimichi, T., 109
Alcock, J., 10
Alcorn, J.B., 101, 142, 155
Alexander, R.D., 11, 12, 13, 14, 174, 289, 290
Alroy, J., 3
Alvard, M., 3, 32, 69, 72, 75, 76, 78, 81, 101, 102, 103, 113, 143, 144, 282
Andamari, R., 116
Anderson, A., 3, 72, 102
Anderson, T.L., 112
Andersson, M., 172
Andrews, G., 264
Appleton, J., 262
Arnhart, L., 289, 290
Arnhem, K., 57
Arnold, J.E.M., 139
Aspelin, L., 57
Atran, S., 77
Auger, J., 286
Aunger, R., 83, 102
Axelrod, R., 11, 25, 37, 111, 140
Aydede, S., 168

Baden, J., 25
Bagnoli, M., 33
Baier, S., 191, 227
Bailey, C., 113, 115, 118, 122
Baines, G., 115, 116
Baksh, M., 57
Baland, J.M., 140
Balee, W., 142
Ball, S.B., 230, 235

Barkan, C.P.L., 38, 177
Barkow, J.H., 2, 287
Barnett, C., 2, 189
Barnett, H.G., 189
Baron, J., 36
Barro, R.J., 202
Bartholomew, K., 102
Basalla, G., 282
Bateman, A.J., 40
Bateman, I., 168
Bates, D.G., 6
Bateson, M., 46
Beauchamp, G., 246
Becker, G.S. (G.), 33, 101, 202
Beckerman, S., 57, 63, 100, 102, 103, 142
Bell, N.J., 47, 174
Bell, R.W., 47, 174
Bellinger, D.C., 287
Belovsky, G., 223
Ben-Ner, A., 33
Benson, K., 46, 143
Bergman, R., 57
Bergstrom, T.C., 33, 101
Berkes, F., 4, 109, 115, 127, 139, 144, 154
Berrigan, D., 101
Betzig, L.L. (L.), 2, 6, 13, 40, 173, 190, 202
Binger, B.R., 36
Binmore, K.G., 33
Bird, M.I., 282
Birdsell, J., 142
Bishop, R.C., 143, 145
Bjorkbom, L., 28
Bliege Bird, R.L., 6
Blomquist, W., 139
Blume, M.E., 174
Blumenshine, R.J., 266
Blurton-Jones, N.G., 246
Boaz, N.T., 284
Bock, J., 6
Bodley, J.H., 9, 68, 72

Bodmer, R., 102
Boehm, C., 289
Bohm, P., 37
Boone, J.L., 6, 37, 41, 176, 183, 191, 196, 219, 221, 228, 229, 230, 233
Borgerhoff Mulder, M., 3, 4, 6, 99, 103, 113, 141, 145, 148, 151, 153, 183, 202, 219, 237, 247
Boswell, J., 13
Boulding, K.E., 15
Bowles, S., 139
Boyce, J., 36
Boyce, M.S., 221
Boyce, R.R., 36
Boyd, R., 2, 10, 25, 99, 103, 110, 111, 112, 139, 162, 194, 213, 241, 242, 246, 289
Bradburd, D., 153
Bradbury, J.W., 186
Bradlow, H.L., 286
Bragg, B.W.E., 48, 175
Brewer, G.D., 28
Broecker, W.S., 139
Bromley, D.W., 139
Brown, D.E., 4, 8
Brown, I.D., 48, 175
Brown, J.L., 64, 103, 116
Brown, J.W., 290
Brown, L.R., 9, 157
Bruening, K., 287
Brunt, P.A., 234, 235
Buchanan, J., 140
Buck, S., 139
Budiansky, S., 67, 68, 101
Bulmer, R.N.H., 114, 115, 144
Bunyard, P., 68, 101
Burger, J., 129, 140
Burke, E., 262
Buss, D.M. (D.), 2, 171, 283
Butt-Colson, A., 61
Byelich, J., 22
Byrne, R., 100

Caldwell, L.K., 28
Callicott, J. B., 78
Campbell, B., 266
Caraco, T., 38, 102, 177
Carlisle, T.R., 193, 195
Carlsen, E., 286
Carlsson, G., 247
Carrier, J., 114, 115
Carroll, J.E., 28

Carson, R.T., 35, 36, 43
Cashdan, E., 91, 102
Cassels, R., 72
Caughley, G., 103
Cavalli-Sforza, L.L., 10, 25
Chagnon, N., 13, 63, 76, 97, 103
Chapman, G.B., 47
Charnov, E.L. (E.), 6, 15, 55, 61, 62, 75, 101, 202
Cheung, S., 103
Chiras, D.D., 9
Christakis, D.A., 6
Christy, F.T., Jr., 115
Ciriacy-Wantrup, S.V., 143, 145
Clark, C.W. (C.), 31, 93, 103, 115, 116, 140, 144, 158, 247
Clarke, A.L., 41, 176
Clay, J.W., 72, 101
Clinton, W.L., 46
Clutton-Brock, T.H., 17, 39, 40, 103, 153
Coale, A.J., 246
Cobb, J.B., Jr., 20
Cockburn, A., 9
Colborn, T., 7, 286, 287
Colwell, R.K., 22
Common, M., 50
Connor, R.C., 75, 151, 193
Conrad, C., 246
Conrad, N., 103
Cook, E.W., 265
Cook, M., 265
Cooper-Marcus, C., 272
Coplan, M.J., 287
Cordain, L., 285
Corning, P.A., 291
Cosgrove, D., 262
Cosmides, L., 2, 32, 34, 37, 101, 140, 287, 288, 290
Costanza, R., 1, 4, 5, 7, 32, 139
Cowhey, P.F., 18
Crawford, E.M., 192
Cronk, L., 9, 246
Cropper, M., 166, 168
Crosby, A.W., 72
Cummings, R.G., 35, 36
Curio, E., 40
Cziko, G., 278

Daly, H.E., 20
Daly, M., 3, 10, 31, 32, 33, 39, 46, 47, 48, 75, 169, 171, 173, 174, 175, 202, 221, 234

Darwin, C., 10, 172, 276, 279
Dasmann, R., 101
Davies, N.B., 10, 75
Davis, D.L., 286
Davis, J.N., 234
Dawes, R.M. (R.), 4, 139
Dawkins, M.S., 194
Dawkins, R., 2, 10, 75, 233
de Graaf, J., 5, 6
de Waal, F.B.M. (F.), 291
Deevey, F.S., 72
Demsetz, H., 140
Denevan, W.M. (W.), 65, 68, 72
Denton, K., 289
Devine, P.G., 289
Dewar, R., 102
Diamond, J.M., 3, 32, 72, 101, 251, 256
Diamond, N., 102
Dietrich, K.N., 287
Dibb, S., 286
Dietz, T., 4
Dimberg, U., 272
Dion, D., 25
Dixon, J.A., 21
Dixon, S., 234, 235
Dove, M.R., 25
Downing, D., 286
Driver, B.L., 272
Drucker, P., 189
Duff, A.M., 235
Dugatkin, L.A., 111, 112, 151, 193
Dunbar, R.I.M. (R.), 222, 246, 285
Dunford, C., 261
Dunlap, T.R., 22, 170
Dupont, D., 36
Durham, W., 2, 63
Durning, A., 5, 6, 142
Dyer, O., 286
Dyson, T., 285
Dyson-Hudson, N., 147
Dyson-Hudson, R., 64, 102, 115, 116, 147

Eaton, S.B., 284, 285
Eckel, C.C., 33, 230, 235
Edgerton, R., 4
Eggan, F., 227
Eggertson, T., 116
Ehrlich, A.H., 14, 15, 281
Ehrlich, P.R., 14, 15, 281, 282
Eibl-Eibesfeldt, I., 2
Ellickson, R., 99, 103

Ellis, J.E., 145
Ellison, P.T., 247
Elster, J., 84, 102
Engen, S., 168
Enquist, M., 48
Ensminger, J., 154
Essock-Vitale, S., 202
Ewald, P.W., 278, 284

Fanshawe, J., 102
Farmer, N., 102
Feeny, D., 112, 139
Fehr, E., 139, 292
Feit, H., 53, 102
Feldman, M.W., 10, 25
Fershtman, C., 189
Fester, R., 291
Field, C.B., 129
Finlayson, A.C., 139
Finn, R., 48, 175
Firth, R., 191, 224, 225, 226
Fischoff, B., 36
Fisher, I., 102
Fisher, R.A. (R.), 46, 96, 103, 161
Fitzgibbon, C., 102
Fitzpatrick, J.W., 11
Flannery, T., 3
Flynn, J., 42
Foley, R., 101, 103
Folke, C., 139
Fox, J.G., 284
Fox, R., 273
Francis, C., 272
Frank, R.H., 5, 6, 32, 33, 230, 231, 235, 291
Fratkin, E., 148, 152, 153, 155
Freeman, D., 4
Freet, J.C., 287
Fretwell, S.D., 202, 246
Friend, I., 174
Frumkin, H., 283

Gadgil, M., 154
Galaty, J.G., 145, 151, 153
Gardner, G.T., 1, 7, 40, 283, 293, 294
Gardner, R., 139
Gardner, W., 49, 175, 176, 177
Garrod, G.D., 35
Gibson, C.C. (C.), 139, 154
Gifford, R., 1
Gigerenzer, G., 34, 37
Gillespie, B., 6, 102

Gillespie, J.H., 221
Gintis, H., 139, 292
Glass, D.C., 273
Gleditsch, N.P., 283
Glick, D., 22
Global Governance, 140
Goodall, J., 11
Goodin, R., 168
Goodwin, B.K., 35
Gordon, H.S., 4, 103, 139, 143
Gordon, S., 3, 31, 169
Gorsline, J., 101
Gove, W.R., 47, 48, 174, 176, 177
Grabowski, R., 155
Grafen, A., 184, 186, 194, 230
Gramlich, E.M., 22
Grant, P., 169
Gray, L.E., 286
Green, L., 46, 102
Greenberg, S., 173
Greene, J., 36
Gregory, R., 36
Groeger, J.A., 48, 175
Gross, D., 63, 64
Grossman, P.J., 33
Guagnano, G.A., 36
Guilford, T., 194

Haas, P., 140
Hackworth, D.H., 25
Hadlock, W.S., 140
Haines, F., 72
Hames, R., 3, 32, 53, 56, 57, 58, 59, 60, 63, 64, 65, 69, 72, 73, 75, 76, 77, 78, 101, 102, 113, 127, 142, 143, 144
Hamilton, W.D. (W.), 11, 25, 75, 103, 110, 111, 127, 160, 186, 288, 291
Hanna, S.S. (S.), 139, 154
Harcourt, A.H., 40
Hardin, G., 4, 13, 25, 73, 98, 102, 105, 107, 109, 130, 136, 139, 157
Hargrove, E.C., 78
Harpending, H.C., 161, 203, 220, 233, 247
Harris, M., 115, 271
Harrison, G.W., 35, 36
Hartig, T., 273
Hartung, J., 289
Harvey, P.H., 186
Hassan, F., 223
Hastjarjo, T., 38, 177
Hauert, C., 7

Hausfater, G., 11
Hawken, P., 283
Hawkes, K., 4, 6, 15, 32, 37, 40, 55, 69, 75, 76, 77, 102
Hawks, R., 6
Hecht, S., 9
Hechter, M., 111, 170
Heerwagen, J.H., 7, 262, 265, 266, 271, 279
Heinen, J.T., 3, 9, 10, 13, 14, 15, 16, 17, 18, 20, 21, 24, 26, 33, 39, 76, 78, 79, 101, 171, 292
Heizer, R.F., 189
Heller, W.P., 157
Henderson, N., 5, 165, 168
Hepburn, R.W., 262
Hermy, C., 115
Herren, U.J., 152
Herrnstein, R.J. (R.), 46, 47, 158, 175
Hess, C., 139
Hewlett, B.S., 40, 47
Heyde, J.M., 35, 36
Hilborn, R., 139
Hildebrand, G., 262
Hildebrandt, W., 102
Hill, J., 219
Hill, K., 13, 32, 40, 47, 76, 86, 96, 97, 101, 102, 103, 127, 143, 144, 155, 173, 202, 246
Hilton, R.M., 140
Hirsch, F., 5
Hirshleifer, J., 101
Hoch, S.J., 47
Hockey, R., 273
Hodes, R.L., 265
Hodgson, G., 101
Hoffman, E., 34, 140
Hoffrage, U., 34, 37
Holdaway, R.N., 3
Holdren, J.P., 281
Holinger, P.C., 47, 49, 174, 176
Holling, C., 102
Holmes, R., 23
Homer-Dixon, T.F., 7, 8, 283
Homewood, K.M., 145
Hooper, A., 113
Hopkins, K., 183, 234
Hough, J.L., 21
House, L., 101
Houston, A.I., 222, 231
Houston, D., 101
Howarth, R.B., 32, 50

Howell, N., 40, 75, 173
Hrdy, S.B., 11, 247
Hsieh, C.C., 38, 177
Hu, H., 287
Hugdahl, K., 265
Hughes, J., 101
Humphrey, C., 139
Humphrey, N.K., 11
Hunn, E., 101, 113, 142
Hurtado, M., 40, 47, 96, 97, 103, 173, 246
Hviding, E., 113, 115
Hygge, S., 265

IIED, 155
IWIGIA, 142
Irons, W.G., 202, 246, 290
Irwin, C.E., 49, 176
Irwin, J.R., 36
Isen, A.M., 273
IUCN, 21
Ives, J.D., 24

Jacomb, C., 3
Jacowitz, K.E., 169
Jakle, J., 262
James, H.F., 3, 72
Jensen, T.K., 7
Johannes, R.E. (R.), 103, 109, 113, 114, 115, 127, 142
Johansson, S.R., 234
Johnson, A., 63, 65, 77
Johnstone, R., 194
Jonah, B.A., 47, 174
Jones, D., 103
Jones, T., 102
Judge, D.S., 247

Kacelnik, A., 46, 47
Kagel, J.H., 38, 102, 140, 177
Kahneman, D., 34, 35, 36, 37, 39, 169, 177
Kaplan, H.S. (H.), 6, 13, 32, 50, 69, 72, 76, 101, 102, 103, 127, 143, 155, 199, 202, 219, 230, 239, 243, 245, 246, 247
Kaplan, S., 7, 272, 273
Karker, A.C., 265
Kattel, B., 21
Kauffman, W., 4
Kay, C., 3, 32, 101
Kay, J., 115, 142
Keckler, C.N.W., 223

Keeley, L.H., 290
Keller, L., 289, 290
Kellert, S.J., 7
Kelley, P., 57, 58, 61
Kendler, K.S., 264
Kenrick, D., 32
Kensinger, K., 60, 65
Kent, S., 284
Keohane, R., 75, 139, 140
Kessler, K.L., 6, 191, 219
Keyfitz, N., 242, 247
Kiltie, R., 60, 65
Kim, R., 287
King, B., 100
Kingsland, S.E., 259
Kirby, K.N., 46, 47
Kirch, P.V., 225, 226
Klein, R.G. (R.), 3, 103
Klepper, S., 103
Klima, G., 152
Kluger, M.J., 284
Knetsch, J.L., 35, 36, 37, 169
Knight, J., 154
Knodel, J., 246
Knopf, R.C., 272
Kollock, P., 4
Kopishke, K., 155
Kracke, W., 60, 65
Krebs, D.L., 289
Krebs, J., 10, 64, 75, 85, 86, 102, 143
Krech, S.I., 3, 4
Kreps, D.M., 140
Krier, J.E., 140
Krupnik, I., 83, 102
Kurland, J., 100

Lack, D., 82, 101, 238, 239, 246
Lahr, M., 103
Lam, W.F., 140
Lamprey, H.F., 145
Lancaster, J.B., 6, 243, 247
Lande, R., 168
Lane, C., 99, 103, 145, 146, 147, 148, 152
Lang, K., 102
Lang, P.J., 265
LaPointe, J., 57
Lappé, M., 284
Latane, B., 25
Lawlor, L., 85, 102
Leacock, E., 53
LeBoeuf, B.J., 46

Lechner, M., 246
Lees, S.H., 6
Leimar, O., 48
Lenton, T.M., 291
Leslie, P., 221
Lesthaeghe, R., 246
Levy, J., 191, 226, 227
Levy, M., 140
Lewis, H.T., 54
Lewontin, R.C., 74
Libecap, G.D., 140
Lima, S.L., 38, 177
Lind, R., 168
Liverman, M., 22
Livi-Bacci, M., 233, 246
Lloyd, W.F., 105, 106, 107
Lodwick, D.G., 170
Loewenstein, G.F. (G.), 34, 35, 36, 37, 47, 168
Logue, A., 168
Loomis, J., 37
Lopes, L.L., 38, 177
Lorberbaum, J., 111
Lovejoy, C.O., 227, 266
Lovejoy, P.E., 191
Lovelock, J., 68
Low, B.S. (B.), 2, 3, 4, 6, 9, 10, 13, 14, 15, 16, 17, 18, 20, 22, 24, 25, 26, 31, 32, 33, 39, 50, 67, 69, 71, 76, 77, 78, 79, 101, 142, 171, 247, 283, 292
Lowdermilk, W.C., 72
Lu, F., 102, 142
Ludwig, D., 139
Lumsden, C.J., 2, 10, 25, 250, 256
Lundén, O., 272
Luttbeg, B., 247
Lutz, W., 7
Lyman, R., 102
Lyng, S., 47, 48, 174, 176

MacArthur, R.H., 213
MacDonald, H., 37
MacDonald, K., 289
Mace, R., 6, 245, 247, 289
MacFarlane, J.W., 115
MacKinnon, J., 21
Maizels, R.M., 284
Mangel, M., 116, 222, 247
Manson, J., 24, 25
Marcus, G.B., 34
Margulis, L., 68, 291
Marks, S.A., 154

Marler, P., 275
Martin, C., 53
Martin, L.L., 75
Martin, P.S. (P.), 3, 72
Martindale, S., 177
Marx, K., 106
Masters, R.D., 287
Mather, C., 278
Matthews, M.L., 48, 175
Maxwell, J.H., 78, 79
Maynard Smith, J., 101, 110, 188, 289, 291
McAlpin, M.B., 192
McCabe, J.T., 145
McCabe, K.A. (K.), 140
McCartney, A., 102
McCartney, A., 102
McCay, B.J., 4, 109, 115, 139
McCulloch, B., 72
McDonald, D., 53, 54, 60, 102
McEvoy, A., 139
McGinley, M.A., 202
McGinnis, M., 75, 139
McInish, T.H., 174
McIntosh, R.P., 25
McCarthy, S., 102
McKean, M.A., 140, 155
McKenney, D.W., 37
McMichael, A.J., 284
McNally, R.J., 264
McNamara, J.M., 222, 231
McNeely, J.A., 21, 25
McWhirter, N., 173
Mealey, L., 202
Mellor, M., 170
Meltzer, D., 103
Merchant, C., 3
Messerli, B., 24
Mesterton-Gibbons, M., 151
Metcalfe, S., 154
Michelman, F., 140
Michod, R.E., 170
Mielke, H.W., 287
Milgrom, P., 140
Milinski, M., 7
Miller, C.H., 3
Miller, G., 6
Millstein, S.G., 49, 176
Mineka, S., 265
Minnis, P.E., 13
Mitchell, R.C., 35, 36
Mogaka, H., 102
Mohai, P., 43

Moran, A.R., 48, 175, 264
Møller, A.P., 40
Molte, K., von, 28
Monbiot, G., 107
Monod, J., 277
Montagu, A., 283
Moore, M., 102
Moore, O., 85, 102
Moran, A.R., 48, 175
Moran, C., 264
Moss, C., 11
Moss, M.E., 287
Mulligan, C.B., 33
Mundkur, B., 250, 256
Munro, G.R., 140
Murdock, G.P., 69
Myers, D.G., 25
Myers, N., 9, 283
Myerson, J., 46
Mysterud, I., 1, 2, 281, 289, 290

Nadel, L., 170
Nagin, D., 103
National Research Council, 139
Needleman, H.L., 287
Neel, J.V., 285
Neiman, F.D., 6, 197
Nelson, R., 53, 101, 115, 142
Nesse, R.M. (R.), 7, 271, 272, 278, 284,
     285, 287, 292
Netting, R.M., 145
Newman, P.C., 73
Ng, Y.K., 32
Nisbett, R.E., 34
Noë, R., 186
Norgaard, R.B., 50, 129, 139
North, D.C., 140
Norton, B., 256
Nowak, M.A., 111, 140
Nunney, L., 289
Nylander, M., 286

Oelschlaeger, M., 68, 101, 256
Öhman, A., 264, 265
Olsen, M.E., 170
Olson, M., 4, 102, 111, 155, 194
Olson, S.L., 3, 72
Orbell, J.M., 139
Organisation for Economic Co-operation
     and Development (OECD), 139
Orians, G.H. (G.), 7, 62, 259, 262, 265,
     266, 271, 279

Ornstein, R., 282
Ostrom, E., 4, 5, 25, 26, 74, 75, 109, 110,
     129, 135, 139, 144, 153, 154
Ostrom, V., 140
OTA, 141
Oye, K.A., 78, 79

Packer, C., 11, 116
Pagel, M., 289
Palmer, C.T., 112
Parker, G., 103, 153
Parkin, T.G., 234
Paul, C.W., 33, 48, 175
Payton, M., 287
Pearce, D., 168
Pearce, F., 101
Penn, D., 1, 2, 3, 4, 6, 7, 281, 284, 286,
     287, 290, 293, 294
Pennington, R., 247
Peoples, J.G., 115
Peres, C., 102
Perrins, C.M., 221
Perusse, D., 241, 247
Peterson, D., 283
Pinker, S., 2, 3
Pinkerton, E., 139
Platt, J., 4
Platteau, J.P., 140
Pleva, J., 286
Plott, C.R., 139
Policansky, D., 129
Pollard, J.H., 216
Polunin, N., 92, 103, 115, 116, 124, 126
Porter, G., 290
Portney, P., 168
Posey, D.A. (D.), 101, 142
Pratt, A., 169
Profet, M., 278
Pugh, M.D., 38, 177
Pugh, G.E., 261
Pulliam, H.R., 261
Puri, R.K., 155
Pusey, A., 11
Putnam, R.D., 28, 75
Putterman, L., 33
Pyke, G., 55, 64

Qiang, R., 7

Rachlin, H., 168
Rappaport, R., 115
Real, L., 38, 39, 177

Redclift, M., 28
Redford, K.H. (K.), 3, 67, 94, 101, 102, 103, 115, 142
Regelmann, K., 40
Reichel-Dolmatoff, G., 68, 101
Remedial Action Plan, 171
Repetto, R., 139
Repton, H., 271
Revelle, C., 9
Revelle, P., 9
Reznick, D., 84, 102
Richerson, P.J., 2, 10, 25, 99, 103, 110, 111, 112, 194, 213, 241, 242, 246, 289
Richter, L.K., 21
Ridley, M., 2, 4, 7, 31, 50, 79
Rightmire, P., 103
Rimland, B., 286
Ritov, I., 169
Roberts, J., 140
Roberts, R.G., 3
Robinson, J.G. (J.), 94, 102, 103
Robinson, W.C., 247
Rodgers, W.A., 145
Rodin, J., 25
Roebroeks, W., 103
Roelofsma, P.H.M.P., 47
Rogers, A.R., 5, 6, 46, 49, 50, 90, 93, 94, 95, 96, 101, 103, 144, 157, 158, 161, 175, 199, 201, 203, 213, 220, 233, 243, 245, 246, 247
Roitberg, B.D., 40
Rolston III, H., 256
Romer, P.M., 33
Roosevelt, A., 65
Rose, C.M., 140
Rose, M.C., 262
Rose, M.R., 6
Ross, E.B. (E.), 60, 61, 62, 63, 65, 66, 102, 115, 271
Ross, L., 34
Roth, A.E., 140
Roth, E.A., 152, 153
Rubin, P.H., 32, 48, 175
Ruddle, K., 109
Runge, C.F., 110, 112
Ruttan, L.M. (L.), 3, 4, 99, 101, 103, 109, 116, 141, 151, 153, 155
Ruud, P., 102

Sachs, C., 42, 43
Sacks, O., 224

Saether, B., 168
Saffirio, J., 56, 57
Sahlins, M., 53, 83, 101
Safina, C., 140
Salisbury, R.F., 73
Samuelson, P.A., 33
Sandford, S., 145
Savelle, J., 102
Schaller, G.B., 14
Schama, S., 260, 262
Schlager, E., 139
Schmid, A., 168
Schneider, S.H., 22
Schoener, T., 102
Schroeder, H.W., 272
Schwartz, J., 287
Scott, A., 143
Seeley, T.D., 11
Seligman, M.E.P., 261
Sellen, D.W., 151, 153
Sen, A., 228
Sessions, G., 3
Sethi, R., 33, 101, 112
Shackelford, T.K., 283
Shafir, E., 34, 35
Sharpe, R.M., 286
Sherman, P.B., 21
Shermer, M., 4
Sherpa, M.N., 21
Shivakoti, G., 135
Sieff, D.F., 152, 153
Sigmund, K., 111, 140
Simmons, R.T., 112
Simms, S.R., 101
Simon, H.A. (H.), 25, 33, 101
Simon, J.L., 202
Simpson, J., 32
Singer, J.E., 273
Singer, P., 257
Singleton, S., 75
Skakkebaek, N.E., 286
Slobodkin, L., 142
Smil, V., 285
Smith, A., 105
Smith, C.C., 246
Smith, D.M., 9
Smith, E.A., 2, 3, 6, 63, 69, 77, 84, 85, 90, 102, 103, 115, 116, 142, 143, 183, 202
Smith, K., 148, 152
Smith, V.K., 35, 43
Smith, V.L., 140

Smuts, B., 290
Sneath, D., 130, 139
Soares, J.J.F., 265
Sober, E., 74, 288, 289
Soltis, J., 110
Somanathan, E., 33, 101, 112
Sosis, R., 6
Spackman, M., 168
Speck, F.G., 101, 140
Sperling, L., 151, 153
Speth, J.D., 72
Spillius, J., 224, 225
Spoehr, A., 78
Starrett, D.A., 157
Steadman, D.W., 3
Stearman, A.M., 3, 77, 143, 144
Stearns, S.C., 221, 284
Stephens, D.W., 46, 85, 86, 102, 143
Stern, P.C., 1, 7, 283, 293, 294
Stevens, S., 155
Stiner, M., 102
Stocks, A., 54, 57
Stone, R., 23
Strassmann, B.I., 6, 285
Strong, D.H., 78
Strotz, R., 168
Stuart-Macadam, P., 284
Sunstein, C.R., 33, 36
Sutherland, W.J., 5, 40, 165
Suttles, W., 187, 189
Swift, D.M., 145
Szathmáry, E., 289, 291

Talbot, J.F., 272, 273
Tang, S.Y., 139
Taylor, K., 65, 75
Taylor, M., 116
Temple, S., 102
Templeton, A., 85, 102
Thaler, R.H., 34, 35, 168
The Economist, 103
Thomas, K., 278
Thompson, N.S., 290
Tier, C., 222
Tietenberg, T., 103
Tiger, L., 273
Todd, J., 101
Tomikawa, M., 145
Tooby, J., 2, 34, 37, 101, 140, 287, 290
Travis, J., 84, 102
Treadway, R., 246
Trevathan, W.R., 7, 284

Trimpop, R.M., 39, 48, 175, 177
Trivers, R.L., 10, 11, 39, 40, 103, 110,
    173, 288, 291
Trotter, M.M., 72
Tuan, Yi-Fu, 4, 68, 72, 76
Tucker, W.T., 189
Tullock, G., 140
Turelli, M., 102
Turke, P.W. (P.), 6, 13, 202, 246
Turnbull, C.M., 13
Turner, B.L., II., 72
Turner, R., 168
Tuthill, R.W., 287
Tutor, B.M., 221
Tversky, A., 34, 35, 39, 177

Ulrich, R.S., 251, 256, 272, 279
United Nations, 107

Valentine P., 103, 142
van de Kragt, A., 139
van Kolfschoten, T., 103
Vander Wall, S., 101
Veblen, T., 6, 184, 186
Vehrencamp, S.L., 111
Viazzo, P.P., 75
Vickers, W., 3, 56, 58, 59, 63, 101, 102,
    143, 155
Vining, D.R., 190, 201, 214, 219, 237,
    245, 246
Viscusi, W., 102
Vogel, G., 7
Voland, E., 6, 246
von Schantz, T., 198

Wade, R., 139
Walker, J.M. (J.), 139
Wallace, A.R., 118
Walters, C., 139
Wang, J., 32
Wang, X.T., 37
Wasser, S.K., 247
Watkins, S.C., 247
Wazir-Jahan, B.K., 71
Weeks, H., 101
Weick, K.E., 20, 21, 79
Weinberg, E.D., 284
Werner, W., 246
Western, D., 146
Wharton, E., 184
Whary, M.T., 284
Whelan, T., 21

White, D., 69
White, A.T., 77
White, L., 3
White, T.A., 110, 112
Whiten, A., 100
Whiteman, L., 140
Whittam, T.S., 177
Wiener, J.B., 140
Williams, G.C., 7, 13, 18, 40, 54, 60, 68,
    75, 84, 102, 110, 112, 143, 271, 278,
    284, 285, 287, 288
Willis, K.G., 35
Wilson, C., 246, 294
Wilson, D.S., 74, 103, 112, 288, 289,
    290
Wilson, E.O., xiii, 1, 2, 3, 7, 8, 9, 10,
    25, 31, 213, 249, 250, 256, 260, 261,
    287, 288
Wilson, J., 91, 103, 114
Wilson, J.Q., 46, 175
Wilson, M., 3, 5, 10, 31, 32, 33, 39, 42,
    44, 46, 47, 48, 75, 169, 171, 173, 174,
    175, 202, 221
Wilson, M.S., 284
Wilson, R., 140
Wilson, T., 34
Winston, G., 168

Winterhalder, B., 2, 55, 63, 84, 99, 101,
    102, 103, 142, 183, 221, 223
Wishnie, M., 3
Wolfson, R., 22
Woodbury, R., 168
Woodwell, D.A., 41
Woolfenden, G.E., 11
Wrangham, R., 24, 25, 283
Wright, R., 2, 291
Wrigley, C., 247
Wynne-Edwards, V., 13, 20, 74, 82, 101, 288

Yen, D., 225, 226
Yochim, M.J., 3
Yost, J., 57, 58, 61
Young, J.F., 157
Young, O., 140
Young, T.P., 222

Zahavi, Amotz, 6, 184, 186, 193, 194,
    195, 230
Zahavi, Avishag, 6, 230
Zala, S.M., 286
Zann, L.P., 113, 142
Zerner, C., 113, 115, 116, 118, 122, 123
Zuckerman, M., 39, 49, 176, 177
Zuk, M., 186

Subject Index

Acacia tortilis, 270
Ache (South America), 90, 96-97, 223
Achuara (South America), 61
acid rain, 19
adaptation, 84
aesthetics (environmental), *see* environmental aesthetics
African elephants, 11
altruism, 12, 32
    asymmetrical, 110, 111, 122
antibiotics, 284
archaic *Homo sapiens*, 98
asymmetrical altruism, 110, 111, 122
Augustus (Roman emperor), 234

Barabaig (Africa), 145-155
Bari (South America), 57
beauty, 262
behavioral depression, 61-62
behavioral ecology, 10, 260
behavioral ecological paradigm, 17
Bengal (famine in 1943), 228-229
bet-hedging strategies, 221-222
binary decisions, 276
bioaccumulation, 286
biodiversity, 252-256, 279
    amount, 252
    loss, 253
biologically prepared learning, 261-262, 264
biophilia, 7, 249-256, 260-261, 283
    definition, 249
biophobia, 264-265
birth order, 41, 176, 238
blank slate, 2
blow gun, 61
boat-tailed grackles, 221
by-product mutualism, 193
bystander effect, 25

Canis lupus, 19, 22

catastrophic population dynamics, 222-223
cereals, 285
chemical pollutants, 7, 285
    DDE, 286
    DDT, 286
    dioxins, 286
    environmental estrogens, 286
    fluoride compounds, 287
    hormone mimics, 286
    lead, 287
    mercury, 285, 286
    PCBs, 285, 286
Chernobyl accident, 22
Chief Seattle, 68
childlessness, 7
chimpanzees, 11
Christmas (European societies), 187
clutch size (optimal), 221
coalitions (as solutions to environmental problems), 26-27
Cocamilla (South America), 57
coercion, 151-152
collared peccaries, 89, 93-95
collective action problem, *see* tragedy of the commons
commitment models, 291-292
common-pool resources, 129-140, 141-155
    also see tragedy of the commons
    definition, 131
    farmer-managed irrigations systems (Nepal), 133-136
    global, 137-139
    grazing system (Barabaig), 145-148
    local and regional, 136-137
    property-rights systems, 132
competition (intrasexual), 40
conceptual integration, 287
conceptual structure of science, 273-276
concorde fallacy (Dawkins), 233

conflicts of interest over resources, 10
conservation hypotheses (response to game depletion), 54
conservation (of resources), 81-100, 124, 141-155, 157-163
  conventional wisdoms, 9, 15, 67
  definition, 54, 84, 112-114, 141, 143
  epiphenomenal, 83, 113
  ethic, 25, 75-76, 113, 126, 254-256
  for children, 157-163
  opportunity costs, 93
  political issues, 155
consilience, 287
conspicuous consumption, 5, 6, 184, 233
consumption
  conspicuous, 5, 6, 184, 233
  of resources, 5
contingent valuation method, 35-37, 42-43
conventional wisdoms for conservation, 9, 15, 67
cooperation, 7, 11
  benefits of, 290-291
  by-product mutualism, 193
  commitment models, 291-292
  human, 7
  reciprocity, 11, 110-111, 122, 133, 291
corporate recycling, 19
costly signals, 184-186, 193
  altruistic behavior, 193
  higher education, 189
cows, 270
creationism, 277-278
Cree (North America), 53, 85, 90
cultural ecology, 53
cultural group selection, 110, 112, 288-291
cultural inheritance models, 240-241
cultural inheritance of wealth, 203, 213
cultural transmission, 25

Darwinian ecology, 7
Darwinian medicine, 7, 278, 284
DDE, 286
DDT, 286
decisions (binary), 276
deep ecology, 3
deer, 270
degree of relatedness, 11

deliberate overexploitation of resources, 16
demographic sink, 233
demographic transition, 233-234, 237-246
  evolutionary hypotheses, 242
  Imperial Roman populations, 234
  psychological mechanisms, 243
demography
  evolutionary, 6
Dendroica kirklandii, 22
dietary prohibitions, 83-84
diet-breadth hypotheses (response to game depletion), 61-62
dioxins, 286
discounting, 5, 20, 22-23, 43-47, 49-50, 86, 93, 96-98, 157-163, 165-168, 174
  definition, 165
  evolutionary, 96-98
  sensory, 22
  species, 23
discount rate (evolutionary), 96
disdain for health risks (sex differences), 41-43
disregard of environmental degradation (sex differences), 43-47
docility, 25
ducks, 270
dynastic utility function, 202

ecofeminism, 3, 170
ecological economics, 1, 33, 294
ecological noble savage, 3, 53-66, 115, 142-145, 282
ecology, 259
  behavioral, 10, 260
  cultural, 53
  Darwinian, 7
  deep, 3
economic incentives (as solutions to environmental problems), 26
economics, 33
  ecological, 1, 33, 294
  experimental, 292
Efe (Africa), 84
effectively polygynous species (humans), 40
efficiency hypotheses (responses to game depletion), 55
egoism, 12, 32
emotions, 261, 265

environmental aesthetics, 7, 62
  flowers, 271
  habitat selection, 265-271
  manipulation of landscapes, 272-273
  postoperative recovery, 272
  restorative responses, 271-272
  savannah hypothesis, 266-271
environmental estrogens, 286
environmental impact, 74, 281
environmental information, 262-263
  time frames, 263
  types of, 262-263
environmental problems
  causes, 1
  solutions, 23-28
environmental psychology, 1
evolutionary approaches to studying human behavior, 2
  evolutionary demography, 6
  evolutionary psychology, 2, 246
  gene-culture co-evolutionary theory, 2, 246, 250
  human behavioral ecology, 2, 183, 246
  human ethology, 2
evolutionary demography, 6
evolutionary discount rate, 96
evolutionary environments (novel), 14-15, 79, 284-285
evolutionary psychology, 2
experimental economics, 292
extinctions, 3
  Late Pleistocene, 72
  Quaternary, 72

family size, 6
famines
  Bengal in 1943, 228-229
  nineteenth-century India and Ireland, 192
farmer-managed irrigations systems (Nepal), 133-136
fertility
  decline/reduction, 6, 219-236
  low, 7
  rate, xiii, 98
fever, 284
fisherman's problem, see tragedy of the commons
flowers, 271
fluoride compounds, 287

food production, 285
food taboos, 83-84
foraging theory, see optimal foraging theory

gain-loss framing effect (Kahneman and Tversky), 35
gap between humans and animals, 277-278
game theory, 144-145, 148-155
gas prices, 19
geese, 270
gene-culture co-evolutionary theory, 2, 246, 250
generalized reciprocity, 122
genetic group selection, 13, 20, 74, 78, 82, 112, 288-291
genetic inheritance of reproductive strategies, 203, 213
ghost in the machine, 2
global population, xiii
global warming, 22
government regulations (as solutions to environmental problems), 27-28, 137
grackles (boat-tailed), 221
grains, 285
grazing system (Barabaig), 145-148
Great Lakes Indians (North America), 73
green corn ceremony (North America), 187
grizzly bears, 22
group selection
  cultural, 110, 112, 288-291
  genetic, 13, 20, 74, 78, 82, 112, 288-291

habitat selection, 265-271
halibut fishery (north Pacific), 132
handicap principle, 6, 184-199, 230
  testability, 188
harvest festivals (Europe), 187
health monitoring, 41
heavy metals, 286-287
  lead, 287
  mercury, 285, 286
Homo economicus, 33
Homo erectus, 98
Homo sapiens (archaic), 98
honest signals, see costly signals
Hopi (North America), 191, 226-228
hormone mimics, 286

horses, 270
household recycling, 19
Hudson's Bay Company, 73
human behavioral ecology, 2, 183, 246
human ethology, 2
human groups
    Ache (South America), 90, 96-97,
        223
    Achuara (South America), 61
    Barabaig (Africa), 145-155
    Bari (South America), 57
    Cocamilla (South America), 57
    Cree (North America), 53, 85, 90
    Efe (Africa), 84
    Great Lakes Indians (North America),
        73
    Hopi (North America), 191, 226-228
    Inuit (North America), 90
    Hutterites (North America), 106
    Kipsigis (Africa), 243-245
    !Kung (Africa), 223
    Kwakiutl (North America), 187, 189
    Lese (Africa), 84
    Machiguenga (South America), 57,
        92, 94
    Maine lobstermen (North America),
        91
    Makuna (South America), 57
    Mamainde (South America), 57
    Mayans (Central America), 72
    Montagnai (North America), 71
    Naskapi (North America), 85
    Piro (South America), 69, 87, 92,
        93-94
    Plains Indians (North America), 72
    Semang (Asia), 71
    Shipibo (South America), 57
    Siona-Secoya (South America), 56,
        58-59, 62
    Sri Lankans, ancient (Asia), 72
    Tikopia (Polynesia), 191, 224-226
    Tuareg (Africa), 191, 227
    Turkana people (Africa), 106
    Waorani (South America), 57-58
    Wana (Asia), 87
    Wayana (South America), 57
    Yanomamö (South America), 56-59,
        63, 69
    Ye'kwana (South America), 57-58,
        60, 63, 69, 73
    Yenewana (South America), 61

humans and animals (gap between),
    277-278
humans versus nature, 1
Hutterites (North America), 106
hypothetical dilemmas (risk), see trade-off
    dilemmas

ideology, 123
incentives (as solutions to environmental
    problems)
    economic, 26
    social, 24-26
inclusive fitness, 127, 160
indirect bias (cultural inheritance models),
    241, 246
indirect reciprocity, 11
individual selection, 13, 78, 82, 288-291
infanticide, 243
information/education (as solutions to
    environmental problems), 23-24
in-groups vs. out-groups, 289
integrating the biological and social sci-
    ences, 1, 287-288
intelligence
    human, 11
intelligent design, 277-278
intrasexual competition, 40
Inuit (North America), 90
irrigations systems, farmer-managed (Ne-
    pal), 133-136
iterated prisoner's dilemma, 111
    moralistic strategies, 111
    Pavlov, 111
    suspicious tit-for-tat, 111
Ismail the Bloodthirsty (Moulai), 173
Ius liberorum (the right of children), 234

Kei Besar Island (Maluku Province, In-
    donesia), 118
K-selection, 98, 191, 213
kin selection, 110, 111-112, 123, 291
Kipsigis (Africa), 243-245
Kirkland's warbler, 22
!Kung (Africa), 223
Kwakiutl (North America), 187, 189

Late Pleistocene extinctions, 72
lead, 287
Lese (Africa), 84
Leslie matrix, 203
levels of selection problem, 13

life-history theory, 82
lions, 11
Loxodonta africana, 11
Macaca tonkeana, 88
macaques, 88
Machiguenga (South America), 57, 92, 94
magnanimity, 193
magnanimous displays, 195-197
Maine lobstermen (North America), 91
Makuna (South America), 57
Mamainde (South America), 57
manipulation of landscapes, 272-273
marginal value (of resource), 92
marginal-value theorem (=patch-choice model), 55, 85
mass extinctions, 3
mating effort, 39
mating vs. parental return curves, 17
maximum sustainable yield models, 93
Mayans (Central America), 72
medicine, 284
    Darwinian, 7, 278, 284
mercury, 285, 286
money (invention of), 34
Montagnai (North America), 71
moralistic strategies, 111
morality, 289
mother-of-pearl shell (trochus), 117-127
Mount Rushmore (USA), 273
multilevel selection, 288-291
multi-person prisoner's dilemma, *see* tragedy of the commons

Naskapi (North America), 85
natural selection vs. sexual selection, 172
negotiations (as solutions to environmental problems), 27
Nepal, 133
novel evolutionary environments, 14-15, 79, 284-285
noble savage, 2, 3, 67
    ecological, 3, 53-66, 115, 142-145, 282
    Rousseau, 2, 67
norms, 25, 123, 132, 294

opportunity costs (conservation), 93
optimal clutch size, 221
optimal foraging theory, 55, 64, 85, 143

optimizing the number of offspring, 6
Origin of Species by Means of Natural Selection (Darwin), 276
overpopulation, 6

Pan troglodytes, 11
Panthera leo, 11
patch hypotheses (response to game depletion), 54, 55, 58-60, 63
PCBs, 285, 286
peccaries (collared), 89, 93-95
people's perceptions and views of the local wetland ecosystem (Lake Ontario, Canada), 171
personal watercraft (USA), 134
pigs, 88
Piro (South America), 69, 87, 92, 93-94
Plains Indians (North America), 72
polar concepts/constructs (in ecology and evolutionary biology), 274-276
policing, 151-152
pollutants (chemical), *see* chemical pollutants
polygynous species (humans), 40
population
    dynamics (catastrophic), 222-223
    global, xiii
postoperative recovery, 272
potlatch (Kwakiutl), 189
prepared learning (biological), 261-262, 264
preventive health care, 41
prices to non-market resources, 35
primogeniture, 243
prisoner's dilemma, 110
    iterated, 111
private ownership, 91, 99, 105, 125
privileged group (Olson), 194
problem of externalities, *see* tragedy of the commons
problems of scale, 21-22, 130
psychology
    environmental, 1
    evolutionary, 2, 246
proximate cues, 14

Quaternary extinctions, 72

rainforest, xiv
rational choice theory, 291
reciprocal altruism, *see* reciprocity

reciprocity, 11, 110-111, 122, 133, 291
    African elephants, 11
    chimpanzees, 11
    generalized, 122
    indirect, 11
    lions, 11
    solution to environmental problems,
        24-26
    strong, 292
    tit-for-tat, 11, 25, 111
recovery programs, 22
recycling
    corporate, 19
    household, 19
rejection of evolution, 277-278
relatedness (degree of), 11
reproduction, 201-217
    successful, 11
reproductive decisions (long-term effects),
    203-216
reproductive interests, 11
reproductive strategy, 203
reproductive success, 11
    effect of wealth, 203, 215
reproductive value, 17
resource
    acquisition, 10
    coalitions (traditional societies),
        73-75
    control, 17
    consumption, see resource use
    deliberate overexploitation, 16
    information or feedback about, 16
    management (definition), 54
    marginal value, 92
    most difficult resource-use problems,
        20
    use, 5, 10, 15-17, 31-33, 281
    use, sex differences, 39-41
    use, traditional societies, 76-79
    use, violence and conflicts, 283
resource holding power, 186-188
resource-related problems (characteris-
    tics), 20
response to game depletion
    conservation hypotheses, 54
    diet-breadth hypotheses, 61-62
    patch hypotheses, 54, 55, 58-60, 63
    taboo hypotheses, 54, 55, 60-61
    time-allocation hypotheses, 54,
        56-58, 63

restorative responses (environmental aes-
    thetics), 271-272
return curves (mating vs. parental), 17
risk, 37-43, 47-50, 174-181
    personal health risks vs. economic/
        status advantages, 178-179
    present profits with environmental
        degradations vs. reduced
        profits with long-term sustainability,
        179-181
    young men as risk takers, 47-50,
        174-176
risk acceptance, 35
    sex differences, 39-41
risk aversion, 34
r-selection, 98, 191, 213

Salmon War (1997), 91
savannah hypothesis, 266-271
Seattle (Chief), 68
scale (problems of), 21-22, 130
science (conceptual structure), 273-276
sea lions (Stellar's), 83, 84
selection
    group (cultural), 110, 112, 288-291
    group (genetic), 13, 20, 74, 78, 82,
        112, 288-291
    habitat, 265-271
    individual, 13, 78, 82, 288-291
    K-, 98, 191, 213
    kin, 110, 111-112, 123, 291
    multilevel, 288-291
    r-, 98, 191, 213
    sexual, 39-41, 169, 172-181
    sexual vs. natural, 172
selfishness, 12, 32
Semang (Asia), 71
sense organs, 275
sensory discounting, 22
serviced provided free by Earth's natural
    ecosystems, xiv
sex differences, 17, 39-41, 169-181
    disdain for health risks, 41-43
    disregard of environmental degrada-
        tion, 43-47
    risk acceptance, 35
    resource use, 39-41
sexual selection, 39-41, 169, 172-181
    vs. natural selection, 172
sheep, 270
Shipibo (South America), 57

Siona-Secoya (South America), 56, 58-59, 62
small wins, 21, 79
snakes, 250-251
social dilemma, *see* tragedy of the commons
social incentives (as solutions to environmental problems), 24-26
social norms, *see* norms
social power, 186-188, 230
social status, 183-199, 219-236
    anxiety, 234
    competition and fertility reduction, 229-235
    definition, 186
    economic benefits, 230
    positional good, 231
social trap, *see* tragedy of the commons
socioeconomic status, 198
solutions to environmental problems, 23-28, 50-51, 293-294
    coalitions, 26-27
    combination of strategies, 293
    government regulations, 27-28, 137
    information/education, 23-24
    economic incentives, 26
    negotiations, 27
    social incentives, 24-26
species decline/extinction, xiv
species discounting, 23
spider monkeys, 93-95
spotted owl, 19, 22
Sri Lankans, ancient (Asia), 72
standard cross-cultural sample, 69
starvation threshold, 205
status (social), *see* social status
Stellar's sea lions, 83, 84
Stone Age Economics (Sahlins), 83
strong reciprocity, 292
successful reproduction, 11
Sus celebensis, 88
sustainable development, 28
synthetic chemicals, 286

taboo hypotheses (response to game depletion), 54, 55, 60-61
technology, 282
territoriality (definition), 91
The Origin of Species by Means of Natural Selection (Darwin), 276
The Wealth of Nations (Smith), 105

Three Mile Island accident, 22
Tikopia (Polynesia), 191, 224-226
time-allocation hypotheses (response to game depletion), 54, 56-58, 63
time preference, 86, 96
Titanic disaster, 229
tit-for-tat, 11, 25, 111
toxicology, 286
trade-off dilemmas
    personal health risks vs. economic/status advantages, 178-179
    present profits with environmental degradations vs. reduced profits with long-term sustainability, 179-181
tragedy of the commons, 4, 13, 26, 53, 90, 105-107, 109-127, 129-140, 141-155, 157, 282
    coercion, 151-152
    commonism, 106
    farmer-managed irrigations systems (Nepal), 133-136
    global, 137-139
    grazing system (Barabaig), 145-148
    halibut fishery (north Pacific), 132
    local and regional, 136-137
    "mutual coercion, mutually agreed upon", 98
    policing, 151-152
    private ownership (privatism), 91, 99, 105, 125
    property-rights systems, 132
    scale effects, 106, 130
    socialism, 106
    solutions, 98-99
    "The tragedy of the commons" (essay), 105, 129
    tragedy of the unmanaged commons, 105-107
    trochus management, 116-127
trochus management, 116-127
Trochus niloticus, 117-127
Tuareg (Africa), 191, 227
Turkana people (Africa), 106

unigeniture, 243
Ursus arctos horribilis, 22

vertical integration, 287
violence and conflicts over access to scarce resources, 283

Waorani (South America), 57-58
Wana (Asia), 87
warfare, 24-25
watersheds, 19
Wayana (South America), 57
wealth differences among individuals, 203
Wealth of Nations (Smith), 105
wetland development, 19
wetland ecosystem, people's perceptions and views (Lake Ontario, Canada), 171
whaling, 93

winter dance (Kwakiutl, North America), 187
wolves, 19, 22

Yanomamö (South America), 56-59, 63, 69
Ye'kwana (South America), 57-58, 60, 63, 69, 73
Yenewana (South America), 61
young men as risk takers, 47-50, 174-176